Biology of Root Formation and Development

BASIC LIFE SCIENCES

Recent Volumes in the series:

A Continuation Order Plan is available for this series. A continuation order will bring delivery of each new volume immediately upon publication. Volumes are billed only upon actual shipment. For further information please contact the publisher.

Biology of Root Formation and Development

Edited by

Arie Altman

The Hebrew University of Jerusalem
Rehovot, Israel

and

Yoav Waisel

Tel Aviv University
Tel Aviv, Israel

Plenum Press • New York and London

Library of Congress Cataloging-in-Publication Data

Biology of root formation and development / edited by Arie Altman and Yoav Waisel.
p. cm. -- (Basic life sciences ; v. 65)
"Proceedings of the Second International Symposium on the Biology of Root Formation and Development, held June 23-28, 1996, in Jerusalem, Israel"--T.p. verso.
Includes bibliographical references and index.
ISBN 0-306-45706-7
1. Roots (Botany)--Formation--Congresses. 2. Roots (Botany)--Development--Congresses. 3. Plant cuttings--Rooting--Congresses. I. Altman, A. II. Waisel, Yoav. III. International Symposium on the Biology of Root Formation and Development (2nd : 1996 : Jerusalem) IV. Series.
QK644.B55 1997
575.5'4--dc21

97-31141
CIP

Proceedings of the Second International Symposium on the Biology of Root Formation and Development, held June 23 – 28, 1996, in Jerusalem, Israel

ISBN 0-306-45706-7

A Division of Plenum Publishing Corporation
233 Spring Street, New York, N.Y. 10013

http://www.plenum.com

10 9 8 7 6 5 4 3 2 1

Printed in the United States of America

PREFACE

This book contains the majority of the presentations of the Second International Symposium on the Biology of Root Formation and Development, that was held in Jerusalem, Israel, June 23–28, 1996. Following the First Symposium on the Biology of Adventitious Root Formation, held in Dallas, USA, 1993, we perceived the need to include all kinds of roots, not only the shoot-borne ones. The endogenous signals that control root formation, and the subsequent growth and development processes, are very much alike, regardless of the sites and sources of origin of the roots. Therefore, we included in the Second Symposium contributions on both shoot-borne (i.e., adventitious) roots and root-borne (i.e., lateral) roots.

Plant roots have remained an exciting and an intriguing field of science. During the years that followed the first symposium, an exceptional proliferation of interest in root biology has developed, associated with the intensive research activity in this field and the contemporary developments in the understanding of root function and development. New methods have been applied, and old ideas and interpretations were reexamined. Altogether, it became necessary to update our viewpoints and to expand them.

The chapters of this book are not intended to provide complete overviews. Our purpose is to present a cross section of the accomplishments of the past and of the direction into the future. The book covers molecular biology, physiology, etc.; passes through topics such as micropropagation, root production, etc.; but also tackles ecological questions and interactions between roots at the field scale. The articles present the opinion of the symposium participants, and their points of views. They were intended to be provocative, to introduce the reader to the pressing current questions of root studies, and to point out new bearings for future research.

The book presents a major source of information for plant root scientists, and designates the frontiers of our knowledge in this field. It points out the pressing questions that should be investigated in the future. It presents a multidisciplinary view of plant roots and the state of the art. It covers various aspects of root structure, development and behavior, the interactions between roots and their environment, and the various uses of roots.

The book contains some 78 contributions. It examines a wide range of critical topics, unexplored areas, potential applications and future directions. It covers the following themes:

- Diversity of roots (Section 1).
- Induction of root formation and root development (Section 2).
- Hormonal control of root induction and development (Section 3).
- Molecular biology of root development (Section 4).
- Ecological aspects of root development (Section 5).
- Applied aspects of root formation and development (Section 6).

- Root products (Section 7).
- Models and methods for root studies (Section 8).

Some of the topics were given more weight than others, especially those where the literature has amply proliferated during the last years, or where new techniques were introduced. Several new themes were included in the Second Symposium:

- Innovative Methods of Root Studies
- Modeling and Simulation
- Ecological Aspects of Root Biology

While we have made every effort to reach uniformity in terminology and style, the presented results, the expressed ideas, and the final shape of the manuscript remained the sole responsibility of the authors.

We wish to express our sincere gratitude to each of the eminent contributors for their scholarly contribution and for their fast and enthusiastic cooperation. We are grateful also to the members of the International Advisory Board and to the Organizing Committee for their unequaled help in organizing the symposium.

Arie Altman
The Hebrew University of Jerusalem

Yoav Waisel
Tel Aviv University

CONTENTS

SECTION 1: DIVERSITY OF ROOTS

Short Papers

SECTION 2: INDUCTION OF ROOTS AND THEIR DEVELOPMENT

Short Papers

SECTION 3: HORMONAL CONTROL OF ROOT INDUCTION AND DEVELOPMENT

Short Papers

SECTION 4: MOLECULAR BIOLOGY OF ROOT DEVELOPMENT

SECTION 8: MODELS AND METHODS FOR ROOT STUDY

Short Papers

SECTION 9: CONCLUDING REMARKS

THE SECOND INTERNATIONAL SYMPOSIUM ON THE BIOLOGY OF ROOT FORMATION AND DEVELOPMENT
(JERUSALEM, ISRAEL, JUNE 23 – 28, 1996)

ORGANIZING COMMITTEE

Arie Altman (Chairman)
Rehovot, Israel

M. Raviv
Newe-Ya'ar, Israel

J. Riov
Rehovot, Israel

T. Sachs
Jerusalem, Israel

Yoav Waisel
Tel Aviv, Israel

INTERNATIONAL ADVISORY BOARD

P. N. Benfrey
New York, New York

Tim D. Davis
Dallas, Texas

A. H. Fitter
York, United Kingdom

H. Flores
University Park, Pennsylvania

Thomas Gaspar
Liège, Belgium

Wesley P. Hackett
St. Paul, Minnesota

Bruce E. Haissig
Rhinelander, Wisconsin

I. M. Sussex
Berkeley, California

D. Tepfer
Versailles, France

1

THE PLACE OF ROOTS IN PLANT DEVELOPMENT

Peter W. Barlow[1] and Beatriz Palma[2]

[1]IACR—Long Ashton Research Station
Department of Agricultural Sciences
University of Bristol
Long Ashton, Bristol BS18 9AF
United Kingdom
[2]Instituto de Biología
Universidad Católica de Valparaíso
Casilla 4059, Valparaíso
Chile

1. INTRODUCTION

A plant is a complex system of branched axes, any of which may be in either a vegetative or a reproductive phase of development. Indeed, given the amazing diversity of plant life, repetitive branching of morphologically distinctive axes is probably the only true generality that can be attributed to plants (see Schultz-Schultzenstein 1861). At a more detailed level, in those axes expressing the vegetative phase, it can be seen that their structure is comprised of a set of reiterated morphological units, a structure which, moreover, tends to be recapitulated in each new axis originating from a branching event. But what exactly is the reiterated structure that is so faithfully reproduced by branching? Where do the type of roots termed "adventitious" fit into this branching scheme, and what contribution do they make to the plant's life cycle? We shall attempt to answer these questions in the following pages but, put briefly, our thesis is that so-called "adventitious roots" comprise a range of shoot-borne roots which are members of a set of reiterated morphological units that, in turn, are integral to plant architecture. Not unexpectedly, these roots are components of a strategem of plant development that is geared to vegetative propagation and nutrient acquisition. These developmental ploys are accomplished in diverse manners, revealing the astonishingly multifarious nature of root growth and development. First, however, it is necessary to clarify what is meant by the two terms already used, "shoot borne-root" and "adventitious root". In doing so, some explanation is required of the reiterative nature of plant construction which, although more evidently of relevance to the shoot system, does nevertheless emphasise the indissoluble link between root and shoot systems in the life of the plant.

Biology of Root Formation and Development, edited by Altman and Waisel.
Plenum Press, New York, 1997.

2. ROOTS IN RELATION TO DEVELOPMENT

2.1. Terminology of the Types of Roots Borne by Shoots

The term "adventitious root" is used widely, but not with any degree of precision. For the horticulturist, an adventitious root is generally one that is caused to form by some artificial means, often with the assistance of growth regulating chemicals (Avery et al. 1947, Blazich 1988), on a plant part (usually of the shoot system) that might never have borne roots if left undisturbed. The roots are "adventitious" in the sense that they can be produced at anatomically unexpected locations. The origin of the root primordia can be traced to the totipotency of cells in the affected part, a property evoked when the integrity of the system is challenged either chemically or physically.

Agronomists and botanists also refer to "adventitious roots", but here the roots are not necessarily artificially induced but are ones which appear during the normal course of plant development. While there is clearly some common ground between this and the situation mentioned in the paragraph above, it does seem that the term "adventitious root" is often used in two distinct senses: in one case, the root in question is a perfectly natural manifestation of the "branching of axes" that characterises plant ontogeny, whereas in the other case, the root has been forced into being by some practice (e.g. severance) which has left the affected part with no option but to take recourse to cellular totipotency if survival and ramification of the plant is to continue.

To accommodate the various temporal and spatial relationships that a root may have with its supporting *shoot* system, we propose the classificatory scheme shown in Table 1 and thereby arrive at eight types of shoot-borne root.

1. Sylleptic shoot-borne root
2. Proleptic shoot-borne root
3. Sylleptic, adventitious shoot-borne root
4. Proleptic, adventitious shoot-borne root
5. Sylleptic, adventive shoot-borne root
6. Proleptic, adventive shoot-borne root
7. Induced, adventitious shoot-borne root
8. Induced, proleptic, adventitious shoot-borne root

Implicit within the scheme is the fact that the respective root primordia may, with reference to the timing of (and response to) a rooting stimulus, natural or otherwise, be either pre-formed (i.e. initiating roots of types 1–4) or post-formed (initiating types 5–8). Fragmentation (or severance) of the shoot may stimulate the emergence of roots from pre-formed, but dormant, primordia on the fragmented part (types 2 and 4), or cause the initia-

Table 1. Eight types of roots (1–8) developing on shoots (or roots) dependent on the timing and location of their initiation and the timing of their outgrowth

	Root primordium initiation			
	Usual timing location		Unusual timing location*	
Root primordium outgrowth	Usual	Unusual	Usual	Unusual
Immediate (sylleptic)	1	3	5	7
Delayed (proleptic)	2	4	6	8

*Delayed or stimulated in relation to the usual temporal sequence of primordium initiation during the development of a particular phytomer.

tion of new root primordia (types 5–8), especially if assisted by root-inducing compounds. The shoot-borne roots induced in horticultural practice are often those of types 5 and 7, but may also be of types 2 and 4. By contrast, roots of type 1 are entirely natural components of the plant, which play an essential role in the life of the plant as a whole. Usually, no specific name is given to them, but here we call them "sylleptic roots".

The terminology associated with the scheme in Table 1 is helped by the adjective "adventive", which conveys the sense of "spontaneous occurrence". In the present context, spontaneity of rooting would probably be perfectly predictable if all the pre-conditions for this event were known. However, if spontaneity is equated with some accidental, or chance, occurrence in time, but not necessarily in location, then "adventive" may appropriately apply to type 5 or type 6 roots borne on a shoot system. For a type 7 or type 8 root, the same spontaneity of induction could also hold true, but here the location is also different and hence we look to "adventitious" to convey this more obviously recognizable trait.

2.2. Metamers, Phytomers, Shoots and Roots

An important notion, emphasised in the above terminology, is that roots arise on shoots at predictable sites. Thus, there are "usual" sites (for roots of types 1, 2, 5 and 6), and these contrast with "unusual" sites (for root types 3, 4, 7 and 8). The predictability with which roots arise endogenously and then emerge at characteristic anatomical (usual) sites indicates that they are ontogenetically pre-determined, and that the primordia from which they arise are just as much an integral component of the developing shoot as are leaf or bud primordia (Kawata et al. 1963, Kurihara et al. 1978). The root primordium probably arises in response to positional information, perhaps in the form of local variations in auxin concentration that are, in turn, related to anatomical features such as a subtending leaf or nodal tissue. But at this point we may justifiably ask whether we understand fully what is meant by shoot development and how it is that root primordia should be part of it?

As mentioned in the Introduction, plant development consists of the reiteration of axes by means of a branching process. The axes themselves are composed of reiterated morphological units known as metamers or phytomers. (Later, we shall attempt to distinguish between these two terms.) Each morphological unit, M, is a portion of shoot that is defined by (1) a leaf, L, at its distal end, and (2) an associated node, N. At its proximal end there is (3) a shoot primordium, or apical dome, D, which often appears to be initiated in the axil of a subtending leaf in an earlier-formed unit; new M units will form from this apical dome. The intervening portion of the stem in M is (4) the internode, I. A fifth component is (5) a root primordium, R, or a site at which R will emerge at some later time. Often, the primordia are dormant ($\overline{D}$ or $\overline{R}$), awaiting a condition that will renew their growth. Thus, the reiterated unit, M, is a quintuplet of cell-groups, each possessing a different developmental state: $M = [L\ N\ I\ D\ R]$. The relationship of these units with the construction of the stem and its phyllotaxy has been considered elsewhere (Barlow 1994a, b). Briefly, there are two concepts of stem construction (Čelakovský 1901): (1) that the stem is subdivided into holocyclic segments or frusta [i.e. they are stem segments whose ends (or transversal boundaries) are bounded by successive nodes, and whose sides (longitudinal boundaries) are the flanks of the segments], or (2) the stem consists of mericyclic sectors (i.e. M is a sliced segment, or sector, whose ends are defined by two successive leaf bases that lie approximately on an orthostichy and one of whose sides is an arc comprising the sector's flank, the other sides being planes lying on two radii that join the edges of the leaf bases to the centre of the stem). In accordance with these sub-divisions of the plant axis, we propose that the holocyclic segments are equivalent to phytomers whereas the mericyclic sectors are metamers. Both types of unit arise from little-understood rhythmic activities within the apical dome of the shoot.

Metamers are units of development which appear to have their origin in cells with similar or common lineages and, being mericyclic, they can, in shoots with spiral phyllotaxy, be ordered in a staggered sequence along the stem. The phytomer is a unit that is sometimes easier to apprehend visually. On occasion, one phytomer may correspond to one metamer (as in shoots with distichous phyllotaxy and encircling leaf bases), but in spiral phyllotaxy the frustum-shaped phytomer would comprise portions of three or more successive metamers. The organizational power of the shoot may be such that, in members of the Gramineae, for example, the spirally arranged metamers are organized into a sequence of phytomers which give the stems an articulated appearance, with the leaf base and associated nodal tissue extending transversely like a disc across the whole stem. Metameric and phytomeric constructions of a notional stem are shown in schematic form in Fig. 1, which also indicates the spatial arrangement of the five components of these reiterated units. The quintuplet of characters (=*M*) can apply equally to both phytomer and metamer.

2.3. Shoots and Roots of Maize and Other Gramineae

Plants are evidently populations of reiterated units, ΣM, from which shoot-borne roots can arise. The branching pattern, or architecture, of the shoot is defined by the sequence, rate and orientation at which their apices grow in response to internal correlative factors within ΣM. The architecture is supplemented by the growth pattern of the root primordia, *R*, which subsequently contribute to the shoot-borne root system. Taking the maize plant (*Zea mays*) as a simple and relatively well studied example, there is a genetically defined number of type 1 (or type 2) root primordia associated with each successive internode along the stem. Roots then grow out from the respective population of primordia on each internode with a rhythm determined by the rate of formation of new *M* units (the

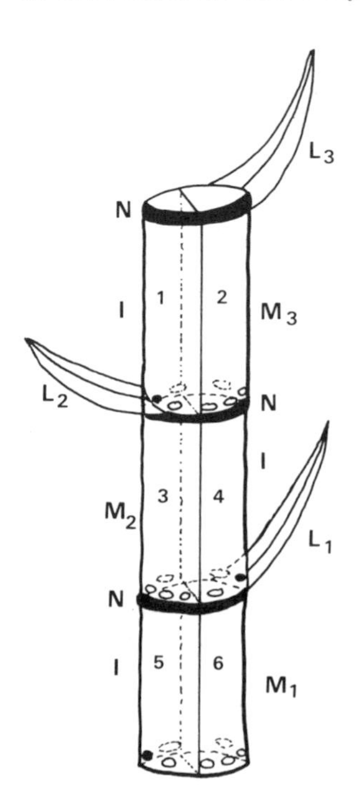

Figure 1. Notional construction of a stem to show the interrelationship between metamers, phytomers and their component subunits, leaves (*L*), nodes (*N*, and filled segments), internodes (*I*, and unfilled segments), bud primordia (filled circle) and root primordia (unfilled ellipses). The present scheme might apply to a shoot with distichous phyllotaxy. The mericyclic sectors (i.e. sectors 2+4 and 3+5) correspond to metamers, whereas the holocyclic segments (i.e. sectors 1+2, 3+4 and 5+6) correspond to phytomers. In a plant whose cellular construction is clearly based on merophytes (e.g. a fern), these latter constructional units also correspond to the metamers (i.e. sectors 2+4 could correspond to both metamer and merophyte).

plastochron) (Picard et al 1985). A detailed consideration of the circumferential arrangement of shoot-borne roots around successive internodes of maize shoots (in the F1 hybrid Dea) led Pellerin et al. (1991) to suggest that there was "some kind of rhizotaxis" (see also Fig. 1). In this variety, internodes 2–5 each tended to bear three roots. Their location around the circumference of a given internode (I_n) was such that they lay between the orthostichies of root sites associated with the previously formed internode (I_{n-1}). This observation suggests that the position of the root primordia is determined by the phyllotaxy of the shoot — which in turn corresponds to the mericyclic arrangement of the metamers.

It is only during the vegetative phase of development that internodes of the maize shoot bear roots (Poethig 1990); internodes of a shoot which has entered the reproductive phase do not form root primordia. Phase transition of the shoot is associated with a reprogramming of phytomer (or metamer) production in the apex: e.g., $[N\,L\,I\,D\,R] \rightarrow [N\,L\,I\,_\,_]$, the "_" indicating a missing member of the usual vegetative phytomer quintuplet.

Consideration of the maize shoot recalls another terminological puzzle that has, from time to time, beset the literature relating to the shoot-borne roots of graminaceous plants (Hoshikawa 1969, Onderdonk and Ketcheson 1972) — that is, what names to give to the roots emerging at morphologically distinct sites along the stem, especially in the early stages of its growth? Favoured positions for roots are (1) just distal to the nodes (here it is important to note that the roots do not emerge from the nodes themselves — they are supra-nodal and emerge from the base of an internode), and, less frequently, (2) at sites along the internodes. In maize and many other Gramineae, the first two whorls of supra-nodal roots arise just distal to the first and second visible nodes of the stem, the scutellar node, N_1, and the coleoptilar node, N_2. These sets of roots are in positions anticipated by the phytomer concept and hence can be considered as shoot-borne roots of type 1. Roots intercalated between the base of I_2, the first emerged internode, and N_2 are designated mesocotylar roots. These roots are not in an expected position (according to the phytomer concept), nor are they always present (Hoshikawa 1969). Hence, they may be considered as spontaneous, adventitious roots of types 4, 7 or 8 (see Table 1). It will be noticed here that the *first* whorl of shoot-borne roots actually lies within I_2 of the *second* phytomer, M_2. By numbering the phytomers in order of their production, M_1, M_2, ..., M_i, the roots, leaves, etc., of which they are composed can also be numbered in a consistent way: $M_1 = [N_1\,L_1\,I_1\,_\,R^*]$, $M_2 = [N_2\,L_2\,I_2\,_\,R_2]$, $M_3 = [N_3\,L_3\,I_3\,_\,R_3]$, $M_4 = [N_4\,L_4\,I_4\,\overline{D}_4\,R_4]$, where L_1 = scutellum (tentative), L_2 = coleoptile, L_3 = first true leaf, I_1 = hypocotyl, I_2 = mesocotyl, _ = absent member, $\overline{D}_4$ = first apical dome or bud primordium (dormant), R_2 = first whorl of supra-nodal, shoot-borne roots (or seminal roots), R_3 = second whorl of supra-nodal, shoot-borne roots (or crown roots). R^* indicates a root whose status we shall discuss in the next section. The intercalated mesocotylar roots on I_2, where present, can be designated $R_{(I)2}$. This terminology does away with the inconsistency in the otherwise perceptive "harmonisation des notations" of Girardin et al (1986) whereby the first internode is associated with the second node, and the first leaf (F_1 in these authors' terminology) appears on the third node.

The fourth phytomer, M_4, of the maize shoot is the first one to be complete, having all five components. By means of D_4, it is also the phytomer which initiates the branching process of the stem that is so characteristic of plant life. Evidently, this phytomer, and others with a full complement of members, are self-sufficient constructional units in the sense that any one of them could form a new plant, even if all other phytomers were destroyed. Here, then, we see the essence of the phytomer (or metamer) in the life of plants generally. The reiterative unit, M, is an autonomous and self-reproducing unit of vegetative plant life.

The above illustration of the construction of the maize shoot *via* phytomers, and the part that shoot-borne roots play in the architecture of the plant can probably be generalized to all species of angiosperms, with the proviso that evolution can supress the property of root development (recall the rootless plants of the *Utriculariaceae*, for example). This leads us now to consider the nature of the *embryonic*, or *pole-borne* root, for it is suggestive that, in listing the phytomeric components of the maize shoot, supra-nodal, shoot-borne roots were not explicitly included in M_1, but were only hinted at with the designation R^*. If a root were to have been included for this unit, where would it have been? We presume that some sort of polarization of the developing phytomer would have caused it to be at or near the proximal end of M_1. Thus, can it be that the embryonic, pole-borne root is, in reality, a shoot-borne root?

3. THE POLE-BORNE ROOT

The logic of the phytomer concept, as applied to the youngest phytomers of maize, inevitably focuses attention on the first of these morphological units, M_1, which initiates the development of a new generation of branching axes. The leafy sub-unit L_1 in this monocot species has already been tentatively equated with the scutellum; in a dicot, L_1 would be equated with a cotyledon (or a pair of them). But what of I_1? This must be a portion of the embryonic axis itself. One end of it should possess the single root, R^*, that is usually termed "pole-borne" or "embryonic". The other end of I_1 would bear an apical shoot primordium or meristem. It would be reasonable to suppose the course of development of the pole-borne maize root to be analogous to that of any of the primordia (R_2, R_3, ...) of the shoot-borne roots present on any other phytomer (M_2, M_3, ...). A root primordium presumably forms at the proximal "pole" of the M_1 phytomer for the same reasons that shoot-borne root primordia form at the proximal end of later phytomers. Accordingly, the physiological stimulus for primordium initiation, which perhaps has its basis in localized auxin content and activity, evokes similar patterns of gene activity at sites where either the embryonic or the shoot-borne root primordia form. Embryonic root and shoot-borne roots may therefore be of qualitatively similar origin. Any subsequent differences between their structures should be largely quantitative (i.e. their size, gravitropic response, etc.) and likely to be due to nutritional or local environmental factors.

It should be recalled that monocots are evolutionarily more advanced than dicots. Therefore, it is conceivable that their newly evolved strategems of development frustrate attempts to homologise their early phytomers with those of dicots. The very ontogeny of monocot embryos indicates such difficulties: viz. the uncertain homology of the scutellum with L_1 of a dicot.

3.1. The Equivalence of the Embryonic, Pole-Borne Root and the Shoot-Borne Root

We here propose the first of two fairly general schemes for early root and shoot development, and of phytomer construction. In this scheme, the embryonic root primordium and shoot-borne root primordia are held to have similar origins; that is, they have originated from within an established, although underdeveloped, phytomer. For the full realization of the scheme, we must trace it from the zygote, Z, and the resultant proembryo. The cell, Z, is the original stem cell from which all other cells are subsequently derived (Barlow 1996a). As the proembryo grows in size and cell number, a discrete stem-cell compartment, Z^S, is retained at one end and a derivative compartment, Z^V, is differentiated at the other. It is the latter compartment that produces M_1. Z^S goes on to generate the apical dome, ${}_ED$, of the primary shoot. The derivative portion, Z^V, polarizes into apical and anti-apical ends, $V+$ and $V-$. The former ($V+$) differentiates into leaf (L_1, scutellum or cotyledon) and node (N_1), the latter ($V-$) differentiates a root primordium, R^*, which we may now term R_1, at the proximal end. The intermediate part of the derivative, lying between $V+$ and $V-$, is regarded as having an indifferent status, o. It differentiates as hypocotyl; it is also equivalent to the first internode, I_1. This accords with the fact that, anatomically, hypocotyls are of embryonic origin and generally show characteristics of both root and shoot. The hypocotyl may even be regarded as an embryonic rhizophore, akin to that well known shoot-borne structure of *Selaginella* (Lycopsida), which seems to develop as an organ *sui generis* before the developmental pattern of its apex "mutates" to form a root.

The following algorithm (1) summarises M_1 development for this first scheme as follows:

$$\begin{aligned}
&Z \rightarrow [Z^S Z^V] \\
&Z^S \rightarrow {}_E D_1 \\
&Z^V \rightarrow [V_1 + oV_1 -] \\
&V_1 + \rightarrow [L_1 N_1] \\
&o \rightarrow I_1 \\
&V_1 - \rightarrow [_ R_1]
\end{aligned} \tag{1}$$

Therefore, $M_1 = [L_1\ N_1\ I_1\ _\ R_1]$. At this stage, the embryo E = $[{}_E D_1\ M_1]$. The fact that mutation (the *dek* mutation in maize, for example) can permit the formation of an embryonic root in the absence of the embryonic shoot could be because of a block to the transition

$$Z^S \rightarrow {}_E D_1$$

Each successive phytomer follows a similar pattern of development by virtue of the retention of the embryonic character at the apical dome, ${}_E D_1$. The apical dome comprises a stem-cell population, as did the apical portion, Z^S, of the proembryo. Concomitant with its growth, the stem-cell compartment, ${}_E D^S{}_1$, is maintained and a derivative portion, V_2, is differentiated. V_2 polarises as V_2+ and V_2-, and is the basis for the development of M_2. The second leaf (or coleoptile), L_2, differentiates together with N_2; the shoot-borne root primordium, R_2, also differentiates. R_2 is thus, in principle, no different from R_1. It is interesting that, in the Gramineae, the embryonic root primordium does not lie on the same axis as the shoot apex, but to one side, as though it were indeed a lateral (rather than a polar) primordium borne on the I_1 internode. In maize, no lateral shoot apex, D_2, develops on M_2, perhaps because the scutellum (which may or may not be equivalent to L_1) does not exert the requisite inductive influence.

The algorithm (2) for M_2 formation is therefore:

$$\begin{aligned}
&{}_E D_1 \rightarrow [{}_E D_1^S V_2] \\
&{}_E D^S{}_1 \rightarrow {}_E D_2 \\
&V_2 \rightarrow [V_2 + oV_2 -] \\
&V_2 + \rightarrow [L_2 N_2] \\
&o \rightarrow I_2 \\
&V_2 - \rightarrow [_ R_2]
\end{aligned} \tag{2}$$

In the context of maize shoot development, it is only later, when V_4 has been derived from ${}_ED_3$, do we obtain D_4 and R_4. At this stage, $M_4 = [N_4\ L_4\ I_4\ D_4\ R_4]$.

3.2. The Distinctiveness of the Embryonic, Pole-Borne Root

The second of our two schemes for the origin of the pole-borne roots confers on this root a rather distinctive status. The scheme is more in keeping with embryological observations (Randolph 1936, Guttenberg et al. 1954) which reveal that, in maize and other grasses, the embryonic axis appears to arise by a process of histogenetic compartmentation from within the body of the embryonic group of cells, a large number of cells not directly contributing to the proembryo. It might do this as a result of a cell-lineage-based program of differentiation (see Barlow 1996b). The same could also occur in dicots, though here the embryogenetic compartmentation process prevails within nearly the entire proembryonic cell group. In both monocots and dicots, only the suspensor differentiates as a terminal-state tissue, distinct from the embryo proper.

It is quite clear that, in the Gramineae, both the primary root and primary shoot meristems arise simultaneously in two distinct locations, denoted $E+$ and $E-$, within the enlarging embryonic cell cluster. The first metamer, M_1, appears to be intercalated between these two developing meristems. It is as though there is a zone within the embryo that is pre-determined to become polarized; a distinctive primary shoot apex, ${}_ED$, forms at $E+$ and an equally distinctive, but different, primary root apex, ${}_ER$, forms at the $E-$ end. The intervening portion, o, becomes hypocotyl. There is no leaf; the scutellum, S, differentiates from embryonic cells external to the differentiating embryonic axis. In this scheme, S is not part of M_1. Thus, in this second scheme, the algorithm (3) for M_1 development is as follows:

$$
\begin{aligned}
&Z \rightarrow [SE + oE-] \\
&S \rightarrow S \\
&E+ \rightarrow {}_ED_1 \\
&o \rightarrow [_N_1 I_1_] \\
&E- \rightarrow {}_ER
\end{aligned}
\qquad (3)
$$

Therefore, $M_1 = [_\ N_1\ I_1\ _]$. At this stage, the embryo, $E = [S\ {}_ED_1\ M_1\ {}_ER]$.

The two distinctive apical meristems contain stem-cell compartments, ${}_ED^S{}_1$ or ${}_ER^S$, and then ${}_ED^S{}_1$ behaves similarly, according to algorithm (2), as described for the first-mentioned scheme of embryonic development.

4. SOME CONSEQUENCES OF PHYTOMERISM

In the first of the two schemes, the pole-borne, or embryonic, root appears to be subordinate to the shoot apex since the progenitors of the root primordium have descended from the embryonic shoot's apical dome in the form of a derivative compartment that later developed as the first phytomer, M_1. Leaf and root have equivalent status as meristematic members of the quintuplet of phytomer sub-units. A distinctive behaviour of the shoot apex is also to be suspected from the very concept of the phytomer since this morphological unit is a rhythmically-produced derivative of shoot apex activity. No similar rhythmic production of visibly distinct morphological sub-units seems to occur in roots, indicating that root and shoot apices, whatever their origin, have patterns of organogenetic behaviour that differ in this respect. However, it is possible that, if a root apex did possess a rhythmic behaviour which resulted in a distinctive morphological sub-division (as "rhizomer segments") over and above that associated with root branching, this might have some negative survival value, whereas in shoots this type of behaviour could have beneficial consequences. Evolution would ensure the minimizing of overt rhythmicity in roots and enhance it in shoots. One benefit of a morphologically sub-divided shoot is that each unit (whether phytomer or metamer) is a potential unit of vegetative propagation, in which the natural shoot-borne roots serve as both anchor and feeder during the dispersal and establishment of the phytomer/metamer propagules. On the other hand, what is required of the root is that it should be a ramifying foraging organ that efficiently transports solutes without any interruption due to its segmentation.

Despite a negative survival value, some trace of innate rhythmicity may still persist in root apices. In this respect, root and shoot meristems may be considered equivalent in their structural and functional attributes, as proposed in Scheme 2 above, and that one organ is not dependent on, or subordinate to, the other. Both are of equal status in their contribution to the plant. Moreover, in some circumstances, roots can give rise to shoots (Bonnett and Torrey 1965). The contrary situation of shoots giving rise to roots is well known. Thus, both organs have equivalent morphogenetic potentials.

Evidence of an internal rhythm is seen from longitudinal sections of roots of the sea grass, *Thalassia testudinum* (Tomlinson 1969). Files of cells in the cortex, at the proximal end of the meristem, simultaneously undergo unequal elongation so that segments, or bands, of two small cells alternate with bands of four long cells. This pattern of growth is indicative of a rhythmic event at the base of the cortical portion of the meristem that generates the equivalent of node (N', short cells) and internode (I', long cells). It is as though the base of the meristem behaves according to algorithm (4):

$$\begin{aligned} &R \rightarrow [R^S V] \\ &R^S \rightarrow R \\ &V \rightarrow [V + oV-] \end{aligned} \tag{4}$$

In this case R^S would comprise most of the meristem, not just a small stem-cell zone. The derivative, $V+$, of a division represents a progenitor of the non-growing cells, $V-$ that of the growing cells. With one or two additional divisions, respectively, $V+ \rightarrow 2\ N'$ cells and $V- \rightarrow 4\ I''$ cells.

Other roots, particularly of ferns, show metameric, rather than phytomeric, segmentation [as defined earlier (see Fig. 1); in these cases the metamers correspond to merophytes]. Here, too, there may be segregation of + and – sites in the derivative cells which is expressed in different morphogenetic outcomes. For example, in both *Ceratopteris thalictroides* (Chiang and Gifford 1971) and *Marsilea quadrifolia* (Lin and Raghavan 1991), a certain cell within a metamer or merophyte, and within the endodermal lineage, grows isotropically; it forms the apical cell for a new lateral root primordium. Other cells in this same lineage within the merophyte grow anisotropically and do not develop in this way. Each merophyte is a derivative, V, of the apical cell, ${}_{E}R^{S}$. The new apical cell of the lateral root could represent most or all of the $V+$ site, while the other cells represent the $V-$ site. This clearly has echoes of the rhythms of leaf production (see algorithm 2) and could conform with arguments put forward that leaves and roots are at least comparable categories of plant organs (Arber 1941). However, it is possible that, in other ferns there may be more than one lateral root per merophyte. It would be interesting to determine the timing of their development: that is, whether one lateral root (the most distal) forms first, being located at the *V*+ sites, whereas those that occur more proximally are initiated later, "filling-in" the space between two successive *V*+ sites along an orthostichy of spirally arranged merophytes. Published evidence suggests that this does not occur and that lateral primodium initiation occurs in a strictly acropetal sequence (Charlton 1983), as expected of a one-lateral-per-merophyte origin. In-filling does, however, occur after auxin treatments (Charlton 1983), the neo-formed roots being type 7 or 8 adventitious root-borne roots. Similar arguments have been put forward for a regularity of lateral initiation at predetermined sites in cultured roots of a dicot (*Lycopersicon esculentum*) by Barlow and Adam (1988), though the rather short spacing between laterals may be an indication that the production of the organogenetic derivative, *V*, occurs towards the base of the meristem and is not an early event, as it is in fern merophytes.

5. CONCLUSION

Adventitious root is a term which, when examined closely, actually comprises a number of different categories of root types that may, in turn, have different anatomical origins determined by different organogenetic conditions. Often they are shoot-borne roots, but they could also be root-borne roots; in theory, there may be eight types of roots depending on the time and position of their origin. The problem of whether or not the embryonic or pole-borne root represents a category of root distinct from the shoot-borne roots is still unresolved. From a phylogenetic point of view, the embryonic root could well be shoot-borne in certain taxa (viz. the embryology and rhizogenesis of club mosses and ferns) and the same argument can be applied in the Angiosperms.

Emphasis is placed on the root as part of a phytomeric unit, which is deemed to be an autonomous modular unit of plant construction, phytomers being particularly well manifested in Gramineae and Equisitaceae, for example. Phytomers are often not so evident in dicotyledonous Angiosperms, but this may depend on the extent to which the leaf base encircles the stem. Just as the inherent shoot-borne nature of the root has become incorporated into the embryonic stages of dicot development, so the segmented phytomer is likewise incorporated into the over-arching development of a more completely integrated shoot organ. At other stages of phylogeny, the incorporation of one morphological unit into another is also evident from the equivalence of merophytes and metamers of mosses, for example, and, at a later stage, with the incoporation of the metamer into the phytomer. In all cases, the one common principle of plant development — branching — is retained.

The plant is thus viewed as a set of branching, caulescent and radicant axes, the roots usually being located at predictable sites on a phytomerically sub-divided stem. Because of the inherent plasticity of primordia, roots may also lead back, developmentally speaking, to the shoot via a root-borne bud. On occasions when the integrity of the plant is broken, cellular totipotency is the fundamental condition that leads to a renewed cycle of plant development and branching. It is then that an adventitious origin of roots is most strongly marked.

ACKNOWLEDGMENTS

IACR receives grant-aided support from the Biotechnology and Biological Sciences Research Council of the United Kingdom. Support from Fundación Andes/CONICYT and Universidad Católica de Valparaíso, Chile, is also gratefully acknowledged.

REFERENCES

Arber, A., 1941, The interpretation of leaf and root in Angiosperms, Biol. Rev. 16:81.

Avery, G.S., Jr, Johnson, E.B., Addoms, R.M., and Thomson, B.F., 1947, "Hormones and Horticulture. The Use of Special Chemicals in the Control of Plant Growth," McGraw-Hill Book Co, New York and London.

Barlow, P.W., 1994a, From cell to system: repetitive elements in shoot and root development, in: "Growth Patterns in Vascular Plants," M. Iqbal, ed., Dioscorides Press, Portland, Oregon, p. 19.

Barlow, P.W., 1994b, Rhythm, periodicity and polarity as bases for morphogenesis in plants, Biol. Rev. 69: 475.

Barlow, P.W., 1996a, Stem cells and founder zones in plants - root organs, in: "Stem Cells," C.S. Potten, ed., Academic Press, London, p. 29.

Barlow, P.W., 1996b, Cellular patterning in root meristems: Its origin and significance, in: "Plant Roots. The Hidden Half," Y. Waisel, A. Eshel, and U. Kafkafi, eds, Marcel Dekker, New York, p. 77.

Barlow, P.W., and Adam, J.S., 1988, The position and growth of lateral roots on cultured root axes of tomato, *Lycopersicon esculentum* (*Solanaceae*), Plant Syst. Evol. 158:141.

Blazich, F.A., 1988, Chemicals and formulations used to promote adventitious rooting, in: "Adventitious Root Formation in Cuttings," T.D. Davis, B.E. Haissig, and N. Sankhla, Dioscorides Press, Portland, Oregon, p. 132.

Bonnett, H.T.,Jr, and Torrey, J.G., 1965, Chemical control of organ formation in root segments of Convolvulus cultured *in vitro*, Plant Physiol. 40:1228.

Čelakovský, L.J., 1901, Die Gliederung der Kaulome, Bot. Z. 59:70.

Charlton, W.A., 1983, Patterns and control of lateral root initiation, in: "Growth Regulators and Root Development, Monograph 10," M.B. Jackson, and A.D. Stead, eds, British Plant Growth Regulator Group, Wantage, p. 1.

Chiang, S.-H., and Gifford, E.M., Jr, 1971, Development of the root of *Ceratopteris thalictroides* with special reference to apical segmentation, J. Indian Bot. Soc. 50A:96.

Girardin, P., Jordan, M.-O., Picard, D., and Trendel, R., 1986, Harmonisation des notations concernant la description morphologique d'un pied de maïs (*Zea mays* L.), Agronomie 6: 873.

Guttenberg, H. von, Heydal, H.-R., and Pankow, H., 1954, Embryologische Studien an Monokotyledonen I. Die Enstehung der Primarwurzel bei *Poa annua* L., Flora 141:298.

Hoshikawa, K., 1969, Underground organs of the seedlings and the systematics of Gramineae, Bot. Gaz. 130:192.

Kawata, S., Yamazaki, K., Ishihara, K., Shibayama, H., and Lai, K.-L., 1963, Studies on root system formation in rice plants in a paddy, Proc. Crop Sci. Soc. Japan 32:163 (in Japanese with English summary).

Kurihara, H., Kuroda, T., and Kinoshita, O., 1978, Morphological bases of shoot growth to estimate tuber yields with special reference to phytomer concept in potato plant, Jap. J. Crop Sci. 47:690.

Lin, B.-L., and Raghavan, V., 1991, Lateral root initiation in *Marsilea quadrifolia*. I. Origin and histogenesis of lateral roots, Can. J. Bot. 69:123.

Onderdonk, J.J., and Ketcheson, J.W., 1972, A standardization of terminology for the morphological description of corn seedlings, Canad. J. Plant Sci. 52:1003.

Pellerin, S., Tardieu, F., and Tricot, F., 1991, Modelling root system architecture: Experimental data on maize root system geometry, in: "Plant Roots and their Environment," B.L. McMichael, and H. Persson, eds, Elsevier Science Publishers BV, Amsterdam, p. 620.

Picard, D., Jordan, M.-O., and Trendel, R., 1985, Rythme d'apparition des racines primaires du maïs (*Zea mays* L.) I. — Etude détaillée pour une variété en un lieu donné, Agronomie 5:667.

Poethig, R.S., 1990, Phase change and the regulation of shoot morphogenesis in plants, Science 250:923.

Randolph, L.F., 1936, Developmental morphology of the caryopsis of maize, J. Agric. Res. 53:881.

Schultz-Schultzenstein, K.H. von, 1861, Die Bedeutung der Verzweigung im Pflanzenreich, Flora (Jena) NR 19 (GR 44):273, 297.

Tomlinson, P.B., 1969, On the morphology and anatomy of turtle grass, *Thalassia testudinum* (Hydrocharitaceae). II. Anatomy and development of the root in relation to function, Bull. Mar. Sci. 19:57.

2

DETERMINATE PRIMARY ROOT GROWTH IN *Stenocereus gummosus* (Cactaceae)

Its Organization and Role in Lateral Root Development

Joseph G. Dubrovsky

Division of Experimental Biology
The Center for Biological
Research (CIBNOR), A. P. 128
La Paz, B.C.S., Mexico 23000

INTRODUCTION

The biology of root formation and development and plant adaptation to the natural environment are two closely related aspects of plant biology and have to be considered jointly. The environmental conditions, highly drastic for some species, may be common for others. Drought is one such, though complex, condition. Effects of high temperature (Taylor and Clowes, 1978; Barlow, 1987; Clowes and Wadekar, 1988, 1989; Francis and Barlow, 1988; Gladish and Rost, 1993) and water deficit (Sharp et al., 1988; Spollen and Sharp, 1991; Tomos and Pritchard, 1994) on meristem function and root growth have been studied mainly in mesophytes. How root formation and growth, in species experiencing drought most of the year, are organized and how the root meristem of these species responds to environmental conditions are fundamental questions. Such questions are important for better understanding of the cellular basis of root growth and development in plants and of adaptive strategies elaborated in the root meristem during plant evolution.

In preliminary experiments, I found some Sonoran Desert Cactaceae, after seed germination, form a short primary root with a growth pattern resembling determinate root growth. The nature of determinate growth and its significance have not been widely studied. This type of growth was described for some Pteridophytes (reviewed by MacLeod, 1991), in *Zea mays* (Varney and McCully, 1991), and in *Opuntia arenaria* (Boke, 1979). The determinate root growth represents a unique development path when meristematic cells divide only for a limited period of time and then differentiate. The study of the determinate meristem and its functioning can be useful for comprehension of apical meristem organization and of the maintenance of meristem integrity in seed plants.

In the Sonoran Desert, seedling establishment *a priori* has to be a rapid process. One can hypothesize determinate growth of a primary root in a desert plant may have an adaptive feature. It may have an ecological significance for rapid induction of lateral roots and for development of a branched root system important for seedling anchorage and for water and mineral absorption. This possibility will be considered here.

One interesting plant where features resembling determinate growth can be found is *Stenocereus gummosus* (Engelm.) Gibson & Horak. This is a widely distributed plant in the Sonoran Desert (Bravo-Hollis and Sánchez-Mejorada, 1978). It is a clonal plant that produces many branches rooted in the ground because of development of adventitious roots. The adventitious roots are well developed, can reach 4 cm in diameter, and penetrate the soil deeper than 60 cm (personal observations). However, the primary roots grow during a very short time. The primary root in this cactus is determinate in its growth pattern. My research was focused on the analysis of determinate root growth, on the timing of meristem activity, and on the analysis of the possible role of determinate growth in induction of lateral roots and in early seedling establishment.

DETERMINATE ROOT GROWTH AND ITS ORGANIZATION

Pattern of Determinate Growth

Primary roots of *S. gummosus* grow only for two days after the start of radicle protrusion (ASRP) (Fig. 1). On subsequent days, there were no changes in root length ($P>0.05$, Student's t-test). The rate of root elongation was at a maximum during the second day ASRP and then declined, being on average 3.4, 0.7, 0.4, and 0 mm d^{-1} during the 2nd, 3rd, 4th, and 5th days. Maximum root length averaged 5.9±0.5 mm (Mean±SE, $n=16$). This pattern of growth was observed under various experimental conditions: in Petri plates on filter paper and in soil, both moistened with 20% Hoagland's mineral solution, and at two temperatures (22° and 30°C). This growth pattern was not induced by lack of water and is a normal development path.

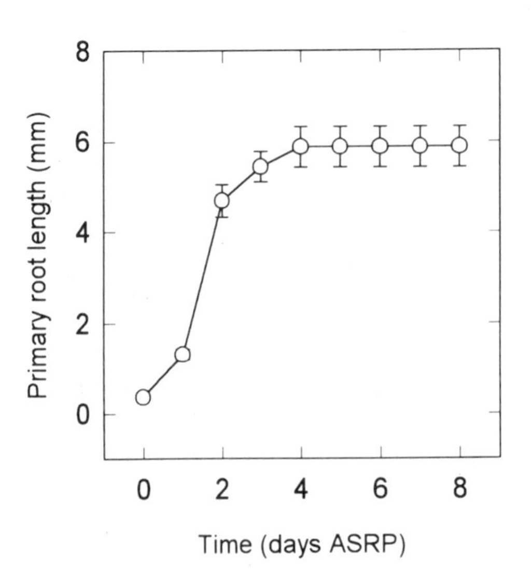

Figure 1. Determinate primary root growth in rapidly growing seedlings of *S. gummosus*. Seedlings were maintained in inclined Petri plates on filter paper moistened with 20% Hoagland mineral solution. Data are means ± SE (n=16).

Behavior of Meristematic and Nonmeristematic Elongation Zones

The distance from the youngest root hairs to the cap and root body junction is the length of the growing part of the root (the length of the meristem and the nonmeristematic elongation zone together). For the first 12 h ASRP, the length of the growing part of the root increased, did not change in the next 24 h, and then sharply declined (Fig. 2). The root-hair zone approached the tip of the root.

The relative meristem height (Rost and Baum, 1988) was determined on Feulgen-stained roots embedded in balsam. The relative meristem height was determined for the epidermis as the distance from the cap and root body junction to the proximal root portion where the epidermis nuclei can be seen in a file with the interval between neighboring cells in a cell file approximately equal or less than the diameter of the nuclei. During the 12 to 36 h period, the relative meristem height (Fig. 2) and the number of epidermal cells in the meristem (not shown) were unchanged, indicating this period can be considered as steady-state growth. On average, 13 meristematic cells in the epidermis were counted during this growth period. After 36 h ASRP, meristematic cells in some roots stopped division and started elongation. Because of this process, 32% and 62% of the roots had no typical meristem at 36 and 48 h ASRP.

When the meristem was not present, the distal cells of the root apex, exmeristematic cells, were on average twice as long as typical meristematic cells. Some of these cells occasionally were found in division. Sixty hours ASRP, all roots lacked the meristem.

Duration of Cell Division Cycle in the Root Apical Meristem

Twelve to 24 h ASRP in roots 1 to 2 mm long, the duration of the cell division cycle duration was estimated by the method of accumulation of metaphases (0.025% colchicine was applied), assuming exponential kinetics of cell proliferation (Webster and Macleod, 1980). The duration of the cell division cycle T was calculated as $T=\ln 2/k$, where k is the rate of accumulation of cells in metaphase. The calculated T was equal to 14.5 h.

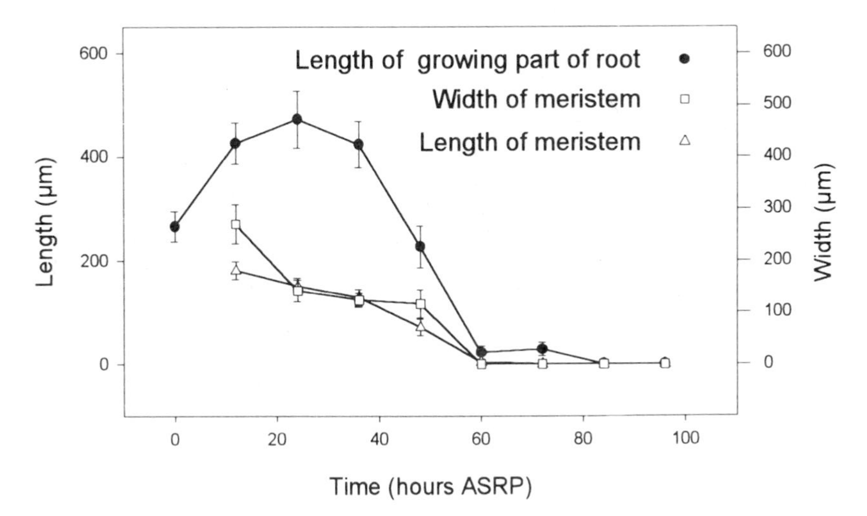

Figure 2. Changes in length and width of meristem and of the growing part of the root (meristem and nonmeristematic elongation zone) over time ASRP. The relative height (length) of meristem (Rost and Baum, 1988) was measured in epidermis on fixed, Feulgen-stained material embedded in balsam. Data are means ± SE (n=6 to 21).

Dynamics of Meristem Exhaustion

The root development described represents a case where meristem integrity was lost. In roots growing steadily, the number of cells in the meristem is relatively constant (Ivanov, 1981). In *S. gummosus*, the number of meristematic cells in the meristem was maintained at the same level up to 36 h, and then began to decrease sharply. In maize roots irradiated with X-rays in doses arresting cell division, the basal half of meristematic cells have left the meristem during a time approximately equal to the duration of one cell division cycle (Ivanov, 1981, 1994). If we assume at 36 h ASRP meristematic activity started to decline in *S. gummosus* roots, and at this time the basal half of meristematic cells began leaving the meristem, then, one can conclude the time, when the number of cells in a meristematic cell file has decreased to half of that at 36 h, is equal to the duration of the cell division cycle. Based on this analysis, the cell cycle duration has to be equal to 12 h (Fig. 3). In our case, the number of epidermis cells in the meristem during its exhaustion was close to that predicted by Ivanov's model of the life-span of cells in the meristem (Ivanov, 1994) when T was taken as 14.5 h (Fig. 3).

After cells leave the quiescent center, they go through a few cell division cycles in the meristem before they cease to divide (Barlow, 1976; Ivanov, 1981, 1994; Webster and Macleod, 1980). Meristem integrity is maintained both by the meristematic activity of cells in the meristem and by the presence and activity of a quiescent center or the initial cells. In the roots studied, meristematic activity was maintained for 48 h. Thus, on average, three cell division cycles in the meristem preceded its exhaustion. It is unclear whether the mechanism for the switchoff control in the meristem is related to a critical number of cell divisions in the meristem. Hypothetically, a limited period of meristematic activity during determinate growth can be explained by early exhaustion of the quiescent center, the initial cells, or their absence in the root. This question requires further investigation.

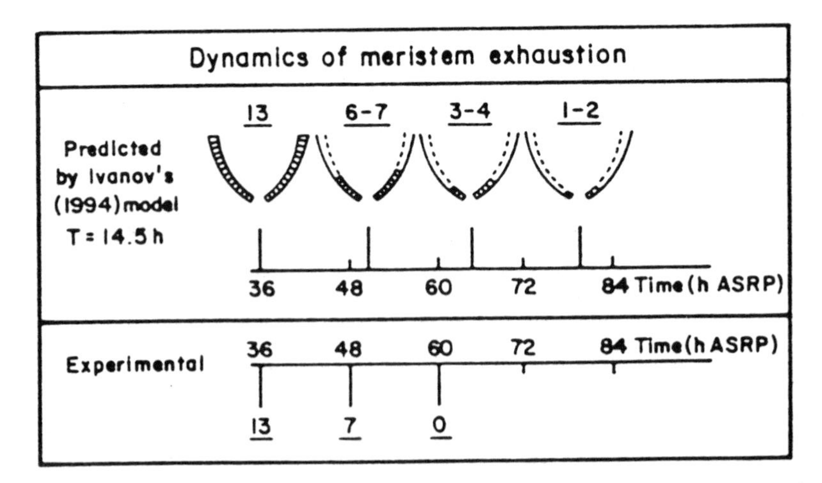

Figure 3. Theoretical and experimental dynamics of meristem exhaustion. The time when a decrease in the meristem length began was on average 36 h ASRP. Ivanov's model (Ivanov, 1994) of the life span of cells in the meristem suggests when division is arrested, during the time equal to one cell division cycle (*T*), the basal half of the meristematic cells leave the meristem. Upper box: dynamics of meristem exhaustion in accordance with the Ivanov's model when *T* was taken 14.5 h as determined. Lower box: dynamics of meristem exhaustion as obtained in the experiment. Underlined numbers are numbers of epidermal cells in a cell file predicted by the model and experimental data (means are given).

LATERAL ROOT FORMATION

Timing of Lateral Root Formation

The first lateral root primordia (LRPs) were recorded 24 to 36 h ASRP. Subsequently, there was an increase in the number of LRPs (Fig. 4). The first lateral roots appeared at 96 h. The time from primordia initiation to lateral root emergence in *S. gummosus* was 60 to 72 h, which was close to that of *Raphanus sativus* (Blakely et al., 1982), *Zea mays*, and *Phaseolus vulgaris* (Thompson and Macleod, 1981a), and shorter than that of *Pisum sativum* (Thompson and Macleod, 1981b) and *Vicia faba* (MacLeod, 1976). This comparison shows the ability of the xerophyte under study to have rapid root branching similar or faster than the mesophytes.

Relationship between Determinate Root Growth and Lateral Root Initiation

There was an inverse correlation between the presence of the meristem in a root and the initiation of LRPs in the primary root; the lower the percent of roots with apical meristem, the higher the percent of roots where LRP initiation started (Fig. 5). The number of LRPs increased with time. Two to four LRPs can be found in roots 72 to 96 h ASRP. At this time, there was no apical meristem found in the primary root, indicating a possible relationship between induction of LRP initiation and exhaustion of the meristem.

It is known the root tip affects lateral root formation in primary roots, possibly through production of an inhibitor that moves basipetally (Thimann, 1936; Böttger, 1974; Wightman and Thimann, 1980; Hinchee and Rost, 1986). The cessation of meristematic activity in the primary root apical meristem of *S. gummosus* apparently was directly related to the induction of lateral root primordia formation because: a) the time of lateral root initiation (Fig. 4) and of the beginning of the decrease in size of the meristem (Fig. 2) was concurrent and was 36 h ASRP and b) there was a correlation between the percentage of roots in the population that at the moment of fixation did not have a meristem because of the determinate growth, and the percentage of roots where lateral root primordia initiation commenced (Fig 5). Actively dividing meristematic cells may be a source of an

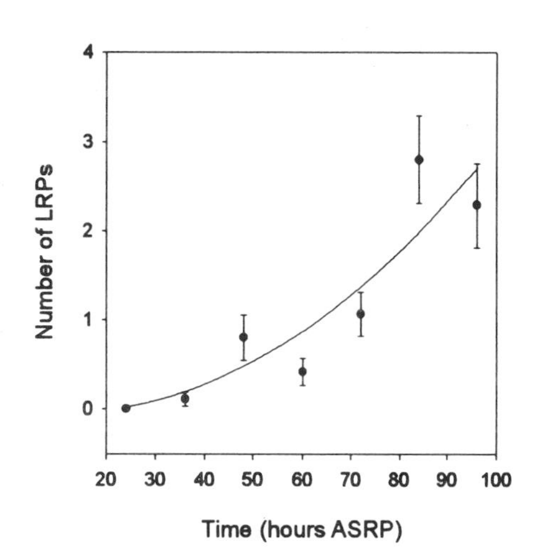

Figure 4. Number of LRPs initiated in primary root over time. LRPs were counted on fixed, Feulgen-stained material embedded in balsam. Data are means ± SE (n=10 to 20).

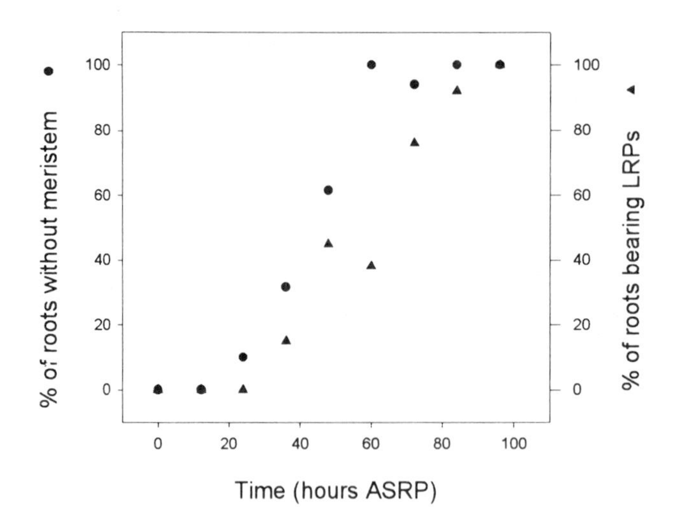

Figure 5. Changes with time in percentages of roots lacking the apical meristem because of differentiation of meristematic cells and of roots bearing LRPs.

inhibitor of lateral root initiation. After 36 h ASRP, when meristematic activity declined, lateral root primordia initiation was apparently promoted. Determinate primary root growth in this cactus species can be thus viewed as a physiological root tip decapitation which stops production of a signal inhibiting lateral root primordia initiation.

ECOLOGICAL SIGNIFICANCE OF THE DETERMINATE ROOT GROWTH

The determinate pattern of root growth demonstrated meristematic cells are programmed to pass through a few division cycles and then to be differentiated. The rapidity of the expression of this developmental program in the primary root suggests, under desert environmental conditions, roots, at any time, are prepared to stop their growth when conditions are inappropriate and to continue growth when conditions permit. The fast completion of growth even under optimal conditions can be an important energy saving mechanism. It is known growing roots have high respiratory costs (Nobel, 1988, 1991). Lateral roots have determinate growth and thus sympodial branching in a root system was found to be typical. The high rate of lateral root formation and early induction of lateral root primordia development associated with determinate growth of the primary root indicated high rates of root system development. Seed germination in an arid environment is usually associated with short wet periods. In *S. gummosus*, rapid formation of the sympodially branched system may be an important factor for successful seedling establishment. Interestingly, a similar feature was found in another cactus species *Opuntia arenaria* Engelm. Root branching was found also to be sympodial, and many short, less than 1 mm long, determinate lateral rootlets were found (Boke, 1979). It might be determinate root growth is a general feature of plants experiencing water deficits. It was shown that mesophytes, such as *Arabidopsis thaliana*, under drought stress are able to produce short roots with limited meristematic activity typical for determinate roots (Vartanian et al., 1994). Thus, this developmental program can be expressed in root meristem constitutively in xerophytes as in *S. gummosus*, or in mesophytes, can be induced by drought, as in *Arabidopsis thaliana* (Vartanian et al., 1994), or expressed under normal field conditions as in *Zea mays* (Varney and McCully, 1991).

CONCLUSIONS

1. Determinate primary root growth in *S. gummosus* is a normal developmental path implying meristematic cells are programmed to undergo a few cycles of division and to be differentiated;
2. The meristem exhaustion was directly related to the induction of lateral root primordia formation and can be considered as a physiological decapitation;
3. High rates of the expression of the determinate developmental program and of lateral root formation suggest the ecological importance of this growth pattern for early seedling establishment in an arid environment.

ACKNOWLEDGMENTS

The author is thankful to Dr. E. Glazier for his careful editing of the English manuscript, to Mr. O. Armendáriz-Ruíz for his help with the preparation of Fig. 3. The research was partially supported by Mexican Council for Science and Technology (CONACyT, Project 5277-N9407).

REFERENCES

Barlow, P. W., 1987, The cellular organization of roots and its response to the physical environment, *in*: "Root Development and Function", P.J. Gregory, J.V. Lake and D.A. Rose, eds., Cambridge University Press, Cambridge, p. 2.

Barlow, P. W., 1976, Towards an understanding of the behaviour of root meristem, *J. Theor. Biol.* 57:433.

Blakely, L. M., Durham, M., Evans, T. A. and Blakely, R. M., 1982, Experimental studies on lateral root formation in radish seedling roots. I. General methods, developmental stages, and spontaneous formation of laterals, *Bot. Gaz.* 143:341.

Boke, N. H., 1979, Root glochids and root spurs of *Opuntia arenaria* (Cactaceae), *Amer. J. Bot.* 69:1085.

Böttger, M., 1974, Apical dominance in roots of *Pisum sativum* L., *Planta* 121: 253.

Bravo-Hollis, H. and Sánchez-Mejorada, R., 1978, Las Cactáceas de México, Vol I. Universidad Nacional Autónoma de México, Mexico City.

Clowes, F. A. L. and Wadekar, R., 1988, Modelling of the root cap of *Zea mays* L. in relation to temperature, *New Phytol.* 108:259.

Clowes, F. A. L. and Wadekar, R., 1989, Instability in the root meristem of *Zea mays* L. during growth, *New Phytol.* 111:19.

Francis, D. and Barlow, P. W., 1988, Temperature and the cell cycle, *in*: "Symposia of the Society for Experimental Biology", p. 181.

Gladish, D. K. and Rost, T. L., 1993, The effects of temperature on primary root growth dynamics and lateral root distribution in garden pea (*Pisum sativum* L., cv. "Alaska"), *Env. Exp. Bot.* 33:243.

Hinchee, M. A. W. and Rost, T. L., 1986, The control of lateral root development in cultured pea seedlings. I. The role of seedling organs and plant growth regulators, *Bot. Gaz.* 147:137.

Ivanov, V. B., 1981, Cellular basis of root growth, *Sov. Sci. Rev.* D2:365.

Ivanov, V. B., 1994, Root growth responses to chemicals, *Sov. Sci. Rev.* D. Physicochem. Biol., 13: 1.

MacLeod, R. D., 1976, The development of lateral root primordia in *Vicia faba* L. and their response to colchicine, *Ann. Bot.* 40:551.

MacLeod, R. D., 1991, The root apical meristem and its margins, *in*: "Plant Roots. The Hidden Half", Y. Waisel, A. Eshel and U. Kafkafi, eds., Marcel Dekker, Inc., New York, p. 75.

Nobel, P. S., 1988, Environmental Biology of Agaves and Cacti. Cambridge University Press, New York, NY.

Nobel P. S., 1991, Ecophysiology of roots of desert plants, with special emphasis on agaves and cacti, *in*: "Plant Roots. The Hidden Half", Y. Waisel, A. Eshel and U. Kafkafi (Eds) Marcel Dekker, Inc., New York, p. 839.

Rost, T. L. and Baum, S., 1988, On the correlation of primary root length, meristem size and protoxylem tracheary element position in pea seedlings, *Amer. J. Bot.* 75:414.

Sharp, R. E., Silk, W. K. and Hsiao, T. C., 1988, Growth of the maize primary root at low water potentials. I.Spatial distribution of expansive growth, *Plant Physiol.* 87:50.

Spollen, W. G. and Sharp, R. E., 1991, Spatial distribution of turgor and root growth at low water potentials, *Plant Physiol.* 96:438.

Taylor, A. T. and Clowes, F. A. L., 1978, Temperature and the coordination of cell cycles within the root meristem of *Allium sativum* L., *New Phytol.* 81:671.

Thimann, K. V., 1936, Auxins and the growth of roots, *Amer. J. Bot.* 23:561.

Thompson, A. and Macleod, R. D., 1981a, Lateral root anlage development in excised roots of *Vicia faba* L., *Pisum sativum* L., *Zea mays* L. and *Phaseolus vulgaris* , L., *Ann. Bot.* 47:583.

Thompson, A. and Macleod, R. D., 1981b, Increase in size and cell number of lateral root primordia in the primary of intact plants and in excised roots of *Pisum sativum* and *Vicia faba*, *Amer. J. Bot.* 68:955.

Tomos, D. and Pritchard, J., 1994, Biophysical and biochemical control of cell expansion in roots and leaves, *J. Exp. Bot.* 45:1721.

Varney, G. T. and McCully, M. E., 1991, The branch roots of *Zea*. II. Developmental loss of the apical meristem in field-grown roots, *New Phytol.* 118:535.

Vartanian, N., Marcotte, L. and Giraudat, J., 1994, Drought rhizogenesis in *Arabidopsis thaliana*, *Plant Physiol.* 104:761.

Webster, P. L. and Macleod, R. D., 1980, Characteristics of root apical meristem cell population kinetics: a review of analyses and concepts, *Env. Exp. Bot.* 20: 335.

Wightman, F. and Thimann, K. V., 1980, Hormonal factors controlling the initiation and development of lateral roots. I. Sources of primordia-inducing substances in the primary root of pea seedlings, *Physiol. Plant.* 49:13.

3

THE ACQUISITION OF CELL FATE IN THE *Arabidopsis thaliana* ROOT MERISTEM

Claudia van den Berg,[1] Willem Hage ,[2] Viola Willemsen ,[1]
Nicole van der Werff ,[1] Harald Wolkenfelt,[1] Heather McKhann,[1]
Peter Weisbeek,[1] and Ben Scheres[1]

[1]Department of Molecular Cell Biology
University of Utrecht
Padualaan 8, 3584 CH Utrecht
The Netherlands
[2]Netherlands Institute for Developmental Biology
Uppsalalaan 8, 3584 CT Utrecht
The Netherlands

INTRODUCTION

During plant embryogenesis an embryo with cotyledons, a shoot apical meristem, a hypocotyl and a root apical meristem, is formed. The primary root and shoot meristems initiate post-embryonic growth generating all plant organs. The root meristem forms the primary root, and the shoot meristem forms the aerial portion of the plant including secondary meristems. Histological and fate map data have shown that there is no precise correlation between the shoot meristem cells and their descendants (Steeves and Sussex, 1989). This indicates that cell fate is flexible. In contrast, in the root a more strict relationship between differentiated cells and their meristematic ancestors is seen. Little is known about the mechanisms specifying cell fate in meristems.

Here, we focus on the cellular communication that is critical for the formation and functioning of the *Arabidopsis* root meristem. Due to its simple cellular pattern, the *Arabidopsis* root is a suitable system to study cell specification and communication. We have used laser ablations to study the flexibility of cells in the root meristem. Furthermore, we have analysed a number of mutations involved in embryonic as well as secondary and adventitious root formation. Taken together, these results show that root meristem initials learn their fate by positional information and that genes involved in cell specification first act early during embryogenesis.

Biology of Root Formation and Development, edited by Altman and Waisel.
Plenum Press, New York, 1997.

EXPERIMENTAL

Laser ablations were performed as described (van den Berg et al., 1995). Living cells could be visualised by the outlining of the cells by the fluorescent dye propidium iodide. Ablation of cells was performed by parking the unfiltered laserbeam (25 mW argon-ion laser, mrc-600, BioRad, Zeiss Axiovert) for 1 sec. on each cell. Successful ablations could be seen by the entering of the propidium iodide into the dead cells (see Fig. 2).

Root mutants were isolated by screening single siliques from 18,000 M1 plants resulting from EMS mutagenesis of dry seeds. Root meristem mutations were selected in the M2 generation. In this way lines were isolated with distortions in root growth or development. Mutants were grouped based on seedling phenotype and subjected to complementation analysis to test allelism.

Histological analysis was performed as described (Dolan et al., 1993; Scheres et al., 1994).

RESULTS AND DISCUSSION

Development of the Root during Embryogenesis

The *Arabidopsis* primary root has a highly regular cellular organisation. It consists of four main tissue types: an inner stele composed of vascular and surrounding pericycle cells, a ground layer of endodermal and cortical parenchyma cells, an outermost layer of epidermal and lateral root cap cells (Fig. 1b, 2a) (Dolan et al., 1993). The basal ends of each cell file are called the initial cells. They are highly constant in number and display regular division patterns. The initial cells divide, generating a new initial and a daughter, which can undergo further divisions and subsequent differentiation. The initials abut four nondividing cells, the quiescent centre, the function of which is unknown.

During the first division of the zygote an apical and a basal cell are generated. The apical cell forms the embryo proper, whereas the hypophysis and the suspensor are generated by the basal cell. The proximal initials are formed from the lowest tier of the embryo proper. The quiescent centre and columella root cap are generated by the hypophyseal cell.

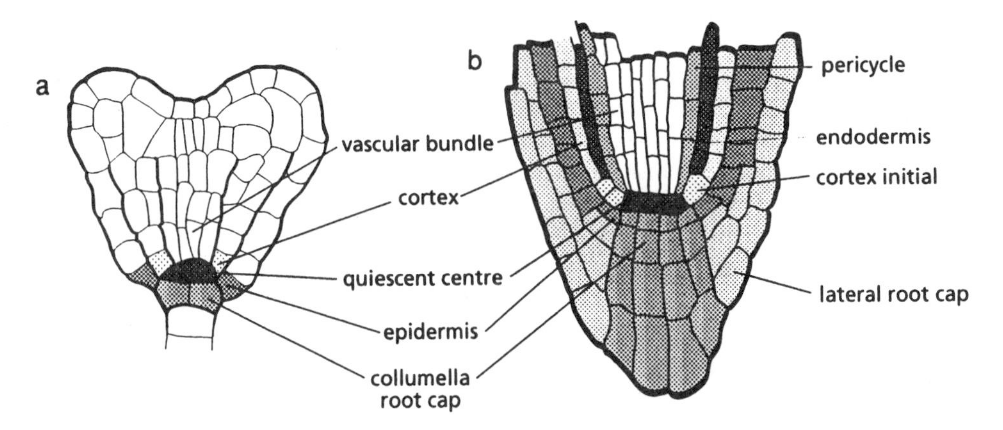

Figure 1. Fate map of the *Arabidopsis thaliana* root; heart stage embryo and seedling.

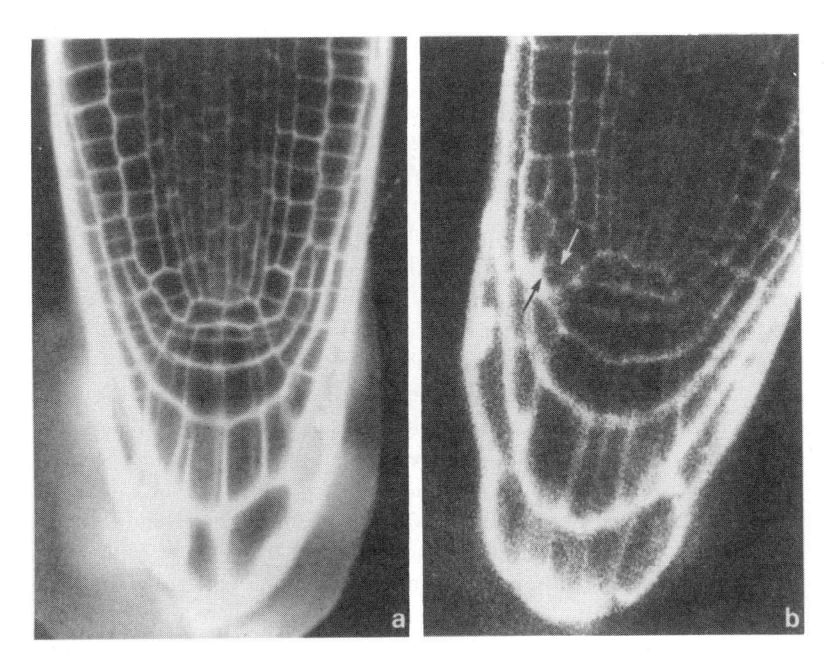

Figure 2. (a) Confocal laserscan of a seedling root stained with propidium iodine. (b) Ablation of an epidermal cell. The dead cell is pressed towards the outside of the root (black arrow) and a cortical cell invades its position (white arrow).

At the heart stage of embryogenesis the cellular organisation of the root apical meristem is completed (Fig. 1) (Scheres et al., 1994).

Communication of Cells in the Root Meristem: The Role of Positional Information

Fate map analysis of cells derived from the shoot apical meristem predicts that cell fate is highly flexible. Positional information rather than clonal origin is thought to regulate cell fate. In contrast, in the root there is a more rigid lineage relationship between meristematic cells and their descendants. This can be taken to suggest that clonal origin makes a contribution to fate determination. To test this in *Arabidopsis* roots, laser ablation can be applied. Cells killed in this way become rapidly flattened and their position is taken up by neighbouring cells (Fig. 2b). The fate of cells in a new position was subsequently studied.

Cell fate decisions in the radial dimension in the *Arabidopsis* root meristem were studied by ablating several proximal initials. After ablation of an epidermal initial, a cortical initial cell invaded (Fig. 2b). This cell formed lateral root cap cells, normally exclusively generated by epidermal initials indicating a switch in fate. Similarly, after ablation of the cortical initial, pericycle-derived cells switched fate; the invading pericycle cell performed the cortex-specific division pattern and generated an endodermis with a casparian strip (van den Berg et al., 1995).

From these experiments, we conclude that the fate of these cells is determined by their position. Furthermore, the signal(s) guiding the functioning of cells is (are) continuously present because switching of cell fates can be observed throughout early seedling development.

What Are the Source and Nature of Positional Information?

To determine the direction of positional signals, we examined further the development of cortical cells. The cortical initial divides first anticlinally generating a cortical daughter. This daughter then undergoes an asymmetric periclinal division forming an inner, smaller endodermal cell and an outer cortical cell (Fig. 3a).

If the cortical initial cell was isolated, by ablating all above daughter cells contacting this initial, the asymmetric division of the isolated cortical cell was prevented (Fig. 3b). This shows that, at least for cortical cells, signals derived from more mature cells of the same tissue type are responsible for fate determination of their initials.

It is clear that positional information acts to reinforce the generated pattern but can lineage still play a role? After all, in the absence of positional information, lineage might determine cell fate. When a single cortical initial was isolated from its daughters, its specification was hindered. This indicates that lineage is of minor importance in cell specification. We concluded that the fate of the proximal initials is instructed by their more mature daughters (Fig. 4a) (van den Berg et al., 1995). Thus, the root meristem initials cannot be seen as the creators of a specific pattern in the plant, whereby the generating cells "know" what to produce. They only conform to an already existing pattern, acting as a copying machine. More apical cells provide the information to copy this pattern.

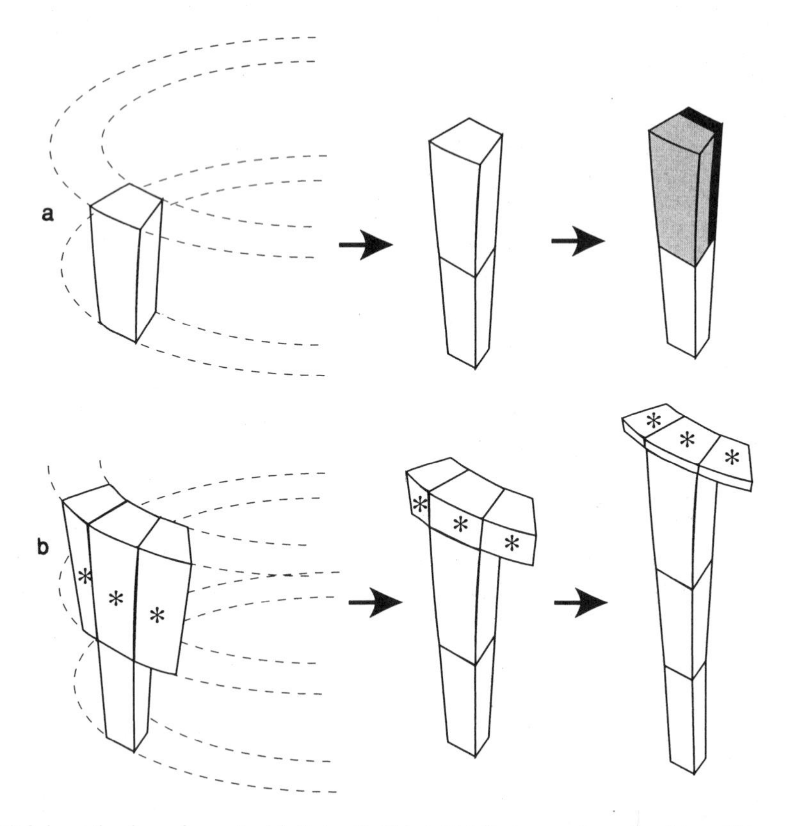

Figure 3. (a) Schematic view of a cortical initial cell. This initial first generates a daughter which performs an asymmetric division forming a cortical (light grey) and an endodermal (dark grey) cell. (b) Schematic representation of ablation of 3 cortical daughters (*) isolating the underlying initial cell. The isolated cell is then unable to perform periclinal divisions, and only anticlinal divisions are seen.

This signal can either act over a long range, providing information along the whole apical-basal axis, or locally in which case a specific group of cells acts as the information source. Ablation of cortical and endodermal cells, 3 cells above the initial, shows that invading pericycle cells still switch fate, indicating that the immediate cortical daughters directly abutting the initials are not the direct source of this positional information.

The most distal cells of the root (quiescent centre and columella) are programmed differently. In these cells, either the signals that determine the radial cell fates are not perceived, or overruling signals are present (Fig. 4b). By ablating the quiescent centre vascular-derived cells move into the former columella position. These cells switched fate and expressed a columella specific marker (a 35S B2-subdomain GUS fusion). In conclusion, the fate of cells along the apical-basal axis is also determined by positional signalling. The source of this signal is currently under investigation. Furthermore, mutants lacking parts of the apical-basal axis are providing more information on how these cell types are established and maintained.

The chemical nature and the method of transport of positional signals are unknown. Cellular components generated in more mature cells could be allocated to the initials via plasmodesmata. It has been demonstrated that at least one transcription factor, *KNOTTED-1*, can be transported from one cell to another (reviewed in Lucas, 1995). KNOTTED-1 is involved in keeping cells in an indeterminate state. Fluorescently labeled injected KNOTTED-1 rapidly moved out of the injected cell into surrounding cells thereby increasing the plasmodesmata size exclusion limit. It is possible that in the root meristem, cell fates are also determined by signals travelling via plasmodesmata. It has been shown that in the root meristem, epidermal cells can transport fluorescent dye into other cells of the epider-

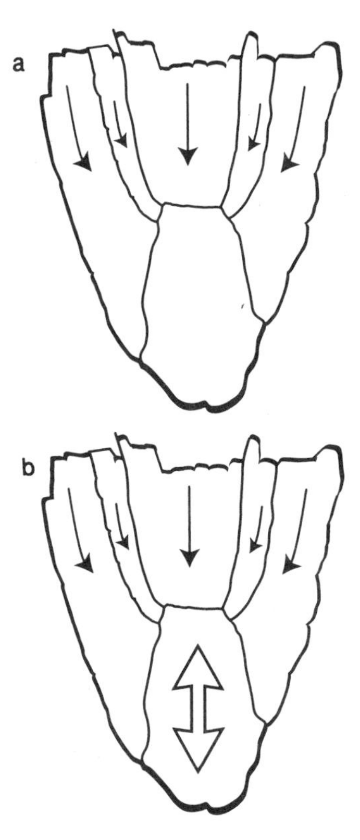

Figure 4. (a) Long range signals determining the fate of cells in the proximal meristem (arrows). (b) Short range positional signals which determine quiescent centre and columella fates (large arrow).

mal cell layer but not into cells in adjacent layers (Duckett et al., 1994). The distribution of plasmodesmata in cell layers could thus be used to restrict the transport of inductive signalling molecules.

Alternatively, laser ablation experiments in the *Fucus* embryo have indicated that the cell wall is involved in cell fate determination (Berger et al., 1994). Contact of one cell with a cell wall of a different cell type causes the cell to switch fate. The basal rhizoid cell in the 2 cell stage embryo was ablated, isolating the thallus cell. Rhizoid cell redifferentiation from the thallus cell was only observed when this cell was contacting the former rhizoid cell wall. If wall-to-wall contact was prevented, isolated thallus cells did not develop a rhizoid. In our ablation experiments the cell wall was not destroyed. This means either that the cell wall is not involved in cell specification, or that the signalling molecules located in the cell wall have a short half life and need to be continuously replenished.

The Function of the Quiescent Centre

The quiescent centre was first identified in maize roots as a group of cells in the centre of the root meristem showing a low rate of cell division (Clowes, 1958). In *Arabidopsis thaliana*, the quiescent centre consists of only four cells. The function of this group of cells has not been demonstrated directly. Barlow (1976) proposed that the quiescent centre cells serve as replacements for initial cells, thereby acting as a stem cell reservoir. Damaging of initial cells should cause the quiescent centre cells to reenter mitosis. It was shown that excision of the root cap in maize roots causes reactivation of quiescent centre cells to reform a complete root cap (Barlow, 1974).

In our experiments, after ablation of cortical or epidermal initials, the dead cells were always replaced by neighbouring initials and not by quiescent centre cells. Thus, apparently, the major function of the quiescent centre is not replacing damaged cells. However, when columella cells were ablated, quiescent centre cells did replace the dead cells. Furthermore, in older roots, quiescent centre cells have been reported to divide occasionally, generating additional cortical cell layers (Rost et al., 1996). Thus quiescent centre cells can replace neighbouring cells, as has been suggested, but no exclusive function for the quiescent centre can be proposed.

A Genetic Approach to Understand Root Development

We used a second, genetic, approach to obtain more information concerning signals involved in root formation and functioning. A number of mutants have been isolated which lack pattern elements in the apical-basal axis of the embryo (Mayer et al., 1991). For example, mutations in the *monopteros* (*mp*) gene interfere with the formation of both the root and hypocotyl during embryogenesis (Berleth and Jürgens, 1993). Cells of the lowest tier of the embryo proper and of the hypophysis display abnormal division patterns from the octant stage onwards. This leads, in the strong mutant phenotype, to a seedling lacking all basal structures.

During wild type embryogenesis, the basal part of the embryo is subdivided into a root and hypocotyl. The root is further subdivided into a prospective embryonic and a meristematic root. We have isolated mutant seedlings which contain an embryonic root but which are directly impaired in the formation of a functional root meristem. Here we will discuss two of them, *hobbit* and *bombadil.* Both show their primary defect in the hypophyseal cell region, the prospective quiescent centre and columella.

Hobbit. Seedlings homozygous for strong *hobbit* (*hbt*) alleles display no root meristem activity at all. They show abnormal root meristem anatomy with regard to both cell shapes and number. Based on anatomy, the quiescent centre and columella root cap region is predominantly affected. We investigated whether these cell types are still present by using cell type specific markers. In wild type plants the more mature cells of the columella contain starch granules. In strong *hbt* mutants these are absent. Moreover, a specific GUS fusion, normally expressed in the root cap, shows no expression in *hbt* seedlings. These results show that the specification of columella cell fate is altered (Fig. 5b). In contrast, a root meristem-specific marker (em101; Topping et al., 1994) is expressed in *hbt* seedlings, showing normal basal cell fate specification (Fig. 5b). We conclude that although the identity of this region is not changed, its cellular specification is lost.

Anatomical analysis of *hbt* embryos shows that the hypophyseal cell undergoes aberrant divisions from early globular stage onwards (Fig. 5b). Furthermore, the periclinal divisions in the epidermal initials that generate the lateral root cap cell are mostly absent. Marker gene expression in *hbt* indeed confirmed that no functional lateral root cap is formed. The *HBT* gene is not only involved in root formation during embryogenesis. Roots generated from homozygous *hbt* seedlings show a similar phenotype and quickly arrest development, showing that *HBT* is involved in all developmental pathways of root formation. *HBT* maps on chromosome 2 and we are currently isolating the corresponding gene.

Bombadil. Besides *HBT*, other genes are involved in root formation including *BOMBADIL* (*BBL*). *BBL* maps on the upper arm of chromosome 3 and we have isolated 2

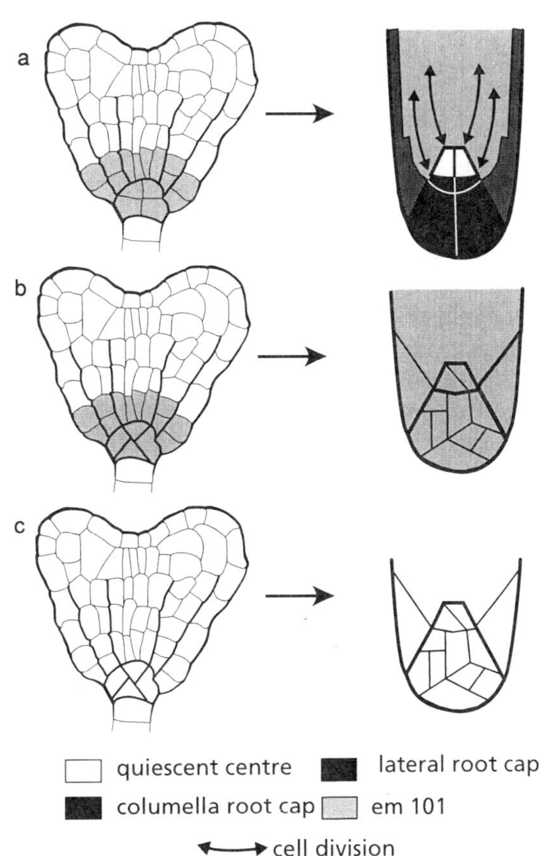

Figure 5. Schematic representations of heart stage embryos and seedling of (a) wild type; (b) *hobbit* and (c) *bombadil*.

alleles to date. Within each allele, the seedling phenotype of *bbl* is variable and two phenotypic classes can be distinguished. The strongest phenotype shows a very short root, a short hypocotyl and small closed cotyledons. The weaker phenotype has a longer root and hypocotyl and larger open cotyledons. The root mostly shows a hook. *bbl* seedlings show a similar phenotype as *hbt* seedlings in that they are also mainly affected in the specification of quiescent centre and columella root cap. *bbl* mutant embryos show more divisions in cells derived from the hypophyseal cell from early stages onwards (Fig. 5c). Later during development, aberrant divisions are seen in other parts of the embryo. In contrast to *hbt*, *bbl* seedlings show no em101 expression (Fig. 5c). This shows that both the regional identity and the cellular specification are lost.

CONCLUSIONS

Based on anatomical data and fate map analysis, the *Arabidopsis* root meristem shows a very rigid cellular organisation. Laser ablation studies show, that cells in the meristem are highly flexible and fate is most likely completely determined by positional information. Furthermore we showed that initial cells learn their fate from more mature cells of the same tissue type. Where exactly this signal is generated is not known. It is likely that all cells in the meristem are able to respond to positional cues. This leads to a model in which a long range signal with an apical to basal direction determines the fates of the different cell layers in the radial dimension (Fig 4a). However, in the most distal part of the root, other cell types are present (quiescent centre and columella). In these cells, fates are also determined by position. We think that these signals act locally, overruling the radial signals (Fig. 4b). We are currently performing laser ablation experiments of cells in this region to determine the direction and source of signals. It will be interesting to see what the fate of the columella and quiescent centre will be if this local signal is removed. It remains to be seen whether the mutants we study lack the same signals. If this is the case, we could conceive that similar positional signals are important for cell fate both during embryogenesis, to pattern the root, and at seedling stage to maintain this pattern. Answering the question whether meristems utilise one or more systems to confer positional information will be a major future challenge.

REFERENCES

Barlow, P., 1974, Regeneration of the cap of primary roots of *Zea mays*, *New Phytol.* 73: 937.

Barlow, P., 1976, Towards an understanding of the behaviour of root meristems, *Theor. Biol.* 57: 433.

Berger, F., Taylor, A., and Brownlee, C., 1994, Cell fate determination by the cell wall in early *Fucus* development, *Science* 263: 1421.

Berleth, T., and Jürgens, G., 1993, The role of the monopteros gene in organising the basal body region of the *Arabidopsis* embryo, *Development* 118: 575.

Clowes, F. A. L., 1958, Development of quiescent centres in root meristems, *New Phytol.* 57: 85.

Dolan, L., Janmaat, K., Willemsen, V., Linstead, P., Poethig, S., Roberts, K., and Scheres, B., 1993, Cellular organisation of the *Arabidopsis thaliana* root, *Development* 119: 71.

Duckett, C., Oparka, K., Prior, D., Dolan, L., and Roberts, K., 1994, Dye-coupling in the root epidermis of *Arabidopsis* is progressively reduced during development, *Development* 120: 3247.

Lucas, W. J., 1995, Plasmodesmata: intercellular channels for macromolecular transport in plants, *Current Opinion in Cell Biol.* 7: 673.

Mayer, U., Torres Ruiz, R., Berleth, T., Misera, S., and Jürgens, G., 1991, Mutations affecting body organisation in the *Arabidopsis* embryo, *Nature* 353: 402.

Rost, T.L., Baum, S.F., Nichol, S., 1996, Root apical organization in *Arabidopsis thaliana* ecotype 'ws' and a comment on root cap structure, *Plant and Soil* 187: 91.

Scheres, B., Wolkenfelt, H., Willemsen, V., Terlou, M., Lawson, E., Dean, C., Weisbeek, P., 1994, Embryonic origin of the *Arabidopsis* primary root and root meristem initials, *Development* 120: 2475.

Scheres, B., DiLaurenzio, L., Willemsen, V., Hauser, M.-T., Janmaat, K., Weisbeek, P., and Benfey, P.N., 1995, Mutations affecting the radial organisation of the *Arabidopsis* root display specific defects throughout the embryonic axis, *Development* 121: 53.

Steeves, T.A. and Sussex, I.M., 1989, *in* "Patterns of plant development" 2nd edn, Cambridge University Press, New York, p. 76.

Topping, J.F., Agyeman, F., Henricot, B., Lindsey, K., 1994, Identification of molecular markers of embryogenesis in *Arabidopsis thaliana* by promoter trapping, *Plant J.* 5: 895.

van den Berg, C., Willemsen, V., Hage, W., Weisbeek, P., and Scheres, B., 1995, Cell fate in the *Arabidopsis* root meristem determined by directional signalling, *Nature* 378: 62.

4

ROOT SYSTEM RESTORATION FOLLOWING ROOT PRUNING OF *Acacia senegal* AND ITS ANALYSIS BY MEANS OF AN ELEMENTARY PETRI NET

Beatriz Palma[1] and Peter W. Barlow[2]

[1]Instituto de Biología
Universidad Católica de Valparaíso
Casilla 4059, Valparaíso
Chile
[2]IACR—Long Ashton Research Station
Department of Agricultural Sciences
University of Bristol
Long Ashton, Bristol BS18 9AF
United Kingdom

1. INTRODUCTION

Acacia senegal is an economically valuable tree species growing in the arid and semi-arid Sahel grassland of northern Africa. Not only is it a protein-rich forage crop, but it is also the sole source of gum arabic, a commercially valuable water-soluble polysaccharide. Natural regeneration of *A. senegal* is compromised by over-grazing, fires, drought, *etc.*, and so it has been necessary to implement re-forestation programmes in order to ward off the continual threat of desertification. In this context, it is fortunate that *A. senegal* has a root system that (a) stabilizes soil structure and (b) enriches the soil with symbiotically-fixed nitrogen. However, during the replanting of young trees, the main root easily becomes damaged. Although a new root system does eventually become re-established, this takes time, leaving the plantlets vulnerable to unfavourable conditions and hence making recovery of the ecosystem less certain.

The regenerative potential of the root systems of seedlings and plantlets of *A. senegal*, the two main sources of new trees in re-forestation programmes (Palma 1990), has been evaluated by subjecting them to experimental amputation (Palma et al. 1994). The morphology of the initial root system and that of the new systems formed in response to amputation were sufficiently varied to permit the application of a formal approach — in this case by means of Petri nets — to provide a coherent framework for the experimental

Biology of Root Formation and Development, edited by Altman and Waisel.
Plenum Press, New York, 1997.

results obtained. This approach has been shown to work well in other regenerating plant systems (Lück et al. 1983, Lück and Lück 1991).

2. MATERIALS AND METHODS

The plant material was of two types, seedlings and plantlets. Seedlings of *Acacia senegal* (L.) Willd. (Papilionaceae) were obtained following scarification of the seed testa in conc. H_2SO_4 for 14 min, as described by Vogt and Palma (1991), with subsequent germination at 30°C in a greenhouse. Plantlets were raised following rooting of excised branchlets in "Woody Plant Medium" (Lloyd and McCown 1980) supplemented with 3% sucrose and 1.5 mg/l indole butyric acid. They were then transplanted to vermiculite/soil mixture and placed in a greenhouse.

When the tap roots of either seedlings or plantlets were 10–12 cm long, the distal portion was amputated either 4 cm from the root base or 1.5 cm from the root tip. The remaining plants were then transferred to minirhizotrons (Riedacker 1974) where the regrowth of the root systems was observed over the next 24 days.

3. OBSERVATIONS ON THE ROOT SYSTEM AND ITS RESTORATION

3.1. The Seedling Root System

3.1.1. The Normal Condition. During the immediate post-germination period, there was growth and development of first and second order axes (tap root and lateral roots). At the end of this period, it was evident that, although all tap roots had extended similarly

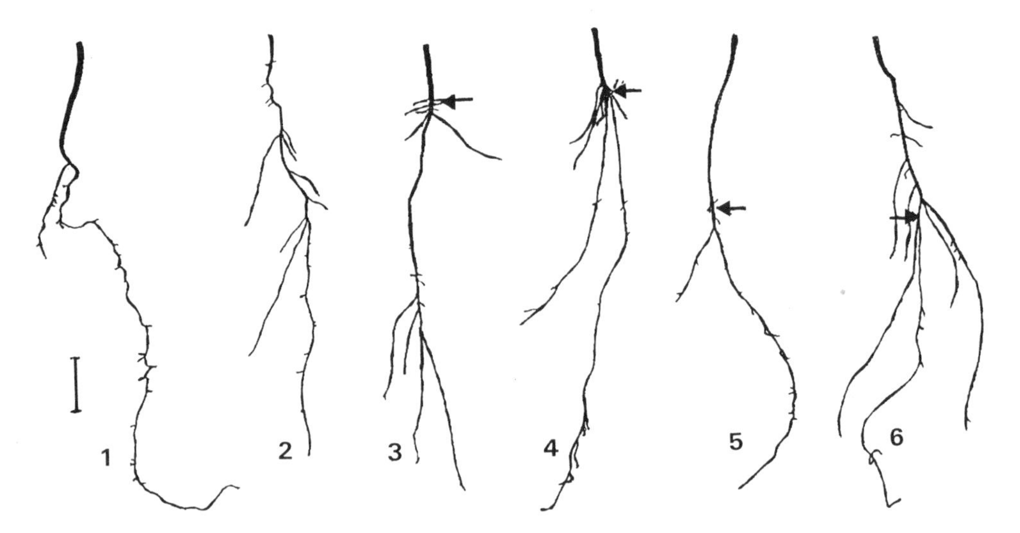

Figures 1–6. Drawings of root systems of *Acacia senegal* developed by seedlings. The systems are either intact (Figs 1 and 2) or have recovered following an amputation at a position in either a pre-formed (Figs 3 and 4) or a post-formed (Figs 5 and 6) region of root. Further descriptions of the root systems are found in the text. Scale bar = 10 cm.

and were 50 cm or more long, the root systems had developed in one of two ways; i.e. the root system at this stage was dimorphic. The first type of system, System 1 (Fig. 1), was found to consist of a pre-formed portion of tap root (i.e. the portion of radical that was formed in the ungerminated seed) which lacked lateral roots, and a post-formed portion of root (i.e. formed after germination) which bore short, diagravitropic laterals. In the second type of system, System 2 (Fig. 2), the pre-formed tap root bore short laterals and the post-formed portion supported longer, plagiogravitropic laterals.

3.1.2. Restoration of the Root System after Amputation near the Root Base. When the tap root of System 1 was amputated at 4 cm, the excision was made into the pre-formed portion. The system which regenerated over the 24-d period conformed to System 1. That is, the new root which grew from the excised stump bore short laterals and the original, pre-formed portion of root continued not to form any lateral roots. More frequently, however, the new roots which developed conformed to System 2 (Fig. 3). The new root growing from the excised stump of the tap root bore lateral roots of the longer, plagiogravitropic type. Moreover, lateral roots emerged from the formerly unlateralized, pre-formed portion of tap root.

When a tap root of System 2 was amputated in the same way, the new system which developed was either of System 1 (Fig. 4) or of System 2, though more frequently it was the latter system that was restored.

3.1.3. Restoration after Amputation Near the Root Tip. When tap roots of either System 1 or System 2 were amputated near the tip, the regenerated portion of root always conformed to System 1 (Figs 5 and 6).

3.2. The Plantlet Root System

3.2.1. The Normal Condition. All root systems which developed on plantlets appeared to conform to System 2, although the laterals still tended to be quite short at the end of the 24-d growth period (not shown), but they were plagiogravitropic, nevertheless.

3.2.2. Restoration of the Root System after Amputation. After either type of amputation (near the base or near the tip), the restored portions of root seemed to conform to System 2. In some cases, however, the restored system after amputation near the tip was comprised of short horizontal laterals which would indicate that it was System 1; it is possible that these could be laterals that had not yet made much growth during the 24-d period.

4. FORMALIZATION OF ROOT SYSTEM DEVELOPMENT AND RESTORATION

In order to help understand the response of the root system of *A. senegal* to pruning, it is useful to have some conceptual framework wherein the experimental findings can be given some coherence. An elementary Petri net (Fig. 7) seems suitable for this purpose. Such a device, moreover, may bring insights into the conditions necessary for the development of the two types of root system.

Details of Petri net construction and application have been presented elsewhere (Lück & Lück 1991, Lück et al. 1983, Barlow 1994). In brief, a Petri net charts the sequence of events within a deterministically developing system, the enablement of each event being

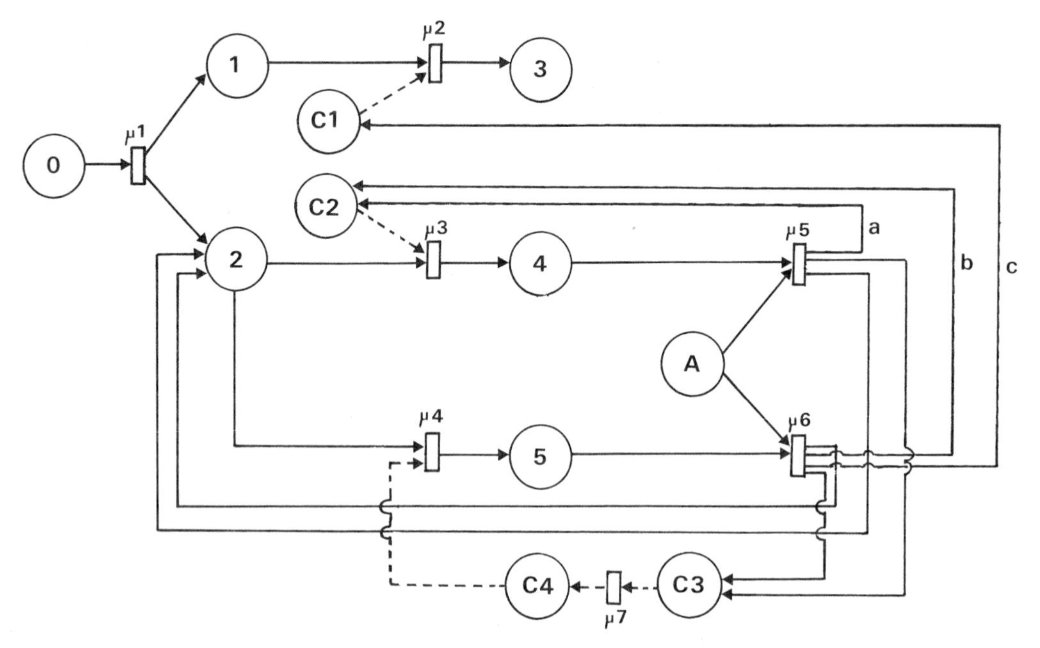

Figure 7. An elementary Petri net representing a parsimonious means of developing the various morphological configurations of *A. senegal* seedling root systems recovering or not from amputation in the pre-formed portion of root (see text). Details of steps in the sequential filling of places (circles) are presented in Table 1. An example of an initial marking of the net with "filled" places (c0, C4, A, each with ●) is shown (see Table 1.4) (though for clarity these places are more simply denoted as 0, 1,...,5 and A, respectively). Arcs denoted by continuous lines indicate connections to and from events (rectangles) and places with root morphological correlates; arcs denoted by broken lines reflect connections which may involve undefined, possibly physiological, correlates. Arcs a, b and c are invoked selectively (see Tables 1.3–1.8).

dependent upon the fulfillment of specific pre-conditions. The set of pre-conditions that enables one or more simultaneously occurring events defines the current "state" of the system. In the present context, each pre-condition corresponds to some identifiable morphological configurations of the root system (c0,c1,...,ci) and to some so-far unrecognised (perhaps physiological) properties (C1,C2,...,Ci). Pre-conditions are represented as "places" (or circles) in the Petri net shown in Fig. 7. An "event" (rectangles in Fig. 7, μ1,μ2,...,μi) is enabled only when the requisite pre-conditions are fulfilled. All the places, which represent these pre-conditions, are then "filled". Places and events are linked by "arcs".

Upon completion of an event, μ, new post-conditions arise which then become pre-conditions for a subsequent event. Hence, the system as a whole is represented as transforming from one morphological state to another, e.g. {c1,c2} transforms to {c3,c4}. The behaviour of the net is sequential: that is, when pre-conditions for an event are fulfilled, the event is then enabled and the post-conditional places become filled. In the present case, events are enabled immediately they are validated, but it is also acceptable to introduce delays before this enabling process occurs.

4.1. An Elementary Petri Net

It is necessary to identify a minimal set of morphological or physiological elements (which correspond to places or pre-conditions) within the root system of *A. senegal* with which to construct an appropriate Petri net. These are as follows:

c0. A pre-formed germinating radicle
c1. A 6-cm-long pre-formed portion of root lacking lateral roots
c2. An immature post-formed vertical root
c3. As c1, but now bearing short lateral roots
c4. A post-formed root bearing long lateral roots
c5. A post-formed root bearing short lateral roots
cA. Readiness for amputation
C1, C2, C3, C4. Physiological conditions which may (but not exclusively) accompany certain morphological conditions, such as those that follow amputation
a, b, c. Arcs invoked in certain circumstances

A parsimonious Petri net that accomodates all the developmental transformations of the *A. senegal* root system, with or without the amputations described below, is shown in Fig. 7.

4.2. Seedling Root Systems

At the stage of growth studied, the seedling root system of *A. senegal* (Figs 1 and 2) consisted of a pre-formed portion (places c1 or c3) that was derived from the embryonic radicle (c0), and a post-formed portion (c4 or c5) developed, *via* c2, from the pre-formed portion (Fig. 7). The two types of seedling root system are easily reached as end-points having the following places: System 1 = {c1,c5} and System 2 = {c3,c4}. Thus, to reach System 1 the pathway of state transformation is, for the root system, c0→{c1,c2}→{c1,c5}, and to reach System 2 the pathway is c0→{c1,c2}→{c3,c4}. However, the derivation of each System, following germination of the radicle, requires two different sets of initial conditions: {c1,c2, C4} for System 1 and {c1,c2,C1,C2} for System 2. This circumstance might suggest that these seedlings of *A. senegal* were of two types, each having slightly different rhizogenetic potentialities, perhaps as a result of different physiological correlates (implied by the necessity for the filling of places C1 and C2 in order for System 2 to be reached, and for the filling of C4 for the reaching of System 1). The complete sets of steps in the development of Systems 1 and 2, as required by the Petri net of Fig. 7, are given in Table 1.1 and 1.2.

4.3. Amputation within the Pre-Formed Portion of Root

Once the seedling root systems were developed, they were subjected to pruning. Their subsequent development was also explored and formalized by means of the Petri net shown in Fig. 7. Following an amputation (cA) in the pre-formed portion of the root, Systems that orginally conformed to types 1 or 2 tended to recover as System 2 (Table 1.4 and 1.6; Fig. 3); more rarely was there the reformation of System 1 (Table 1.3; Fig. 4). [Restoration of an entire System 1 from amputated System 2 is not possible, though see Table 1.6.] The Petri net indicates that, in order to derive an entire root System 2 from one that was originally System 1, amputation supplies a pre-condition which permits the refilling of places C1 and C2 following the invocation of arcs b and c from event μ6 (Table 1.4). In order to recover System 1 from the starting point of System 1, it is necessary either to invoke a different pre-condition supplied by amputation [e.g. to create a new place cA′ (not shown)], or to choose that arcs b and c are not invoked [b-, c-]. The latter option for net construction was preferred (Table 1.3) on the grounds of economy of places. In all, it is possible to reach six different end-points following amputation in the pre-formed portions

Table 1. The following parts of the Table show the workings of a Petri net (see Fig. 7) that permits development of the dimorphic root system derived from seedlings of *Acacia senegal* (Tables 1.1 and 1.2) and the subsequent restoration of root systems following amputation of the post-formed portion of tap root. The Petri net contains supplementary arcs labelled a, b and c (broken lines). Even when the requisite pre-conditions are fulfilled, these arcs are not always invoked. Each part of the Table therefore contains an indication of whether these arcs are invoked (+) or not (–) through the notation a+, a–, b+, b–, c+, or c–. There are circumstances where there is no opportunity for the use of these arcs because the necessary pre-conditions are not met. Here the notation is a0, b0 or c0

Step	Pre-conditions	Events	Post-conditions
1.1. Development of Root System 1 (Fig.1). [a0, b0, c0]			
1	0, C4	μ1	1, 2, C4
2	1, 2, C4	μ4	1, 5
1.2. Development of Root System 2 (Fig. 2). [a0, b0, c0]			
1	0, C1, C2	μ1	1, 2, C1, C2
2	1, 2, C1, C2	μ2, μ3	3, 4
1.3. Restoration of a complete Root System 1 following amputation in the Pre-formed portion of Root System 1. [a0, b–, c–]			
1	0, C4, A	μ1	1, 2, C4, A
2	1, 2, C4, A	μ4	1, 5, A
3	1, 5, A	μ6	1, 2, C3
4	1, 2, C3	μ7	1, 2, C4
5	1, 2, C4	μ4	1, 5

For the five other root system restorations, details of the Petri net are abbreviated to the initial pre-conditions (M_1) and final post-conditions (M_n) after n = S steps.

1.4. Restoration of a complete Root System 2 following amputation in the Pre-formed portion of Root System 1. M_1 = {0, C4, A}, M_4 = {3, 4, C4}, S = 4. [a0, b+, c+].
1.5. Restoration of a complete Root System 2 following amputation in the Pre-formed portion of Root System 2. M_1 = {0, C1, C2, A}, M_4 = {3, 4, C4}, S = 4. [a+, b0, c0].
1.6. Regeneration of a Post-formed portion of Root System 1 following amputation in the Pre-formed portion of Root System 2 (Fig. 4). M_1 = {0, C1, C2, A}, M_5 = {3, 5}, S = 5. [a–, b0, c0].
1.7. Regeneration of a Post-formed portion of Root System 2 following amputation in the Pre-formed portion of Root System 1 (Fig. 3). M_1 = {0, C4, A}, M_4 = {1, 4, C4}, S = 4. [a0, b+, c–].
1.8. Regeneration of a Post-formed portion of Root System 1 and then the continued development of the Pre-formed portion as Root System 2 following amputation in the Pre-formed portion of Root System 1. M_1 = {1, 2, C4, A}, M_5 = {3, 5}, S = 5. [a0, b+, c+].

of either Systems 1 or 2. Complete details can be shown for only one of them (Table 1.3); the remainder are presented in abbreviated form here (Table 1.4–1.8).

4.4. Amputation within the Post-Formed Portion of Root

The Petri net in a slightly more elaborate form (not shown) which maintains pre-existing post-formed portions by means of regenerative events, can also account for regeneration of new types of root systems when amputation is close to the tip of the post-formed tap root. The predominant type regenerated is System 1 (Figs 5 and 6). This occurs irrespective of the original type of System that was amputated and may indicate that amputation at a more distal level no longer supplies information which can modify the type of regenerative response.

5. CONCLUSIONS

It is easy to over-interpret the root restoration schemes depicted by the Petri net. For example, it is by no means certain that the C3 or C4 post-conditions that follow amputation and the enabling of event μ6 are the same as the pre-conditions that favour development of root System 1 (Table 1.1) or that, in certain cases, additional post-conditions (C1 or C2, or both) are similarly created by amputation (Table 1.4 and 1.5). However, the logic established by the net does suggest that some events analogous to those proposed may be occurring.

One assumption underlying the present Petri net is that root Systems 1 and 2 are distinct categories, and that System 1 is not simply an immature form of System 2. The latter seems unlikely, given that all seedlings examined had developed over a long period and that intermediate types of root systems were absent. It is possible that non-optimal environmental conditions (e.g., the root medium) had favoured the dimorphic root system and that dimorphism need not, therefore, be a characteristic inherent to the stock of seeds used (i.e there was some physiological or genetic divergence in the seeds). However, even if environmental effects were at work, there are clearly differences in the responses of roots to them (threshold effects). This again justifies different places (C1, C2, C4) from which to develop the two types of root system by means of the Petri net.

As in all modelling procedures, the process of modelling clarifies an understanding of the system under investigation (see Barlow and Zieschang 1994). The Petri net, because of its rigorous and deterministic structure, reinforces this view. The possibility of choice for arc utilization is, from the formal point of view, a weakness in the present net. It could possibly be over come by constructing an entirely different net, but this would require additional places. The flexibility of arc invocation may be thought to point to certain stochastic events in the root system during regeneration, but it is more likely that such events are reflections of unrecognised, additional pre-conditions.

The net of Fig. 7 is based on morphological conditions, though the places C1–C4 may also reflect physiological or cellular attributes of the system. These latter places would, in this case, represent a different level of biological organization. In principle, it is undesirable to mix levels in modelling procedures. The present Petri net may thus represent a point of cross-over in the modelling process, anticipating the passing from one level (that of organs) to the next lower level (that of cells) (see Barlow and Zieschang 1994, for further discussion of this aspect of modelling). A Petri net constructed at the cellular or physiological level would be feasible but might require a sub-net to represent the higher organ level. Already our net begins to suggest this extension of the modelling procedure.

ACKNOWLEDGMENTS

Support from Fundación Andes/CONICYT and Universidad Católica de Valparaíso, Chile, is gratefully acknowledged. IACR receives grant-aided support from the Biotechnology and Biological Sciences Research Council of the United Kingdom. We also thank Drs J. and H.B. Lück for their helpful remarks.

REFERENCES

Barlow, P.W., 1994, From cell to system: Repetitive elements in shoot and root development, *in*: "Growth Patterns in Vascular Plants," M. Iqbal, ed., Dioscorides Press, Portland, Oregon, p. 19.

Barlow, P.W., and Zieschang, H.E., 1994, Root movements: towards an understanding through attempts to model the processes involved. *Plant and Soil* 165:293.

Lloyd, G., and McCown, B., 1980, Commercially feasible micropropagation of mountain laurel, *Kalmia latifolia*, by use of shoot-tip culture, *Comb. Proc. Internat. Plant Propag. Soc.* 30:421.

Lück, J., and Lück, H.B., 1991, Petri nets applied to experimental plant morphogenesis, *Acta Biotheoret.*, 39:235.

Lück, J., Raoul, F., and Lück, H.B., 1983, Le déterminism de la ramification chez *Tradescantia fluminensis* à la lumière des réseaux de Petri. *in*: "Actes du Troisième Séminaire de l'Ecole de Biologie Théorique du CNRS, Arcachon, 30 May - 4 June 1983," J.-L. Gallis, ed., Université de Bordeaux, p.27.

Palma, B., 1990, "Contribution à l'étude de certains aspects de la multiplication de l'*Acacia senegal* (L.) Willd.," Thèse de Doctorat de l'Université Aix-Marseille III, France.

Palma, B., Vogt, G., and Neville, P., 1994, Comparison of root pruning systems of seedlings and plantlets of *Acacia senegal* (L.) Willd., *ΦYTON (Buenos Aires)* 55:137.

Riedacker, A., 1974, Un nouvel outil pour l'etude des racines et de le rhizosphère: le minirhizotron, *Ann. Sci. Forest.* 31:129.

Vogt, G.-F., and Palma, B., 1991, Influence de quelques produits désinfectants sur le pourvoir d'imbibition des graines d'*Acacia senegal* — Rôle des différents parties du tégument, *Phyton (Horn, Austria)* 31:97.

5

ROOT ARCHITECTURE EFFECTS ON NUTRIENT UPTAKE

Asher Bar-Tal,[1] Ruth Ganmore-Neumann,[1*] and Gozal Ben-Hayyim[2]

[1]Institute of Soils and Water
The Volcani Center
Agriculture Research Organization
Bet-Dagan 50250, P.O.B. 6, Israel
[2]Institute of Horticulture
The Volcani Center
Agriculture Research Organization
Bet-Dagan 50250, P.O.B. 6, Israel

INTRODUCTION

Root size and architecture are considered to form a major factor in nutrient uptake efficiency of plants (Fitter, 1991). Simulation of root growth and architecture has shown that root architecture affects the volume of the soil from which nutrients can be exploited (Fitter et al., 1987; Fitter et al., 1991). Models of nutrient uptake by roots showed that architectural characteristics may affect nutrient uptake. Barber and Silberbush (1984), using an uptake simulation model, demonstrated that nutrient uptake rate is dependent on root radius, length and density. Itoh and Barber (1983) improved the agreement between actual and predicted results by including root hairs in the model. The above information indicates that root architecture may affect nutrient uptake from the rhizosphere. However, there is very little quantitative knowledge on this effect and quantitative experimental data are vague. O'Toole and Bland (1987) found that cotton genotypes showing tolerance to water stress were characterised by long lateral roots, whereas, Petrie et al. (1992) did not find any significant effect of root architecture on water uptake from soil by mono- and di-cotyledonous species that differed in root branching and density. Eghball and Maranville (1993), comparing N uptake by two varieties of corn, showed that the one with longer and thinner roots was more efficient in N uptake. The difficulties in obtaining such quantitative data are: i. monitoring and measuring root parameters in vivo, ii. discriminating between root and shoot effects, and iii. estimating relative impact of morphological and functional effects on nutrient uptake. Changes in endogenous free polyamines (putrescine, spermidine and spermine) were found to be in correlation with root formation (Friedman et al., 1982; Burtin et al., 1990). Recently

* Deceased, August 1997.

Biology of Root Formation and Development, edited by Altman and Waisel.
Plenum Press, New York, 1997.

it was found that application of an inhibitor of putrescine synthesis, DFMO (α-DL-difluoromethylornithine), caused changes in phenotype development (Burtin et al., 1991) and altered the architecture of excised root system of tobacco (Ben-Hayyim et al., 1993). Using excised root simplifies the experimental system by avoiding shoot/root interaction; the biochemical effect is exerted specifically on root architecture without affecting nutrient demand and uptake characteristics of root unit, thus enabling one to quantify the effects of architectural parameters on nutrient uptake. However, tobacco roots grow very slowly, and so, successful application of the method to fast growing roots of another species, e.g., tomato, would improve the experimental system. There has been no study of the effect of DFMO on either the nutrient uptake or the development of excised tomato roots.

OBJECTIVES

In the present study we tested the usefulness of DFMO as a means for studying the effect of root architecture on nutrient uptake efficiency. The objective of the present research was to develop an experimental system for evaluating and quantifying the effect of the root architecture on nutrient uptake by tomato roots. The specific objectives were:

1. to study the effect of DFMO and N and P concentrations on growth and architecture of excised roots in a solution culture and an agar medium.
2. to quantify nutrient uptake by root unit with and without DFMO.

EXPERIMENTAL

The research included three steps: i. excised tomato roots were grown in continuously shaken solution culture in Petri dishes with and without DFMO; ii. excised tomato roots were grown in continuously shaken solution culture in Petri dishes with and without DFMO combined with various N and P concentrations; iii. excised tomato roots were grown in agar medium in Petri dishes, with and without DFMO and with two P quantities. P was applied to the agar medium by two methods: i. even distribution of P throughout the Petri dish; and ii. application of the same quantity of P to a spot, 8 cm away from the root. Roots growth took place in a controlled growth chamber. Root growth was monitored with a flat-bed scanner (HP IIIC) and analyzed with Delta-T software. Nutrient uptake was determined using chemical analysis of the plant material. Nutrient depletion from the solution was estimated by chemical analysis of the initial and final solution. Statistical analysis was carried out with JMP software of SAS.

RESULTS AND DISCUSSION

DFMO enhanced root elongation in solution cultures, relative to the control (Fig. 1). Since the elongation followed exponential curve (the fitted lines in the figure) the observed difference between treatments increased with time. DFMO also increased the number of lateral root tips relative to the control (Fig. 1), indicating that the effect of DFMO on the elongation was due to formation of more lateral roots. The effects of DFMO on both root length and the number of root tips after 8 days of growth were found to be significant at the 0.05 level according to Student's t test.

Significant interaction (0.05 level) of DFMO and N concentration on root length and number of root tips was obtained. The increase of root length promoted by DFMO application to solution cultures was obtained at relatively low N concentration (7.5 mM)

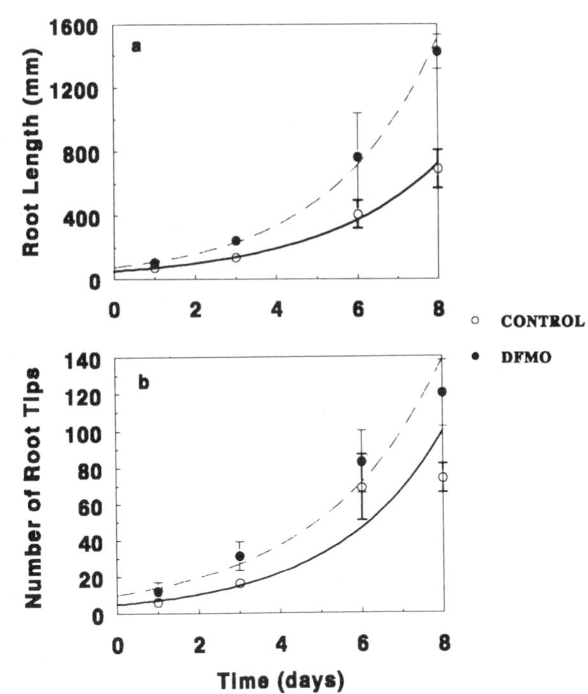

Figure 1. The effect of DFMO on (a) elongation of the root and (b) number of new root tips in a solution culture.

whereas no effect was found at higher N concentrations (Fig. 2). Similarly, an increased number of root tips through DFMO addition was obtained at low N concentrations, whereas an opposite effect was obtained at higher N concentrations (Fig. 2).

A significant interaction of DFMO with P concentration in their effect on root length and number of root tips was also obtained. DFMO application increased root length and

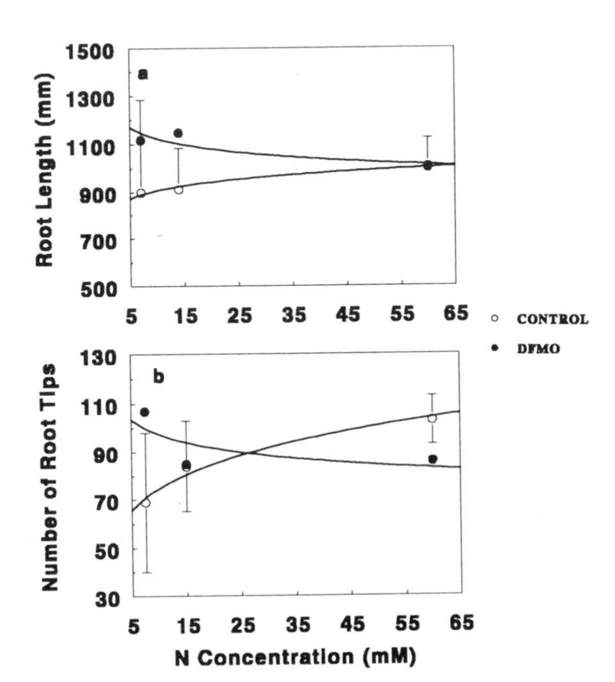

Figure 2. The effect of DFMO in combination with N concentration on (a) root length and (b) number of root tips in 8 days growth in a solution culture.

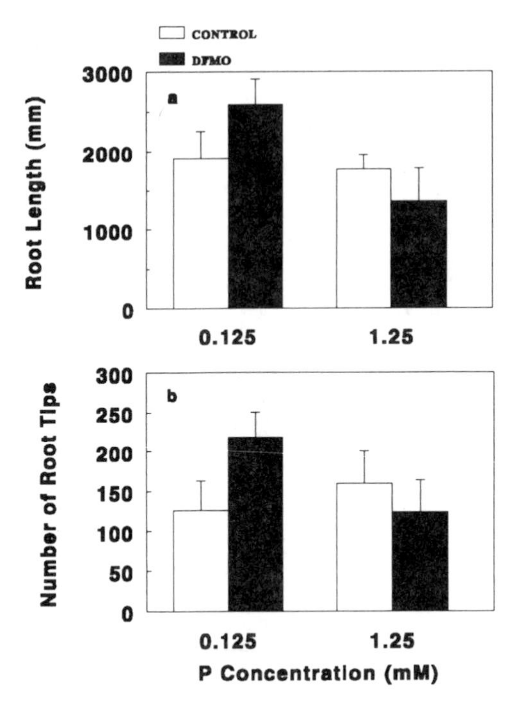

Figure 3. The effect of DFMO in combination with two P concentrations on (a) root length and (b) number of root tips in 8 days growth in a solution culture.

the number of root tips in solution culture containing a relatively low P concentration, 0.125 mM, whereas no effect was found at a higher P concentration, 1.25 mM (Fig. 3). These effects of DFMO in combination with low P on root length and on the number of tips were significants at the 0.05 level.

DFMO in combination with low P increased the ratio of root length to number of tips relative to the control (data not shown), indicating that DFMO also affected root architecture.

Root elongation (expressed as RGR, relative growth rate) was reduced by application of the high P quantity evenly to the agar medium (Fig. 4), similarly to the results in solution cultures. The opposite effect was obtained by concentrated application of the same quantities of P to a single spot (Fig. 4). One can explain this phenomenon by hypothesizing that root elongation as a function of P concentration follows a nonlinear curve with a maximum at a P concentration below 1.25 mM. This is higher than the actual concentration of P near the root when the low level of P was concentrated at a spot far away from it. This hypothesis was indeed supported by results of the analysis of the agar 7 days after the spot application of P. Application of DFMO to the agar medium in combination with the low P (spot treatment), increased root RGR, whereas no effect was obtained with the higher P treatment, similarly to results presented above for solution culture.

So far, we have shown that DFMO can serve as a means to alter root growth and architecture. However, its effect on nutrient uptake from solution culture and from a solid medium is still an open question. In an attempt to answer this question, nitrogen and phosphorus uptakes were estimated from the depletion of the nutrients from the solution or by direct analysis by tissue digestion. On the depletion method we found that DFMO had no significant effect on N and P uptakes from solutions; a good correlation was obtained between root length and N and P uptakes, independently of the presence of DFMO (Figs. 5 and 6, respectively).

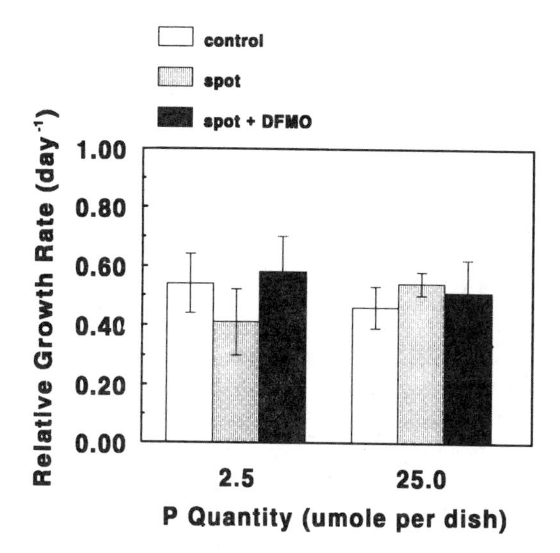

Figure 4. The effect of DFMO in combination with two P quantities and two methods of P application on root RGR (the relative growth rate) in an agar medium.

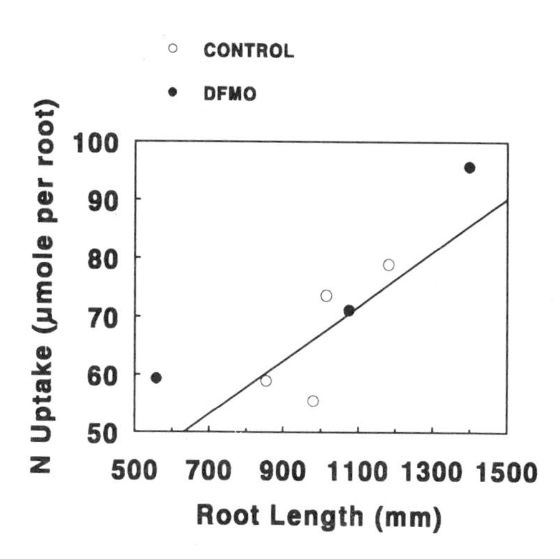

Figure 5. N uptake from 7.5 mM N solution as a function of root length, with and without DFMO application.

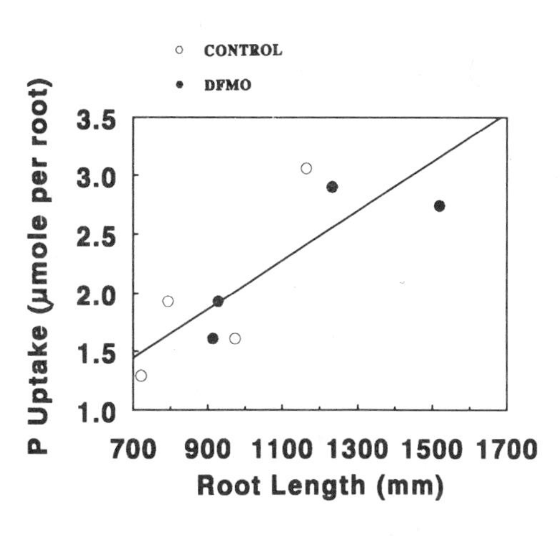

Figure 6. P uptake from 0.125 mM P solution as a function of root length, with and without DFMO application.

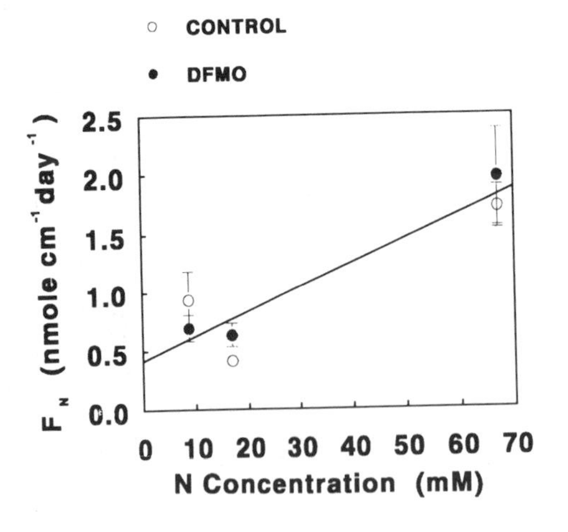

Figure 7. N uptake rate per root unit length, Fn, as a function of N concentration and as affected by DFMO application.

Increasing the N concentration from 7.5 to 65 mM enhanced Fn, N uptake rate per root length unit, from a shaken solution culture. This parameter was not affected by DFMO application, under three N concentrations (Fig. 7). Since DFMO had no effect on root radius, the results of uptake per root unit length or unit weight were similar.

Increasing P concentration from 0.125 to 1.25 mM enhanced Fp, P uptake rate per root unit length, from a shaken solution culture. DFMO application had no significant effect on Fp, with two P concentrations (Fig. 8). Since DFMO had no effect on root radius, the results of uptake per root unit length or root unit weight were similar.

The analyses of the roots were well correlated with the depletion measurements. The analysis of P content in the roots showed that DFMO had no effect on P uptake by the roots, and that P uptake from a 1.25 mM P solution was 9–10 times greater than from 0.125 mM P solution.

CONCLUSIONS

1. DFMO affected the development of excised tomato roots in solution cultures and agar media.

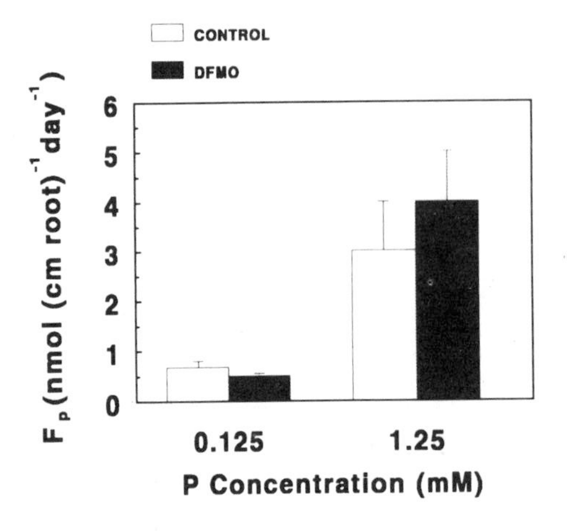

Figure 8. P uptake rate per root unit length, Fp, under high and low P concentration and as affected by DFMO application.

2. An interaction between DFMO and N and P concentrations was revealed. DFMO in combination with high N and P concentrations did not affect root growth.
3. DFMO did not affect the uptake rate of N and P per root unit length or weight of excised tomato roots.

Therefore, excised roots of tomato, with and without DFMO, can be used as a tool for studying the effect of root architecture on nutrient uptake from the soil.

ACKNOWLEDGMENTS

The present study was supported by the Chief Scientist of the Israeli Ministry of Agriculture, Grant 204032494.

REFERENCES

Barber, S.A. and Silberbush, M., 1984, Plant root morphology and nutrient uptake, in: S.A. Barber and D.R. Bouldin, (eds.) "Roots, Nutrients and Water Influx, and Plant Growth", ASA Special Publication Number 49. Madison, WI. p. 65–87.

Ben-Hayyim, G., Leach,F., Damon, J.P., Martin-Tanguy, J., and Tepfer, D., 1993, Biochemical and molecular genetic control of root system architecture, in: First International Symposium on the Biology of Adventitious Root Formation. Dallas, Texas, USA.

Burtin, D., Martin-Tanguy, J., Paynot, M., Carre, M. and Rossin, N., 1990, Polyamines, hydroxycinnamoyl putrescines and root formation in leaf explants of tobacco cultivated in vitro; effects of the suicide inhibitors of putrescines synthesis. Plant Physiol. 93:1398–1404.

Burtin, D., Martin-Tanguy, J., and Tepfer, D., 1991, α-DL-difluoromethylornitine, a specific, irreversible inhibitor of putrescine biosynthesis, induces a phenotype in tobacco similar to that ascribed to the root-inducing, left-hand transferred DNA of Agrobacterium rhizogenes. Plant Physiol. 95:461–468.

Eghball, B., and Maranville, J. W., 1993, Root development and nitrogen influx of corn genotypes grown under combined drought and nitrogen stresses. Agronomy J. 85:147–152.

Fitter, A.H., 1987, An architectural approach to the comparative ecology of plant root systems. New Phytol. 106:61–77.

Fitter, A.H., 1991, Characteristics and function of root systems, in: Y. Waisel, A. Eshel and U. Kafkafi (eds.) Plant Roots: The Hidden Half. Marcel Dekker, Inc., New-York, 3–25.

Fitter, A.H., Stickland, T.R., Harvey, M.L. and Wilson, G.W., 1991, Architectural analysis of plant root systems: 1. Architectural correlates of exploitation efficincy. New Phytol. 118:375–382.

Friedman, R., Altman, A. and Bachrach, U., 1982, Polyamines and root formation in mung bean hypocotyl cuttings. Plant Physiol 70:844–848.

Itoh, S. and Barber, S.A., 1983, Phosphorus uptake by six plant species as related to root hairs, Agron. J. 75:457–461.

O'Toole, J.C., and Bland, W.L., 1987, Genotypic variation in crop plant root systems. Adv. Agron. 41:91–145.

Petrie, C.L., Kabala, Z.J., Hall, E., and Simunek, J., 1992, Water transport in unsaturated medium to root with differing local geometries. Soil Sci. Soci. Am. J. 56:1686–1689.

6

ROLE OF THE ROOT APEX IN LATERAL ROOT DEVELOPMENT IN PINE SEEDLINGS

Nir Atzmon,[1] Oded Reuveni,[2*] and Joseph Riov[3]

[1]Institute of Field Crops
The Volacni Center
Bet-Dagan, 50250 Israel
[2]Institute of Horticulture
The Volacni Center
Bet-Dagan, 50250 Israel
[3]Faculty of Agriculture
The Hebrew University of Jerusalem
Rehovot, 76100 Israel

INTRODUCTION

The inhibitory effect exerted by the apical meristem of a growing root on the development of lateral roots (LR) is a well known phenomenon (Bottger 1974, Wightman and Thimann 1980). LR development consists of three distinct stages: initiation, emergence, and elongation. The pattern of LR development and their distribution along the primary root axis are thought to be controlled by gradients of inhibitory factors originating in the root apex and promoting factors originating in the aerial organs (Wightman et al. 1980). It has been suggested that the inhibitory factors are cytokinins (Torrey 1976, Van Staden et al. 1988) and possibly abscisic acid (Bottger 1974, Torrey 1976) and the promoting factors are auxins (Scott 1972).

The knowledge of the physiology of LR development in woody species characterized by a strong tap root is limited (Young 1990). Many forest tree species are typical tap root plants, in which much effort has been made to induce root branching in the early developmental stages for practical purposes (Geisler and Ferree 1984, Deans et al. 1990, Struve 1990). Previous studies on the control of LR development were performed primarily with excised roots grown in culture (Golaz and Pilet 1987, Blakely et al. 1988). Recently, there has been an increasing interest in root development in woody plants (Young 1990) but the knowledge of LR development in intact plants is still limited.

Atzmon et al. (1994) showed that in the case of the strong tap root system of pine seedlings, numerous LR are formed but only a few of them, about 10–15%, continue to

* Deceased, October 1997.

Biology of Root Formation and Development, edited by Altman and Waisel.
Plenum Press, New York, 1997.

elongate while the others degenerate. The aim of the present study was to determine the relationship between lateral root development and the distribution of endogenous indole-3-acetic acid (IAA) and cytokinin-like substances along an intact tap root and after removal of the tap root tip in *Pinus pinea* seedlings.

EXPERIMENTAL

Young pine (*Pinus pinea* L.) seedlings having fully expanded cotyledons and about 5 cm long tap root were transferred to an aerated nutrient solution culture (0.5 Hoagland nutrient solution) in 7.5-liter containers and were grown there for 4 weeks. The experiments were done in a growth chamber under the following conditions: 16 h light and 25/20°C light/dark. Photon flux density of 130 μ Einstein $m^{-2}s^{-1}$ was measured at the cotyledon level.

Intact seedlings and seedlings whose tap root apex was removed at different periods were examined. The tap root of all seedlings was divided into 4 morphological sections as described in the figures, and each section was analyzed separately.

IAA was extracted and partially purified by a mini-column procedure developed by Thompson et al. (1981). The extract was further purified by HPLC, and IAA was quantified by ELISA according to Sagee (1986). Cytokinin-like activity was determined by the soy-bean callus bioassay (Van Staden 1976).

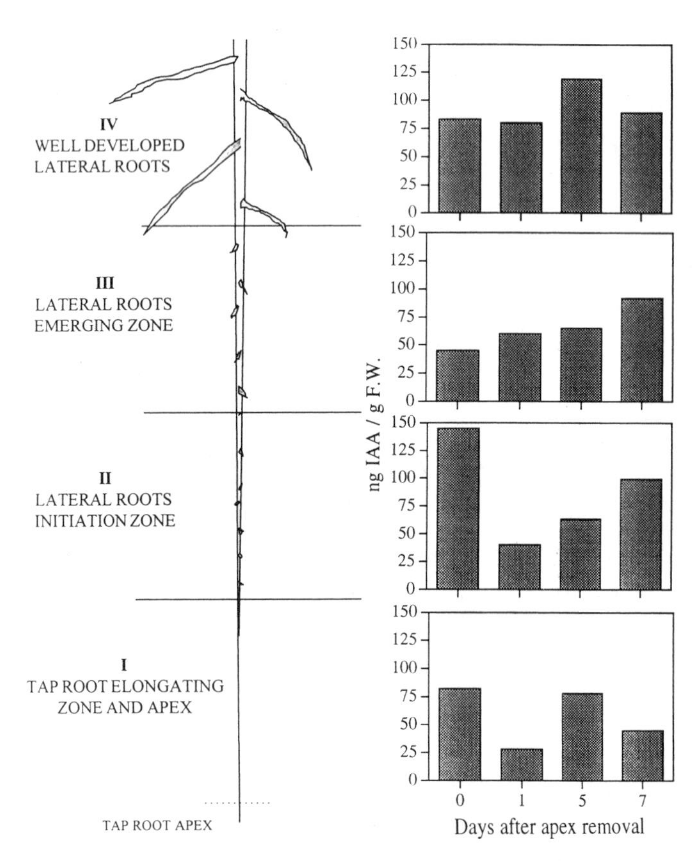

Figure 1. Levels of endogenous IAA in different sections of the intact tap root of *Pinus pinea* at time of decapitation (0) and 1, 5 and 7 days following removal of the tap root apex.

RESULTS AND DISCUSSION

In intact root system, the highest level of IAA was found in the section of LR initiation (II in Fig. 1). The lowest level of IAA was observed in the section where LR emerged (III) while in sections I and IV, IAA level was in between that of the above two sections. These results indicate that the distribution of IAA along the tap root is not a simple gradient from the root base to the apex. It seems that the site and the stage of LR development affect the transport and distribution of IAA. The present observations that IAA accumulates in the zone where LR initiate are in accordance with the view that IAA plays an important role in root initiation and high concentration of the hormone is needed for the process (Blakely et al. 1988; Rowntree and Morris 1979; Singh and Singh 1987).

Root decapitation resulted in a significant reduction in IAA level in the two lower sections while in the upper sections, no significant changes were recorded. The results suggest that the tap root apex also affects the distribution of IAA along the root system. Initiation of LR in section I, following root decapitation, occurred despite the fact that no dramatic increase in IAA was recorded. The reason for this might have been the significant reduction in cytokinins-like activity in this section following the removal of the tap root apex (see below).

The highest cytokinin-like activity in the intact root system was found in section I (Fig. 2), which contained the tap root apex. The lowest cytokinin-like activity was observed in section II, in which LR initiation occurred. Cytokinin-like activity increased as LR continued to develop and the high activity detected in the upper section probably re-

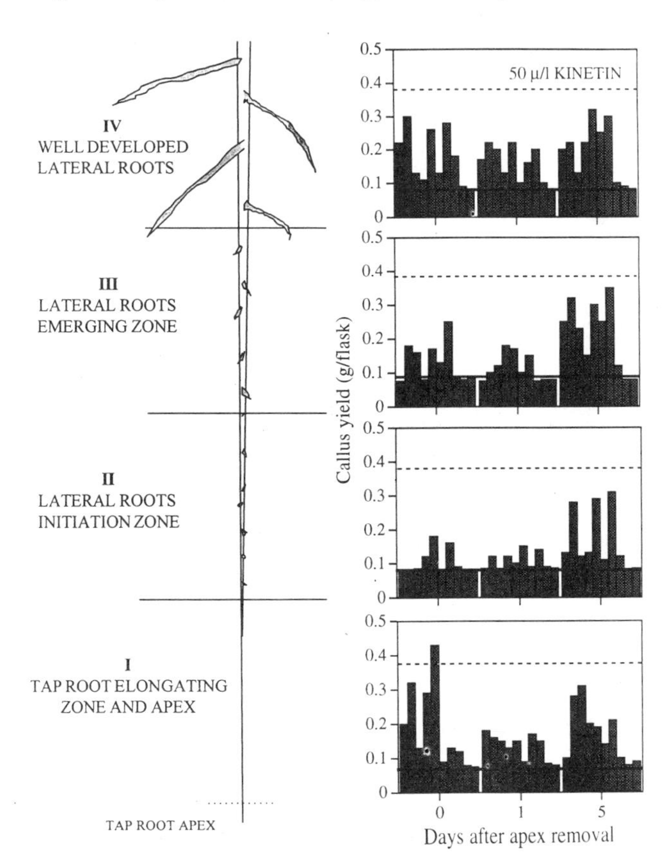

Figure 2. Cytokinin-like activity in different sections of the intact tap root of *Pinus pinea* at time of decapitation (0) or 1 and 5 days following removal of the tap root apex. The lower solid line indicates the callus yield at 0 μl kinetin/ liter.

sulted from the apex activity of newly formed LR in this section. This undoubtedly was the reason why cytokinin-like activity in the upper section was not affected by the tap root apex removal.

Removal of the tap root apex caused a dramatic decrease in cytokinin-like activity in section I, which was accompanied by LR development (see also Atzmon et al., 1994). The results suggest that low concentrations of cytokinins are favorable to the two first stages of LR development. This is in agreement with other studies that show that relatively high concentrations of cytokinins inhibited LR induction and emergence (Wightman et al., 1980).

The results of this study demonstrate that the ratio between stimulators (IAA) and inhibitors (cytokinins) are an important factor in the control of LR development in pine seedlings. The highest IAA/cytokinin ratio in intact seedlings was observed in section II, where LR initiation occurred, whereas the lowest ratio was observed in section I where no LR appeared. Removal of the tap root apex significantly lowered the cytokinin activity in section I, thus creating a favorable IAA/cytokinin ratio which stimulated LR initiation in this section, which normally does not produce LR.

REFERENCES

Atzmon, N., Salomon, E., Reuveni, O., and Riov, J. 1994. Lateral root formation in pine seedlings. I. Sources of stimulating and inhibitory substances. *Trees*. 8: 268–272.

Blakely, L., Blakely, R.B., Colowit, P.M., and Elliott, D.S. 1988. Experimental studies on lateral root formation in radish seedling roots. *Plant Physiol.* 87: 414–419.

Geisler, D. and Ferree, D.C. 1984. The influence of root pruning on water relation, net photosynthesis, and growth of young "Golden Delicious" apple trees. *J. Am. Soc. Hort. Sci.* 109: 827–831.

Rowntree, R.A., and Morris, D.A. 1979. Accumulation of ^{14}C from exogenous labeled auxin in lateral root primordia of intact pea seedlings (*Pisum sativum* L.). *Planta* 144: 463–466.

Sagee, O. 1986. Factors affecting the sensitivity of citrus abscission zone tissues to ethylene with a special reference to auxin metabolism. Ph.D. Thesis, The Hebrew University of Jerusalem, Israel.

Singh, I.S., and Singh, C.P. 1987. Response of excised roots of *Rauvolfia serpentina* benth. to growth regulatory substances. *Acta Hort.* 208: 141–155.

Thompson, D.G., Morris, J.W., Morris, R.O., and Zaerr, J.B. 1981. Rapid isolation and quantification of indol-3-acetic acid. *Plant Physiol.* 67: 98 (Suppl.).

Van Staden, J. 1976. The identification of zeatin glucoside from coconut milk. *Physiol. Plant.* 36: 123–126.

Wightman, F., Schneider, E.A., and Thimann, K.V. 1980. Hormonal factors controlling the initiation and development of lateral roots. II. Effects of exogenous growth factors on lateral root formation in pea roots. *Physiol. Plant.* 49: 304–31.

Wightman, F., and Thimann, K.V. 1980. Hormonal factors controlling the initiation and development of lateral roots. I. Sources of primordia-inducing substances in the primary root of pea seedlings. *Physiol. Plant.* 49: 13–20.

Young, E. 1990. Woody plant root physiology, growth and development. *HortScience* 25: 258–259.

PRELIMINARY EVALUATION OF VARIABILITY IN *MUSA* ROOT SYSTEM DEVELOPMENT

Guy Blomme and Rodomiro Ortiz

Plantain and Banana Improvement Program, International Institute of Tropical Agriculture, Onne Station, PMB 5320, Ibadan, Nigeria.

Studies of *Musa* root systems have generally focused on the high value export dessert bananas. However, a much broader study is required to support the genetic improvement of plantains and cooking bananas. Detailed time course studies of root system development were carried out on 12 genotypes from six diverse *Musa* groups. The performance of tissue culture derived plants was compared with that of suckers taken from field grown plants. Genotypes were assessed during establishment of tissue culture plantlets in the nursery (at 2, 4 and 6 weeks old) and in the field (at 12 and 16 weeks old) (Table 1), and compared to the vegetative growth of suckers in the field (at 6, 8, 12 and 16 weeks). Tetraploid plantain hybrids (TMPx 1658–4, TMPx 548–9 and TMPx 5511–2) were compared with their maternal triploid plantain landrace genotypes (Obino l'Ewai) and their paternal diploid banana genotypes (Pisang Lilin and Calcutta 4). These were also compared with cooking banana cultivars (Cardaba and Fougamou), a cooking banana hybrid (FHIA3), dessert bananas (Valery and Yangambi Km 5) and a plantain landrace (Agbagba). Significant correlations ($p<0.05$) across all genotypes were observed between above ground parameters and root parameters, irrespective of type of planting material or age (Table 2). Associations between leaf area, plant height, circumference at soil level and dry weight of the roots were significant over nearly all the age groups. There were no sig-

Table 1. Total length of the primary and secondary cord roots (cm), across all genotypes, for *in vitro* derived seedlings, *in vitro* derived field established plants and sword suckers at different weeks after planting

	Nursery seedlings			Field established			Sword suckers			
	Weeks after planting									
Genotype	2	4	6	12	16	20	6	8	12	16
Agbagba	54	98	222	923	2234	3015	739	857	2642	5438
Obino l'Ewai	65	98	168	2054	3077	7177	664	731	1519	2841
548-9	64	104	329	2293	4061	5948	177	417	693	3193
1658-4	26	100	140	474	2382	4015	227	999	2499	3819
Calcutta 4	49	218	312	2674	3638	8034	33	105	890	1963
Pisang Lilin	106	315	220	306	1471	1333	351	837	1517	2230
FHIA 3	49	171	277	1543	3068	4100	1217	506	2955	4788
Cardaba	110	161	337	1593	3028	3552	301	854	1546	3373
Fougamou	77	161	224	1148	1518	2153	641	429	1551	4511
Valery	60	125	274	2470	2069	3623	533	1109	1924	3042
Yangambi Km5	37	134	265	2009	5877	7751	47	481	1407	2673
5511-2	35	99	127							
CV	33.43	24.89	18.12	17.6	25.25	28.86	77.03	59.03	45.69	41.55
LSD 0.05	29	53	60	616	1071	1930	NS	NS	1093	2058

Table 2. Significant correlations ($p<0.05$) between above ground and root growth parameters for *in vitro* derived seedlings and field established plants at different weeks after planting

	Nursery (in vitro derived)						Field (in vitro derived)					
	2 week		4 week		6 week		12 weeks		16 weeks		20 weeks	
	Corr	Prob	Corr	Prob	Corr	Prob	Corr	Prob	Corr	Prob	Corr	Prob
Height of plant /dry weight of roots			0.606	0.035	0.726	0.007	0.877	0.000	0.894	0.000	0.861	0.000
Number of primary cord roots							0.683	0.019				
Total length of primary cord roots					0.833	0.001	0.743	0.008	0.821	0.002	0.696	0.016
Average length of primary cord roots					0.676	0.014	0.689	0.017	0.875	0.000		
Length of primary and secondary cord roots							0.820	0.002	0.816	0.002	0.685	0.018
Number of leaves /dry weight roots	0.684	0.013					0.898	0.000	0.828	0.001	0.851	0.001
Number of primary cord roots			0.790	0.002	0.666	0.017	0.781	0.004			0.750	0.007
Total length of primary cord roots			0.856	0.000	0.750	0.004	0.815	0.002	0.894	0.000	0.837	0.001
Average length of primary cord roots							0.701	0.015	0.774	0.004		
Length of primary and secondary cord roots							0.863	0.000	0.870	0.000	0.834	0.001

nificant phenotypic correlations between root parameters of the two types of plant material at 6, 12 and 16 weeks old. This may suggest that different mechanisms affect root development, which may be controlled by different regulatory genetic systems, according to the type of propagule. The use of this type of data to formulate an ideotype, which will direct future breeding strategies at IITA, will be discussed.

ARRANGEMENT OF MICROTUBULES, BUT NOT MICROFILAMENTS, INDICATES DETERMINATION OF CAMBIAL DERIVATIVES

Nigel Chaffey,[1,2] John Barnett,[1] and Peter Barlow[2]

[1]Department of Botany, School of Plant Sciences, University of Reading, Whiteknights, Reading RG6 6AS, United Kingdom. [2]IACR, Long Ashton Research Station, Department of Agricultural Sciences, University of Bristol, Long Ashton, Bristol BS18 9AF, United Kingdom.

The vascular cambium is the sheath of meristematic cells that lies between, and gives rise to, secondary phloem and secondary xylem (the whole comprising the secondary vascular system—SVS—of the plant). The cambium is composed of axially-elongated fusiform cells and cuboid ray cells. Phloem and xylem tissues are each composed of several cell types—axially-elongated fibres, parenchyma, vessel and sieve tube elements, and radially-elongated ray cells—any, or all, of which may influence quality or quantity of the wood produced.

During its active phase, in spring and summer, cambium not only maintains itself, but also produces derivatives which subsequently differentiate as xylem (towards the inside) or phloem (towards the outside). Important questions relating to these activities are: *What determines whether a cell (i) perpetuates the meristematic state or (ii) differentiates? What determines whether, for example, a xylem derivative is to give rise to a fibre cell or a vessel element?* We have attempted to answer these questions by examining the microtubule (MT) and microfilament (MF) components of the cytoskeleton within cells of the SVS of taproots of *Aesculus hippocastanum* L. (horse-chestnut) seedlings. Indirect immunofluorescence microscopy (IIF) using monoclonal antibodies against α-tubulin and F-actin on 6 μm sections, as well as transmission electron microscopy (TEM) of 60 nm sections, have been used (Chaffey *et al.*, 1996, 1997a).

Cambial cells *per se* are practically impossible to identify in hardwood species. Instead, the *cambial zone* is usually referred to; it consists of the cambial initials plus their derivatives at early stages of differentiation to secondary vascular tissue. Fusiform cells are readily-distinguished from ray cells, the latter forming obvious rays which traverse secondary xylem, cambial zone and secondary phloem. Radial files of axially-elongated cells are also apparent, running from the cambial zone into the secondary xylem and comprised of cells at progressively more advanced stages of differentiation which have been derived by periclinal divisions of cambial cells. As seen by TEM, cambial cells contain a prominent single nucleus and vacuome, numerous plastids, mitochondria and active dictyosomes in a ribosome-rich cytoplasm, and thin cell walls, particularly the tangential wall, pierced by plasmodesmata.

The cortical MT cytoskeleton of cambial cells, as seen by IIF, is reticulate with no preferred orientation to the constituent MTs. This reticulate array exists in both ray and fusiform cells. However, a helical array is found in differentiating axial phloem derivatives. A similar difference in MT orientation is observed on the xylem side of the cambial zone within putative fibre cells. TEM of these same cell-types reveals images of the MTs which accord with the two distinct arrays identified by IIF. TEM also reveals images of the MF cytoskeleton associated with vascular differentiation. MF bundles are oriented approximately axially within fusiform cambial cells. This orientation is retained in developing axial phloem cells and xylem fibres, *despite their MT cytoskeleton having undergone rearrangement to a helical array* (Chaffey *et al.* 1997b).

CONCLUSIONS

1. Correlative IIF and TEM offer a powerful method for investigating the dynamics of the cortical MT cytoskeleton during cell differentiation *in situ*.
2. A reticulate arrangement of cortical MTs is present in both ray and fusiform cells of the cambial zone of *Aesculus hippocastanum* taproots.
3. Helical MT arrays are found in developing axial phloem cells and xylem fibres.
4. MT rearrangement, from reticulate to helical, indicates determination of cambial derivatives to the vascular developmental pathway. MFs in these cells, by contrast, continually display an axial orientation. Factors which determine the ultimate fate of an individual cambial derivative remain to be elucidated.

REFERENCES

Chaffey, N.J., Barlow, P.W. and Barnett, J.R. (1996). Microtubular cytoskeleton of vascular cambium and its derivatives in roots of *Aesculus hippocastanum* L. (Hippocastanaceae). In: *Recent Advances in Wood Anat-*

omy (Eds. L.A. Donaldson, A.P. Singh, B.G. Butterfield and L.J. Whitehouse), pp. 171–183. New Zealand Forest Research Institute, Rotorua, New Zealand.

Chaffey, N.J., Barnett, J.R. and Barlow, P.W. (1997a). Cortical microtubules involvement in bordered pit formation in secondary xylem vessel elements of *Aesculus hippocastanum* L. (Hippocastanaceae): A correlative study using electron microscopy and indirect immunofluorescence microscopy. *Protoplasma* 197: 64.

Chaffey, N., Barlow, P. and Barnett, J. (1997b). Cortical microtubules rearrange during differentiation of vascular cambial derivatives, microfilaments do not. *Trees* 11: 333.

7

INDISSOCIABLE CHIEF FACTORS IN THE INDUCTIVE PHASE OF ADVENTITIOUS ROOTING

Thomas Gaspar,[1] Claire Kevers,[1] and Jean-François Hausman [2]

[1]Institute of Botany B 22
Univ. of Liège - Sart Tilman
B-4000, Liège
Belgium
[2]CRP-CU
162 av. de la Faïencerie
L-1511 Luxembourg
Luxembourg

INTRODUCTORY GENERAL CONSIDERATIONS

It happened that the first discovered and identified phytohormone, indolyl-3-acetic acid (IAA), was early shown to promote or favour adventitious rooting (Thimann and Went, 1934). Later identified natural auxins and synthetic compounds of this category had the same effects (Jackson, 1986). With the years, the rooting property of auxins appeared to be specific to this class of growth regulators since no such clear-cut effect could be apparently obtained by exogenous application of other known phytohormones. Some of them, such as cytokinins and gibberellins, for instance, were even classified as rooting inhibitors (Jackson, 1986; Davis et al., 1988; Davis and Haissig, 1994), although there were papers indicating the necessity of cytokinins for rooting (Letham, 1978) and others showing rooting effects of gibberellins under certain circumstances (Gaspar et al., 1977). On the basis of the effects of exogenous application of auxins, a series of wrong concepts as to their roles had arisen: that auxin is the major triggering agent in rooting, that the application of exogenous auxin is needed to augment the endogenous bulk of auxin, that rooting necessitates the maintenance of a "high" amount of endogenous auxin for a certain (unprecise) time, etc. Because there are inductive/adaptative enzymes to regulate the exogenously fed hormones (this is well known for auxins and cytokinins) and because application of a hormone may induce modifications in the metabolism of other hormones, such simplistic conclusions may not be drawn. Another associated error was to consider rooting as a single developmental process.

One of the main achievements in the studies of adventitious root formation in the last years has been the recognition of successive interdependent physiological phases

Biology of Root Formation and Development, edited by Altman and Waisel.
Plenum Press, New York, 1997.

(Mitsuhashi-Kato et al., 1978; Jarvis et al., 1983; Moncousin et al., 1988; Blakesley, 1994). These were generally called induction, initiation and expression (Gaspar et al., 1992, 1994), although some other terminologies can be used (De Klerk et al., 1995). A consensus tacitly emerged to define the rooting inductive phase as the time period covering the necessary biochemical events preceding the initiation of the cell divisions which lead to the formation of root primordia (Jarvis, 1986; Moncousin, 1991). Another definition or a practical way to estimate its duration, is the minimum time required for the presence of the external rooting signal for competent cells (Hand, 1994). Competence itself is defined as a cell reactivity state allowing response to a specific stimulus that finally leads to a specific developmental pathway. Induced cuttings which do not require any more the rooting signals, are also said to be determined, even if other environmental factors are required for the completion of the successive developmental phases (Mohnen, 1994).

The inductive rooting period sometimes appears to be very short, being achieved in less than 24 hours from the time of application of the external auxin (Moncousin et al., 1988; Hausman et al., 1994). It has also appeared that auxin application was made on cuttings which had begun their inductive phase earlier, for instance in shoots at the end of an *in vitro* multiplication cycle where the initial cytokinin/auxin ratio had been progressively reversed and has served as an inductive treatment. Such established facts now explain earlier discrepancies about results of analyses of endogenous compounds.

Taking into account these considerations, a role or several roles for each of the phytohormone types in one or several phases of the rooting process is to be anticipated. The present paper discusses the indissociable involvement of endogenous auxin(s), polyamine(s) and peroxidase(s) in the rooting inductive phase.

IAA LEVEL AND PEROXIDASE CHANGES IN THE INDUCTIVE PHASE OF ADVENTITIOUS ROOTING

Because the different physiological phases had not been determined in many rooting experimental systems, the variations in IAA levels measured in the cuttings before and in the course of the rooting process were quite discrepant (Gaspar and Hofinger, 1988; Blakesley et al., 1991). It appeared in recent years that, whatever the cutting types, elevation of the endogenous level of free IAA always occurs in the inductive phase of rooting (Fig. 1). The early occurrence of the peak (in some cases, in less than 24 hours) and its transient character, may explain why it has been missed by some authors.

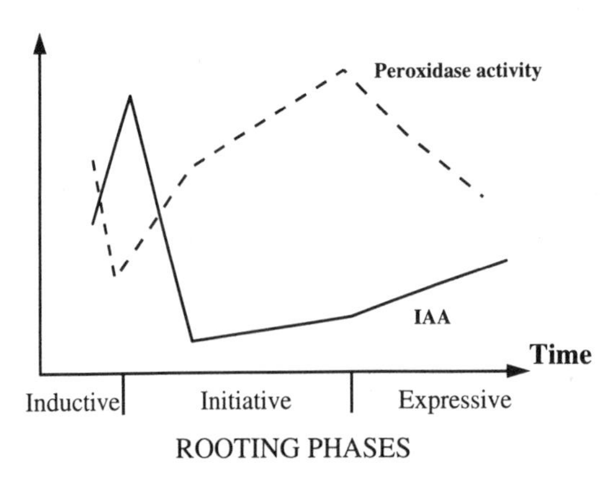

Figure 1. Generally occurring changes of peroxidase activity and of endogenous free IAA level along successive inductive, intiative, and expressive rooting phases.

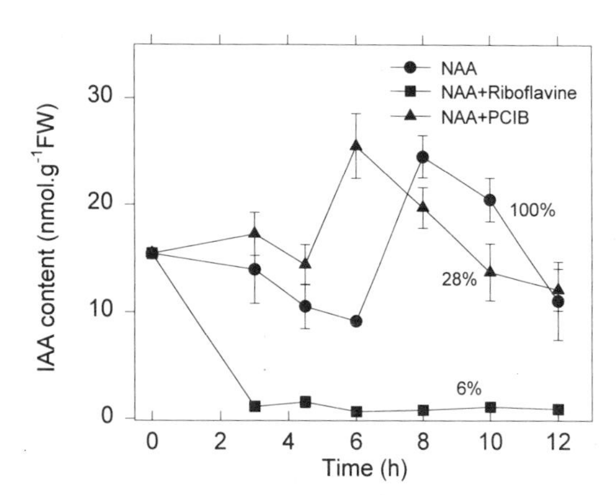

Figure 2. Changes in free IAA level in poplar shoots during their inductive phase in the presence of NAA (0.3 mg l^{-1}; ●) alone, and supplemented either by riboflavine (0.3 mM; ■) or by PCIB (0.1 mM; ▲). The respective rooting percentages are indicated.

This early IAA peaking in the inductive phase has been causally related to rooting:

- it is not observed in non-rooting cuttings (Hausman et al., 1995a);
- it is measurable only in the rooting zone (Hausman et al., 1995a);
- treatments, such as with riboflavine, which inhibit rooting, hinder IAA peaking (Fig. 2);
- PCIB, a competitor of IAA for auxin-receptors or auxin binding proteins (Marumo, 1986), when applied at inductive phase, inhibits rooting without hindering IAA peaking (Fig. 2).

This demonstration of the involvement of free IAA in the rooting inductive phase does not exclude the participation of other auxins such as IAA-aspartate (Blakesley, 1994) and serotonin (Gatineau et al., 1997).

The study of peroxidase activity and isoperoxidases, first as possible enzymes mediating IAA catabolism and second as markers of the successive rooting phases, allowed the establishment of a general time-course of activity with a typical minimum at rooting inductive phase, and a typical maximum at initiative phase (Moncousin, 1991; Gaspar et al., 1992, 1994). Figure 1 shows two curves: the levels of soluble peroxidase activity and endogenous free IAA, which occur in the course of the ongoing successive inductive, initiative, and expressive phases of rooting, as generally measured in a variety of cuttings (Gaspar et al., 1992, 1994). These two curves appear to be approximately the reverse of one another.

The close relations between the peroxidase peak at initiative phase and rooting has been reviewed (Gaspar et al., 1992). The peroxidase minimum at the rooting induction phase has been determined more recently (Moncousin et al., 1988; Hausman et al., 1995b). Its direct relationship with rooting has still to be established.

POLYAMINE CHANGES AT ROOTING INDUCTIVE PHASE

In addition, many studies hypothesized a role for polyamines in the rooting process (Friedman et al., 1982, 1985; Jarvis et al., 1983; Tiburcio et al., 1989; Torrigiani et al., 1989, 1993; Biondi et al., 1990; Altamura, 1994) but very few dealt with precise rooting phases where polyamines were really involved. A series of works on rooting of poplar shoots *in vitro* showed an early typical elevation of putrescine, close to the IAA peak, at the 6–7 h close to termination of the inductive phase (Fig. 3). Spermidine and spermine

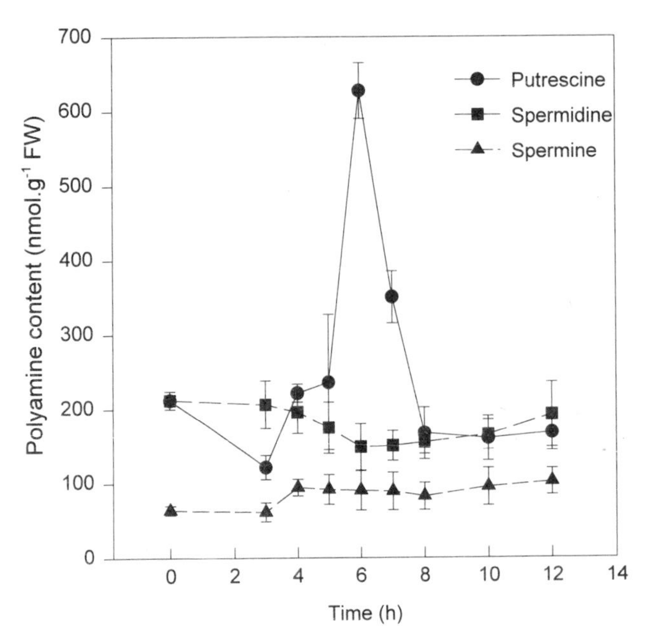

Figure 3 . Changes in free putrescine (●), spermidine (■) and spermine (▲) during the rooting inductive phase of poplar shoots, in the presence of NAA as rooting auxin. In the absence of NAA, no such transient increase of putrescine is measured.

did not change significantly. Other arguments that causally implicated putrescine in the rooting inductive phase of poplar include:

- the transient elevation did not occur in non-rooting cuttings (Hausman et al., 1994);
- it was measurable only in the basal rooting zone (Hausman et al., 1994);
- inhibitors of putrescine biosynthesis, such as DFMO and DFMA (α-difluoromethylornithine and α-difluoromethylarginine, respectively), applied prior to or at the beginning of inductive phase, inhibited rooting (Hausman et al., 1994);
- an inhibitor of putrescine conversion into spermidine and spermine (cyclohexylamine, an inhibitor of spermidine synthase), which promoted the accumulation of endogenous putrescine, favoured rooting, in the absence of exogenously supplied auxin (Hausman et al., 1995a);
- exogenously applied putrescine, prior to or at the beginning of inductive phase, provoked rooting (Hausman et al., 1995a).

Additional results point to a major role played by putrescine catabolism through its Δ^1-pyrroline-GABA (γ-aminobutyric acid) pathway (Hausman et al., 1994, 1995a). Indeed, treatment of poplar cuttings with AG (aminoguanidine, an inhibitor of diamine oxidase, DAO, which converts putrescine to GABA) inhibited rooting (Hausman et al. 1994, 1995a). Similar arguments (Kevers et al., 1997) suggest an involvement of putrescine and its catabolic pathway to GABA, in the rooting inductive phase of walnut shoot cuttings, where endogenous IAA and peroxidase had been also implicated (Ripetti et al., 1994; Heloir et al., 1996).

POSSIBLE INTERRELATIONSHIPS AMONG IAA, PEROXIDASES, AND PUTRESCINE

Because the activity of the so-called IAA-oxidase system has been found to vary in parallel to peroxidase activity, at least at the rooting initiation phase (Gaspar et al., 1992,

1994), it has been tempting to consider the variation of the auxin level as the result of a peroxidase-mediated auxin catabolism. All peroxidases indeed, but some of them with a greater efficiency, can catalyze auxin catabolism (Gaspar et al., 1982). Alternatively, it can be hypothesized that prior changes in IAA levels have modulated peroxidase activity. IAA is known to control the activity of (other ?) peroxidases involved in lignification (Ros Barcelo and Munoz, 1992), an obligatory process in root formation, namely through the differentiation of xylem cells. Other factors involved at different steps of root formation, such as phenolics (Curir et al., 1989; Berthon et al., 1993), cytokinins (Vidal et al., 1994), and ethylene (Moncousin et al., 1989), are also known to influence peroxidase activity. Investigations of the changes of the isoperoxidase spectrum during rooting has not solved the problem of the relationships between IAA and peroxidase, the main difficulty being due to the polyfunctionality of the isoperoxidases (Pedreno et al., 1995). Figures 2 and 3 show that the transient increases of free IAA and free putrescine at rooting inductive phase occur at about the same time. In the same plant material, under the same conditions, a peroxidase increase precedes the typical minimum of the rooting inductive phase at the 6th h (Fig. 4A).

Exogenous application of putrescine or of CHA, in the absence of auxin, promotes rooting (see above) and provokes a peroxidase peak (Fig. 4B and C). It might be tempting to suggest that putrescine controls the IAA level, mediated by some peroxidase(s), but the

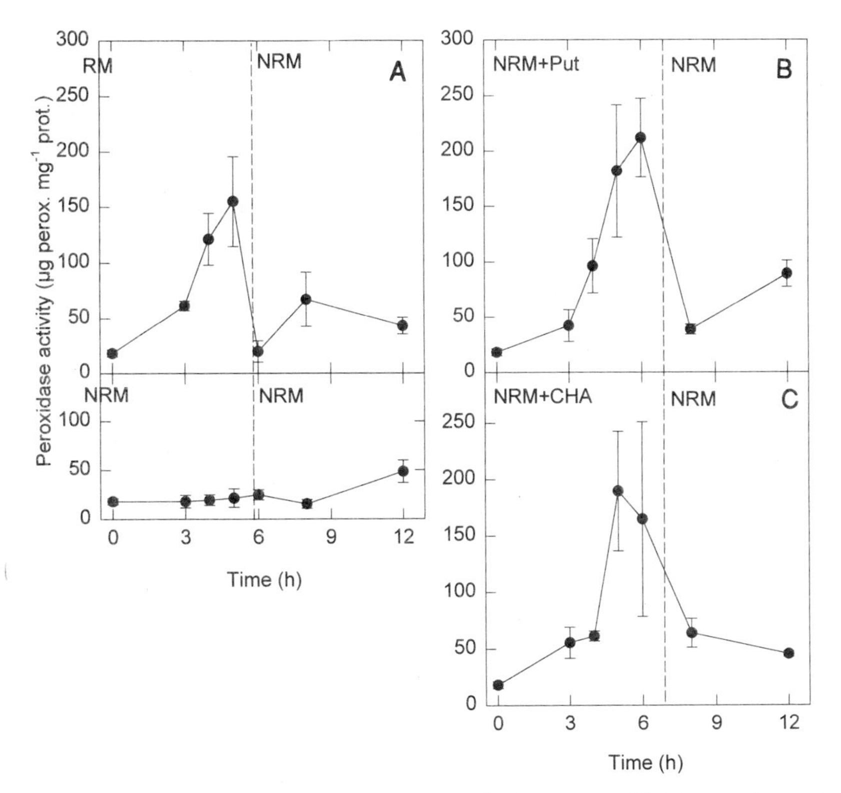

Figure 4. Changes in peroxidase activity of poplar shoots. (A) in the presence of NAA as rooting auxin (RM = rooting medium) or in the absence of NAA (NRM = non rooting medium); (B) in the absence of auxin but in the presence of putrescine 10^{-4}M (NRM + put); (C) in the absence of auxin but in the presence of CHA 10^{-4}M (NRM + CHA).

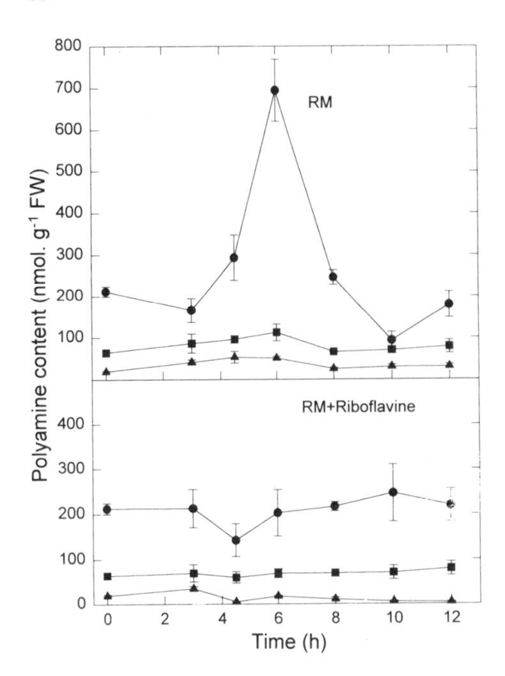

Figure 5. Changes in free putrescine (●), spermidine (■) and spermine (▲) in poplar shoots during their inductive phase in the presence of NAA alone (RM), or supplied with riboflavine 0.3 mM (RM + riboflavine).

reverse is also true. Rooting inhibition by riboflavine (see above) corresponds with a hindrance of the characteristic elevation of the IAA peak at the inductive phase, but also of the temporary putrescine increase (Fig. 5). The transient rise and decline of IAA and putrescine thus seem closely related. It might be hypothesised that IAA and putrescine which are known to control cell division cycles (Del Duca and Serafini-Fracassini, 1993), and are required to initiate cell divisions at the end of the rooting inductive phase, rise simultaneously. However, the considerable peroxidase variation that occurs in the same time period may indicate an eventual role of peroxidase in the relationship between IAA and putrescine.

Putrescine degradation through its Δ^1-pyrroline pathway (Fig. 6) is accompanied by the formation of H_2O_2. Might this H_2O_2 serve some (iso)peroxidases for some of their roles? The following hypotheses can therefore be formulated as to roles for peroxidases in the interplay between IAA and polyamines:

a. putrescine might indirectly inactivate some peroxidase(s) involved in auxin catabolism by providing H_2O_2 through its catabolism, thus favouring an increase in the level of endogenous IAA;
b. an increase in putrescine level might alternatively enhance peroxidase activity and hence bring about a decrease in IAA level; a possible putrescine control of peroxidase activity via the GABA shunt and the Shemin pathway must not be excluded. Indeed a GABA shunt of tricarboxylic acid cycle, from glutamate (Shelp et al., 1995) or from polyamines (Bisbis et al., 1997) to succinate, as illustrated in Figure 7, has been shown to be operative under certain stress conditions, and succinate is a precursor for the biosynthesis of tetrapyrrole-containing compounds such as peroxidase via the Shemin pathway (Castelfranco and Beale, 1983). Rooting might by some way be considered as a response to stress (injury due to cutting plus imposed hormonal imbalance in favour of auxin). GABA alone, in the absence of auxin, is indeed able to promote rooting to some extent in poplar shoots (unpublished data).

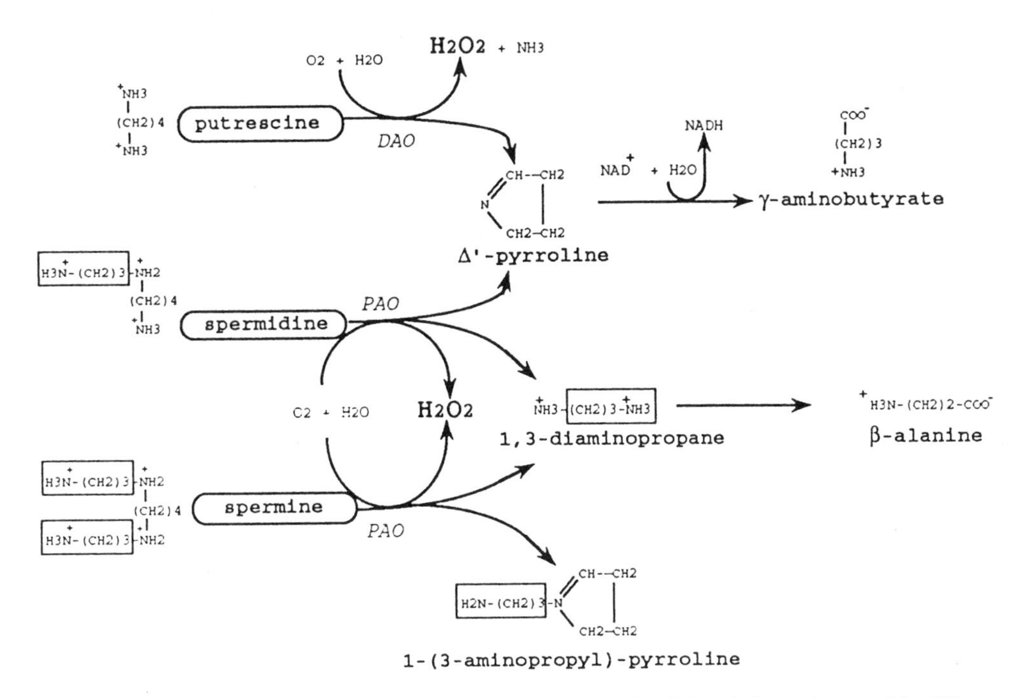

Figure 6. DAO (diamine oxidase) and PAO (polyamine oxidase) mediated degradation pathways of the different polyamines indicating the formation of H_2O_2 and GABA (γ-aminobutyric acid).

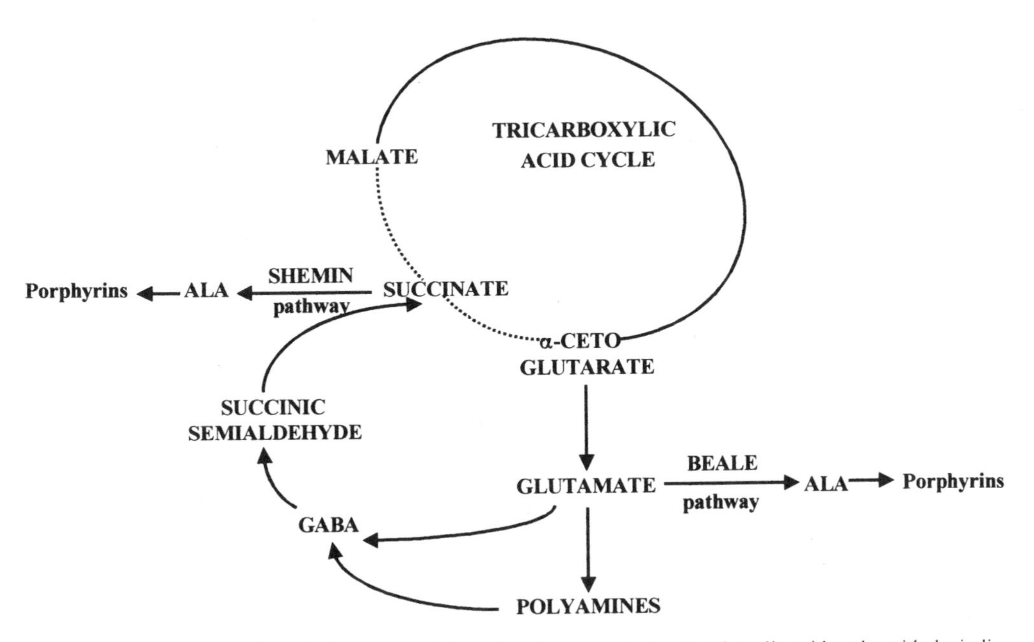

Figure 7. GABA-shunt, through glutamate and through polyamines, in the tricarboxylic acid cycle, with the indications of the Beale (from glutamate) and Shemin (from succinate) pathways of ALA (aminolevulinic acid) and porphyrin (including peroxidase) biosynthesis.

Finally, it becomes evident that many transduction pathways may not be initiated without the participation of the membrane-wall associated peroxidases (and NAD(P)H oxidase) and the generation of active oxygen species. Among the latter, H_2O_2 has been proposed to be involved directly in the regulation of gene expression (Mehdy, 1994). Experiments will be designed to check these hypotheses.

REFERENCES

Altamura, M.M., 1994, Rhizogenesis and polyamines in tobacco thin cell layers, *Adv. Hortic. Sci.* 8: 33.

Berthon, J.Y., Battraw, M.J., Gaspar, Th., and Boyer, N.,1993, Early test using phenolic compounds and peroxidase activity to improve *in vitro* rooting of *Sequoiadendron giganteum* (Lindl.) Buchholz, *Saussurea* 24: 7.

Biondi, S., Diaz, T., Iglesias, I., Gamberini, G., and Bagni, N., 1990, Polyamines and ethylene in relation to adventitious root formation in *Prunus avium* shoot cultures, *Physiol. Plant.* 78: 474.

Bisbis, S., Kevers, C., and Gaspar, Th., 1997, Atypical TCA cycle and replenishment in a non-photosynthetic fully habituated sugarbeet callus overproducing polyamines, *Plant Physiol. Biochem.*, 35: 363.

Blakesley, D., 1994, Auxin metabolism and adventitious root initiation, *in*: "Biology of Adventitious Root Formation,"T.D. Jackson and B.E. Haissig, eds, Plenum Press, New York.

Castelfranco, P.A., and Beale, S.I., 1983, Chlorophyll biosynthesis: recent advances and areas of current interest, *Ann. Rev. Plant Physiol.* 34: 241.

Davis, T.D., and Haissig, B.E., 1994, Biology of Adventitious Root Formation. Plenum Press, New York.

De Klerk, G.J., Kepel, M., Ter Brugge, J., and Meekes, H., 1995, Timing of the phases in adventitious root formation in apple microcuttings, *J. Exp. Bot.* 46: 965.

Del Duca, S., and Serafini-Fracassini, D., 1993, Polyamines and protein modification during the cell cycle, *in*: "Molecular and Cell Biology of the Plant Cell Cycle," J.C. Ormrod and D. Francis, eds, Kluwer Acad. Publ., Dordrecht.

Friedman, R., Altman, A., and Bachrach, U., 1982, Polyamines and root formation in mung bean hypocotyl cuttings. I. Effects of exogenous compounds and changes in endogenous polyamine content, *Plant Physiol.* 70: 844.

Friedman, R., Altman, A., and Bachrach, U., 1985, Polyamines and root formation in mung bean hypocotyl cuttings. II. Incorporation of precursors into polyamines, *Plant Physiol.* 79: 80.

Gaspar, Th., and Hofinger, M., 1988, Auxin metabolism during adventitious rooting, in: "Adventitious Root Formation in Cuttings", B. Haissig and N. Sankhla, eds., Dioscorides Press, Portland.

Gaspar, Th., Kevers, C., Hausman, J.F., Berthon, J.Y., and Ripetti, V., 1992, Practical uses of peroxidase activity as a predictive marker of rooting performance of micropropagated shoots, *Agronomie* 12: 757.

Gaspar, Th., Kevers, C., Hausman, J.F., and Ripetti, V., 1994, Peroxidase activity and endogenous free auxin during adventitious root formation, *in*, "Physiology, Growth and Development of Plants in Culture: "P.J. Lumsden, J.R. Nicholas and W.J. Davis, eds." Kluwer Acad. Publ., Dordrecht.

Gaspar, Th., Penel, C., Thorpe, T. and Greppin, H., 1982. Peroxidases 1970–1980. A Survey of their Biochemical and Physiological Roles in Higher Plants, Univ. Genève - Centre Bot.

Gaspar, Th., Smith, D., and Thorpe, T., 1977, Arguments supplémentaires en faveur d'une variation inverse du niveau auxinique endogène au cours des deux premières phases de la rhizogenèse, *C.R. Acad. Sci. Paris* 285: 327.

Gatineau, F., Fouché, J.G., Kevers, C., Hausman, J.F., and Gaspar, Th., 1997. Quantitative variations of indolic compounds including IAA, IAA-aspartate and serotonin in walnut microcuttings during root induction. *Biol. Plant.* 39: 131.

Hand, P., 1994, Biochemical and molecular markers of cellular competence for adventitious rooting, *in*: "Biology of Adventitious Root Formation", T.D. Davis, and B.E. Haissig, eds., Plenum Press, New York.

Heloir, M.C., Kevers, C., Hausman, J.F. and Gaspar, Th., 1996, Changes in the concentrations of auxins and polyamines during rooting of *in vitro*-propagated walnut shoots, *Tree Physiol.* 16: 515.

Hausman, J.F., Kevers, C., and Gaspar, Th., 1994, Involvement of putrescine in the inductive rooting phase of poplar shoots raised *in vitro*, *Physiol. Plant.* 92: 201.

Hausman, J.F., Kevers, C., Gaspar, Th., 1995a, Auxin-polyamine interaction in the control of the rooting inductive phase of poplar shoots *in vitro*, *Plant Science* 110: 63.

Hausman, J.F., Kevers, C., and Gaspar, Th., 1995b, Putrescine control of peroxidase activity in the inductive phase of rooting in poplar shoots *in vitro*, and the adversary effect of spermidine, *J. Plant Physiol.* 146: 681.

Jarvis, B.C., Shannon, P.R.M., and Yasmin, S., 1983, Involvement of polyamines with adventitious root development in stem cuttings of mung bean, *Plant. Cell. Physiol.* 24: 677.

Kevers, C., Bringaud, C., Hausman, J.F., and Gaspar, Th., 1997, Putrescine involvement in the inductive phase of walnut shoots rooting *in vitro*. *Saussurea* 28, in press.

Letham, D.S., 1978, Cytokinins, *in*: "Phytohormones and Related Compounds: a Comprehensive Treatise", Letham, D.S., Goodwin P.B. and Higgins, T.J.V., eds, Elsevier, Amsterdam.

Marumo, S., 1986, Auxins, *in*: " Chemistry of Plant Hormones", Takahashi N., ed., CRC Press, Boca Raton.

Mehdy, M., 1994, Active oxygen species in plant defense against pathogens, *Plant Physiol.* 105: 467.

Mohnen, D., 1994, Novel experimental systems for determining cellular competence and determination, in: "Biology of Adventitious Root Formation", Davis, T.D. and Haissig, B. eds, Plenum Press, New York.

Moncousin, C., 1991, Rooting of *in vitro* cuttings, *in*: "Biotechnology in Agriculture and Forestry". Vol. 17. High-Tech and Micropropagation I.Y.P.S. Bajaj, ed., Springer-Verlag, Berlin.

Moncousin, C., Favre, F.M., and Gaspar, Th., 1988, Changes in peroxidase activity and endogenous IAA levels during adventitious rooting in cuttings. *in:*" Physiology and Biochemistry of Auxins in Plants", Kutacek, M., Bandurski, R.S. and Krekule J., eds. Academia, Praha.

Moncousin, C., Favre, J.M., and Gaspar, Th., 1989, Early changes in auxin and ethylene production in vine cuttings before adventitious rooting, *Plant Cell Tissue Organ Culture* 19: 235.

Pedreno, M.A., Ferrer, M.A., Gaspar, Th., Munoz, R., and Ros Barcelo, 1995, The polyfunctionality of cell wall peroxidases avoids the neccesity of an independent H_2O_2-generating system for phenolic coupling in the cell wall, *Plant Peroxidase Newsletter* 6: 11.

Ripetti, V., Kevers, C., and Gaspar, Th., 1994, Two successive media for the rooting of walnut shoots *in vitro*. Changes in peroxidase activity and in ethylene production, *Adv. Hortic. Sci.* 8: 29.

Ros Barcelo, A., and Munoz, R., 1992, Peroxidases: their role in the control of plant cell growth, *in* "Plant Peroxidases 1980–1990. Topics and Detailed Literature on Molecular Biochemical, and Physiological Aspects" C. Penel, Th. Gaspar and H. Greppin, eds. Univ. of Genève, Publ.

Shelp, B.J., Walton, C.S., Snedden, W.A., Tuin, L.G., Oresnik, I.J., and Layzell, D.B., 1995, Gaba shunt in developing soybean is associated with hypoxia, *Physiol. Plant.* 94: 219.

Tiburcio, A.F., Gendy, C.A., and Tran Tanh Van, K., 1989, Morphogenesis in tobacco subepidermal cells: putrescine as a marker of root differentiation, *Plant Cell Tissue Organ Cult.* 19: 43.

Torrigiani, P., Altamura, M.M., Scaramagli, S., Capitani, F., Falasca, G., and Bagni, N., 1993, Regulation of rhizogenesis by polyamines in tobacco thin layers, *J. Plant. Physiol.* 142: 81.

Vidal, N., Ballester, A., Vieitez, A.M., Kevers, C., and Gaspar, Th., 1994, Biochemical characteristics of chesnut shoots related to *in vitro* multiplication and rooting capacities, *Adv. Hortic. Sci.* 8: 19.

8

THE EFFECTS OF POLYAMINES AND HYDROGEN PEROXIDE ON ROOT FORMATION IN OLIVE AND THE ROLE OF POLYAMINES AS AN EARLY MARKER FOR ROOTING ABILITY

E. Rugini, G. Di Francesco, M. Muganu, S. Astolfi, and G. Caricato

Dipartimento di Produzione Vegetale
Sez. Ortofloroarboricoltura
Università della Tuscia
Via S. C. de Lellis, 01100 Viterbo
Italy

INTRODUCTION

The polyamines (PAs) such as putrescine (PUT) spermidine (SPD) and spermine (SPM) are involved in numerous processes associated with plant growth by affecting cell division (Bagni et al., 1982) and cell development (Slocum et al., 1984). In explants, treated with auxin, polyamines show increase concentration especially before root emergence (Desai and Mehta, 1984; Friedeman et al., 1982), but the relationship between auxin-induced root formation, polyamine content, polyamine synthesis inhibitors and ethylene is yet unclear (Biondi et al., 1990). It is generally accepted that auxin has a central role in the initiation of adventitious roots (Davis et al., 1988) and many studies have attempted to correlate peroxidase enzyme activity with rooting, the inductive phase being achieved through lowering peroxidase activity (Gaspar et al., 1985). Although for most species, exogenous treatment with PAs does not affect rooting (Biondi et al., 1990), in some species, PAs increase rooting, such as: olive (Rugini and Wang, 1986), azalea (Mirkovic, 1993), hazelnut (Rey et al., 1994) poplar (Hausman et al., 1995). In olive exogenous PUT increases root percentage, promotes earlier rooting and increases the number of roots/explant both when combined with auxin or following treatment with *Agrobacterium rhizogenes* (Rugini, 1992). It has been shown that following exogenous treatment with H_2O_2, a product of PA's catabolism, such as PUT, increases rooting percentage and promotes earlier rooting. It was suggested that the role of PAs in rooting by the H_2O_2 was to increase the peroxidase activity (Rugini et al., 1991) and to see whether they could be used as an early marker for rooting ability. By using olive, which possesses endogenous suboptimal levels of PAs, this work attempts to clarify the role of PAs in the induction of rooting.

Biology of Root Formation and Development, edited by Altman and Waisel.
Plenum Press, New York, 1997.

MATERIALS AND METHODS

Source of Plant Material

Cuttings for in Vivo Experiments. Cuttings were obtained from subapical twigs with 4 leaves, collected from 8-year-old plants of cv Frangivento. Fifty cuttings, subdivided in 5 blocks, were used for each treatment. Cuttings were immediately dipped in auxin by placing two centimeters of their basal end for 10 sec in a solution containing 2000 ppm IBA in alcohol/water (1:1 w/v). Treatment with H_2O_2 at 3.5% volume or in 1 mM acqueous solution of Putrescine-HCl was carried out with the same way for 30 sec and 30 min respectively at 0, 2 and 5 days. The cuttings were placed in perlite and maintained at 23 ± 1°C constant temperature on a bench covered with glassfibre slab.

Explants for in Vitro Experiments. Explants used were derived from olive shoots, cv Moraiolo, which had been maintained for 4 years by subculture every 35 days on OM (Rugini, 1984) plus 18 μM zeatin riboside, 3% sucrose and 0.7% agar. All cultures were maintained under a 16 h photoperiod at 40 μmol $m^{-2}s^{-1}$ provided by cool-white fluorescent lamps at 23 ± 1°C. Bourgin and Nitsch (1967) medium with macroelements reduced to half, supplemented with 2% sucrose, 0.7% agar and 3.2 μM NAA, at pH 5.8 was used as a rooting medium. All media were autoclaved at 121°C for 20 min. The following substances, PUT, SPD, DFMO, DFMA, DCHA were added to the rooting medium at 1 mM concentration. Gibberellic acid at 57, 142 and 285 μM, and AG at 10 μM concentration were also used. Each treatment of the rooting test involved 30 uninodal olive microcuttings placed singly in test a tube (200 × 16 mm) containing 5 ml of medium.

Peroxidase Activity Determination

At the beginning of the rooting experiment 1 gram of uninodal explants, from basal shoots, were taken off the rooting medium with or without PUT, at 0, 6, 31, 48 and 72 h in the dark and under 16 h photoperiod. The explants were homogenized in a cooled mortar and pestle using 4 ml of 0.1 M citrate-phospate buffer (pH 5.0) containing 2 mM ditiotreitol (DTT) and 1% (w/v) polyvinylpyrrolidone. The homogenate was centrifugated at 15,000 g for 20 min and the supernatant was dialized and used as an enzyme extract for peroxidase activity, using guaiacol as a hydrogen donor. Peroxidase activity was measured at 470 nm using a spectrophotometer at 30°C. The assay medium contained 40 mM guaiacol (in citrate-phospate buffer pH 5.0), 5 mM H_2O_2 (in citrate-phosphate buffer, pH 5.0) and 100 μl of extract. The protein content was determined by the Coomassie staining method described by Bradford (1976), using BSA as standard.

Polyamine Analysis

The PAs were analyzed in: 1) One cm apical rapid growing shoots of olive, chestnut and walnut, 2) one cm of basal end of cuttings from easy-to-root genotype (Frangivento) and difficult-to-root genotype (unknown genotype), 3) whole explants placed to root under 16 h photoperiod and in dark conditions, both treated with PUT and untreated ones (as control). Cuttings and whole explants were analyzed at 0, 1, 2 and 5 days. The samples were homogenized in PCA (percloric acid) and distinguished in free PAs (aliquots of the PCA supernatant), conjugated PAs (of the same hydrolyzed supernatant) and bound PAs (of the hydrolyzed pellet); they were then dansylated (Palavan and Galston, 1982), separated by thin layer chromatography in chloroform:triethylamine (5:1) and detected using a spectrofluorimeter.

Determination of Root Elongation and Histological Observation

The explants from rooting medium with or without PUT were transferred immediately after root apex protrusion to the same medium with or without PUT. Every two days, over 15 days, the elongation of the roots was measured by marking the glass with a pen. A sample of ten explants at 7, 10 and 15 days, were removed from the medium and vacuum infiltrated for 10 min with FAA (formalin acetic acid). They were then dehydrated with ethanol at increasing concentrations. Sections (less than 10 μm thick) were cut using a rotary microtome (Microm HM 340 E). Cut paraffin sections were passed through a xylol series, stained with safranin-fast green combination and mounted in Canada balsam before being examined under a light microscope.

RESULTS

Effect of Scalar Treatment with PUT and H_2O_2 in Cuttings

The roots started to appear at the basal end of cuttings after one month. Both H_2O_2 and PUT did not affect rooting when supplied at the 5th day, while both promoted early rooting and increased the final rooting percentage when applied when the cuttings were prepared (0 day); H_2O_2 affected rooting also on the 2nd day (Fig. 1). In addition, PUT alone, without auxin, increased rooting compared to the untreated material (data not shown).

Effect of PUT, SPD, and PA Inhibitors on in Vitro Explants

PUT had the same effect observed in cuttings while SPD caused basal necrosis in explant tissues in contact with the medium in 75% of the explants; however the roots emerged despite the necrosis and the final rooting percentage was no different from the

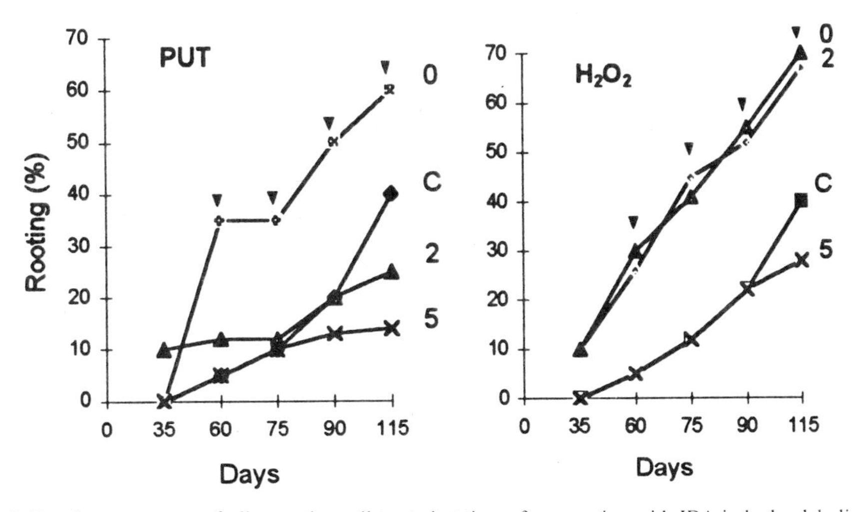

Figure 1. Rooting percentage of olive cuttings all treated at time of preparation with IBA in hydroalcholic solution, followed by treatment with PUT and H_2O_2 at time cuttings were made (0), at 2nd and 5th day. Triangles indicate the differences by X^2 test of homogeneity at P = 0.05 with the corresponding control (C) mean.

control. The necrosis was reduced to 10% by inhibiting PUT catabolism with AG, which doubled the endogenous conjugated of PUT at the first day or to a lesser extent, at 60%, by inhibiting the convertion of PUT to SPD when DCHA was added to rooting medium. The necrosis was completely eliminated by adding AG and DCHA at the same time (data not shown). The AG, applied singly or combined with PUT, increased rooting percentage, but this increase was not seen following addition of DCHA even in the presence of PUT (Fig. 2). However, in those rare experiments in which PUT, did not increase rooting, AG alone did not affect rooting unless added with PUT (data not shown). The inhibitors of PUT synthesis, DFMO and DFMA, both applied to the rooting medium, inhibited rooting; an effect which could be reversed partially by both the addition to the medium of PUT or by dipping the basal explants in H_2O_2 before placing them on the rooting medium (Fig. 2).

Effect of GA_3 and PUT on Rooting and on Peroxidase Activity in Explants. GA_3, at 57 μM added to rooting medium completely inhibited rooting. However, when it was added together with PUT, rooting occurred although at a lower percentage (Fig. 3). Explants derived from shoots grown on proliferation medium with 142 and 285 μM of GA_3 respectively showed decreased rooting ability (17% and 30%) unless PUT was added (data not shown). The peroxidase activity in the explants under a 16 h photoperiod (in which PUT increased rooting) was higher in PUT-treated explants than untreated ones from the 31st

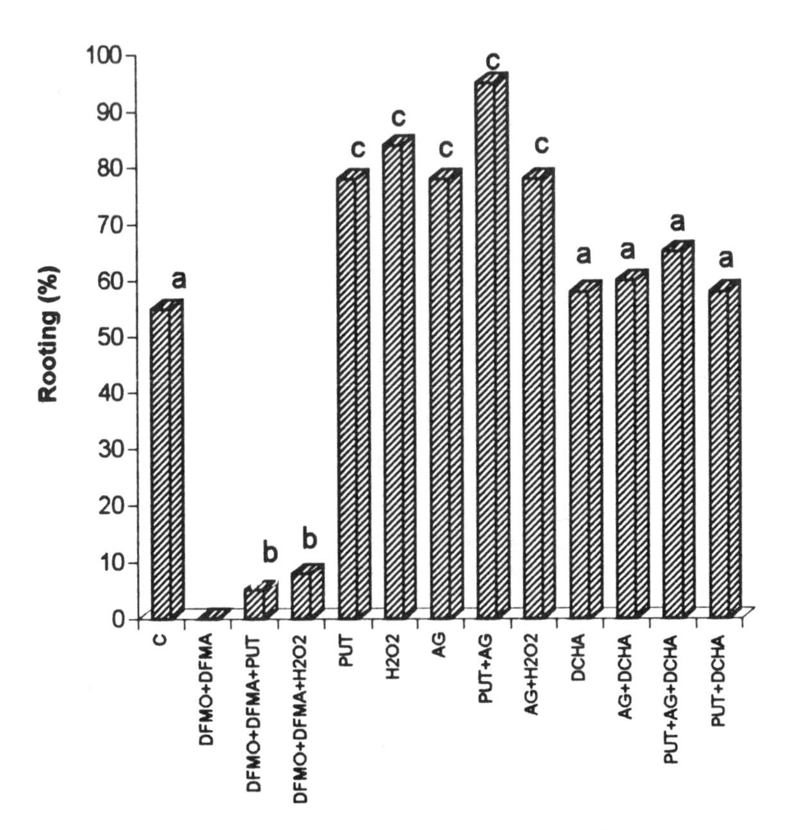

Figure 2. Effect of PUT, H_2O_2 and some inhibitors of synthesis and catabolism of PAs on rooting in *in vitro* olive explants.

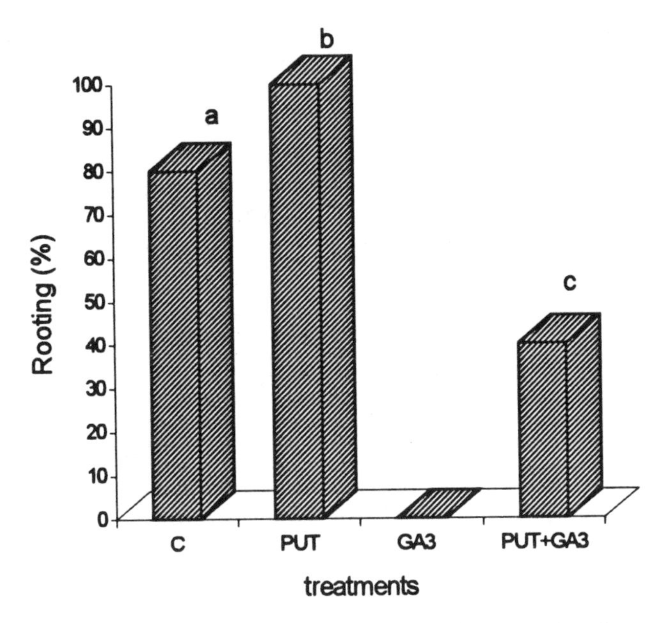

Figure 3. Effect of PUT and GA_3 added to rooting medium on rooting olive explants.

hour onwards (about 120%) (Fig 4). On the contrary, in explants placed in the dark (in which PUT does not increase rooting) the trend was similar but the activity was slighly higher in untreated explants (data not shown).

Effect of PUT on Root Elongation and Histological Observation. During 15 days of observations, the diamine did not increase root elongation in explants initially placed in auxin medium or in explants placed in auxin plus PUT medium. On the contrary, the explants initially treated with PUT showed shorter roots compared with those initiated in auxin only, but those transferred to a medium with auxin only showed more roots/explant (Tab. 1). The histological analysis showed a precocious formation of initials in explants treated with PUT. At 7 days only 33.3% of the untreated explants had formed initials, while 70% of the PUT-treated ones showed well developed root primordia.

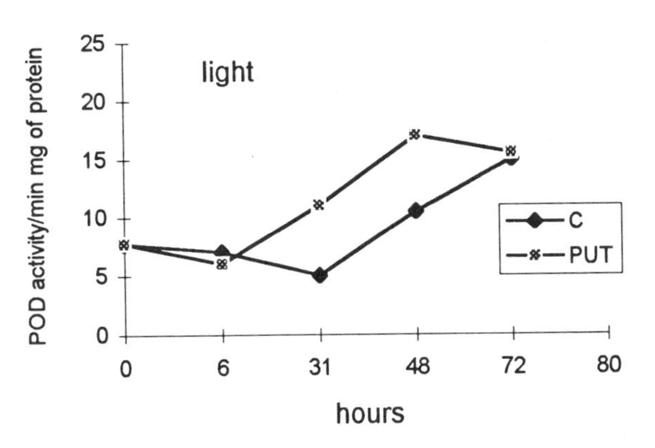

Figure 4. Total peroxidase activity in olive uninodal explants on rooting medium with or without PUT in 16 h light photoperiod. Data are means from an experiment run in triplicate representative of four independent experiments. SD did not exceed 5% of the means.

Table 1. Root length and number of roots/explant in *in vitro* olive explants detected after15th day from root protrusion in fresh NAA or NAA + PUT media

Media			Roots	
Initial	→	Elongation	Length (mm)	n°x explant
NAA	→	NAA	19.2a*	2.00a
NAA	→	NAA + PUT	19.0a	2.33a
NAA + PUT	→	NAA	15.2b	2.90b
NAA + PUT	→	NAA + PUT	15.1b	2.12a

* Mean separation by Duncan's test (P = 0.05)

Endogenous PA Variation during Early Rooting Stage

The total PA content in field grown shoots of olive was about half the content (172 nmol/g.FW) of PAs compared to chestnut and walnut (346 and 367 nmol/g.FW respectively). The content of PUT, SPD and SPM was more or less the same. Both, one centimeter basal end of the cuttings and *in vitro* whole explants, showed similar PA form (free, conjugated or bound) variation within 5 days of observation. However in those which would not root (difficult to root genotype, explants placed under 16 h photoperiod, untreated with PUT) showed, on the 2nd day and sometimes also on the 1st day, a significantly higher accumulation of bound PAs compared to those which would root easier (easy to root genotype, explants in darkness and explants treated with PUT) (Fig. 5).

DISCUSSION

Contrary to the observations with other species, in olive, exogenous treatment with PUT increases rooting percentage, promotes early rooting and increases root number. This is due mainly to the low endogenous content of PAs, both in field grown shoots, as demonstrated here and, as previously observed in *in vitro* shoots at the end of the proliferation phase (Rugini et al., 1993) if compared with species which do not respond to exogenous PUT. In this work 5 experiments were aimed to demonstrate the role of PAs in the early stage of rooting, by means of H_2O_2; other exsperiments demonstrated that PAs can be considered even as a marker of rooting: 1) H_2O_2 promotes earlier rooting and increases rooting percentage in a manner analogous to PUT. Since H_2O_2 treatment affects rooting both at preparation time and at the 2nd day, while PUT is only as affective at the time the cuttings are taken (Fig. 1), this indicates that the cells, to differentiate into roots, require H_2O_2 at an early stage. The H_2O_2, from exogenous PUT treatment at the 2nd day, might become available too late, because it seems that H_2O_2 does not come directly from PUT degradation rather from other PAs. In fact, the inhibition of PUT catabolism with AG does not reduce rooting when added alone or in combination with PUT, but on the contrary increases it. This increase was avoided by the inhibition of conversion of PUT to SPD by DCHA, indicating that the increase of rooting caused by exogenous application of PUT was mainly dependent on the conversion of PUT to SPD. This is in contrast with poplar, in which AG inhibited rooting while DCHA promoted it (Hausman et al., 1995). 2) Like PUT, H_2O_2 can partially remove the inhibition of rooting by the two inhibitors of PUT biosynthesis, DFMA and DFMO. 3) The increase of peroxidase activity in explants treated with exogenous PUT can be due to the H_2O_2, produced from PA degradation, H_2O_2 being a substrate

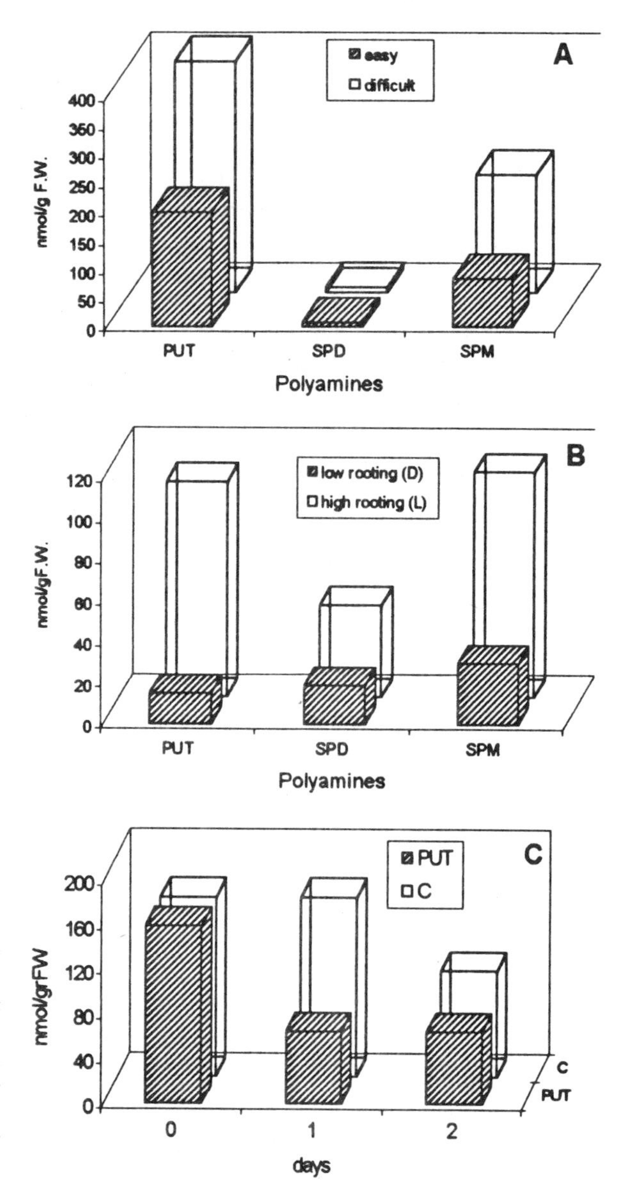

Figure 5. Bound PAs at 2nd day detected in: A) two cm basal cuttings in easy-to-root and difficult-to-root genotypes; B) in *in vitro* whole olive explants placed in 16 h light photoperiod (L), which resulted difficult-to-root and in the dark (D) which resulted easy-to-root; and C) total bound PAs detected in whole explants at 0, 1 and 2 day in rooting medium, with PUT (C) (high rooting %) and without PUT (low rooting %).

for peroxidases. This effect may contribute to the decrease of auxins in the cells which seem essential for rooting induction (Gaspar et al., 1985). This increase was observed only in the 16 h photoperiod, where exogenous application of PUT affected rooting. While, in the dark, where PUT does not affect rooting, it was similar or higher in PUT untreated explants. 4) PUT can avoid the inhibition of rooting caused by the treatment with GA_3,which allows the increase of auxins by protecting them from degradation. PUT re-established rooting both when GA_3 was added to the rooting and to the proliferation medium. Supplying shoots, during the proliferation phase with GA_3 at a concentration greater than 57 μM, may allow for accumulation in the cells. In fact, relatively low concentrations in the proliferation medium results useful because it reduces the lag phase of the shoots and consequently speeds up shoot growth without subsequently affecting rooting. 5) Finally, since PUT does not speed up root cell division (Tab. 1), the exogenously applied PUT acts in the very early stages of rooting phase; so the early root protrusion in PUT-treated explants observed was not due to a higher rate of cell division but to the earlier root initial formation, as the histological observation also demonstrated.

The other successful aim of this work was to consider PAs as a precocious marker of rooting. All of the three experiments aimed to compare the PA variation between explants placed in conditions ranging from easy-to-root to difficult-to-root (such as: 16 h photoperiod vs dark, with vs without PUT treatment and difficult-to-root vs easy-to-root genotypes) behaved in the same manner. All experiments showed an accumulation of PAs in bound form, within the 1st and 2nd day, only in explants which will root with difficult or will not root at all.

In conclusion, this work confirmed our previous observations that exogenous PUT and H_2O_2 treatments increased rooting percentage, root number/cutting or explant and promoted earlier rooting in both olive cuttings and *in vitro* explants. These marked effects are due to the low endogenous PA content compared to other species, such as walnut and chestnut, which are not affected by PA treatment; this suggests that olive could be used as a model plant in the study of the role of PAs. PUT does not affect root elongation at a concentration able to increase rooting, indicating that the earlier protrusion of root apicies exclusively depends on the early primordia formation. PUT affects rooting only if applied at an early stage, probably in the induction phase mainly through H_2O_2 from SPD or SPM catabolism by increasing peroxidase activity. Finally, determination of bound PA content could permit the identification of explants which will not root even at a very early stage by observing the accumulation of bound PAs.

REFERENCES

Bagni N., Serafini-Fracassini D. and Torrigiani P., 1982. Polyamines and cellular growth processes in higher plants. In: Plant growth substances. (P. F. Wareing ed.) Academic Press, London p. 473–482.

Biondi S., Diaz T., Iglesias I., Gamberini G. and Bagni N., 1990. Polyamines and ethylene in relation to adventitious root formation in *Prunus avium* shoot cultures. *Physiol. Plant.* 78: 474–483.

Bourgin, J.P. and Nitsch J.P., 1967. Obtention de Nicotiana haploides à partir d'ètamines cultivèes *in vitro. Annal Physiologie Vegetal* 9, 377–383.

Bradford M.M., 1976. A rapid and sensitive method for quantitation of microgram quantities of protein utilizing the principle of protein dye-binding. Annual. Bichem. 72:248–254.

Davis T.D., Haissing B.E., Sankhlan N., 1988. Adventitious root formation in cuttings. Discorides Press, Portland , pp 150:161.

Desai H.V. and Mehta A.R., 1984. Changes in polyamine levels during shoots formation, root formation and callus induction in cultured *Passiflora* leaf discs. *J. Plant Physiol.* 119:45–53.

Federico R., Angelini R., 1988. Distribution of polyamine and their related enzymes in etiolated and light-grown *Leguminosae* seedlings. *Planta* 173:317.

Friedman R., Altman A. and Bachrach U., 1982. Polyamine and root formation in mung bean hypocotyl cuttings. *Plant Physiol.* 70:844–848.

Gaspar T., Penel C., Castillo F.J. and Greppin H., 1985. A two-step control of basic and acidic peroxidase and its significance for growth and development. *Physiol. Plant.* 64:418

Hausman J.F., Kevers C. and Gaspar T., 1995. Putrescine control of peroxidase activity in the inductive phase of rooting in poplar shoots in vitro, and the adversary effect of spermidine. *J. Plant Physiol.* 146:681–685.

Mirkovic K., 1993. Effect of Polyamines on rooting of hardy deciduous (*Rhododendron* sp) *in vitro* (Master Thesis, Mediterranean Agronomic Institute of Chania, Crete, Greece)

Palavan N. and Galston A.W., 1982. Polyamine biosynthesis and titer during various developmental stage of *Phaseolus vulgaris*. *Physiol. Plant.*, 55:438–444.

Rey M., Diaz-Sala C., Rodriguez R., 1994. Exogenous polyamines improve rooting of Hazel microshoots. *Plant Cell Tissue and Organ Culture* 36:303–308.

Rugini E., 1984. *In vitro* propagation of some olive (*Olea europea* L.) cultivars with different root-ability. and medium development using analytical data from developing shoot and embryos. *Sci. Hortic.* 24:123–134.

Rugini E. and Wang X.S. 1986. Effect of Polyamines, 5-azacytidine and Growth Regulators on rooting in vitro of fruit trees, treated and untreated with *Agrobacterium rhizogenes* Proc. Int. Congress of Plant Tissue and Cell Culture. Minnesota, p 374.

Rugini E., Lupino M., De Agazio M. and Grego S. 1991. Endogenous polyamine and root morphogenesis variations under different treatments in cutting and *in vitro* explants of olive. *Acta Horticulture* 330: 225–232.

Rugini, E. 1992. Involvement of polyamines in auxin and *Agrobacterium rhizogenes*-induced rooting of fruit trees *in vitro*. *J. Amer. Hort. Sci.* 117:532–536.

Rugini, E., Jacoboni, A. and Luppino, M. 1993. Role of basal shoot darkening and exogenous putrescine treatments on in vitro rooting and on endogenous polyamines changes in difficult-to-root woody species. *Scientia Horticulture*. 53:63–72.

Slocum R.D., Kaur-Sawhney R.,and Galston A.W., 1984. The physiology and biochemistry of polyamine in plants. *Arch. biochem. biophys.* 235(2):283–303.

9

HISTOLOGICAL AND BIOCHEMICAL CHARACTERIZATION OF ADVENTITIOUS ROOT FORMATION IN WALNUT COTYLEDON FRAGMENTS

Laurent Duroux, Fabienne Fontaine, Christian Breton, Jean-Paul Charpentier, Patrick Doumas, and Christian Jay-Allemand

Station d'Amélioration des Arbres Forestiers
INRA-Orléans, F-45160 Ardon
France

INTRODUCTION

Studies conducted on the plant model *Arabidopsis thaliana* demonstrate that primary and lateral root meristem formation are controlled by different factors including the number, nature and environment of the original cells as well as hormonal and genetic controls (Aeschbacher et al., 1994, Celenza et al., 1995). The primary root meristem derives from one cell of the embryo and the hypophysis whereas the lateral root meristem rises from pericycle cells of primary root system. Adventitious root formation (ARF) is a multistep developmental process in which competent cells are induced to form a root meristem (Mohnen, 1994). Thus, it can be hypothesized that adventitious root meristem (ARM) formation has its own characteristics. This specificity could reside in 1) the induction process since ARM can be formed from differenciated cells of variable origins (Chriqui, 1985) and 2) the ARM organization process which depends on the cellular neighbouring of the pre-meristematic cells.

An *in vitro* plant model using mature cotyledon fragments cultured without phytohormone was developed in our laboratory to study ARF in walnut (Jay-Allemand et al., 1991). The first part of the work was to describe morphological and histological changes going on during this process leading always toward ARF after 4–5 days of culture. It was shown that induction occurred within the 2 first days of culture giving rise to an unorganized cell cluster with meristematic characteristics (Gutmann et al., 1996). These observations were in good accordance with those of Nougarede and Rondet (1983) in dwarf pea epicotyls. Then, the whole or only part of these pre-meristematic cells are committed to form an ARM (Gutmann et al., 1996). The work presented here emphasizes on the bio-

Biology of Root Formation and Development, edited by Altman and Waisel.
Plenum Press, New York, 1997.

chemical events involved in ARF induction and ARM formation in walnut cotyledons: indolic compounds, β-glucosidase activities and naphtoquinones were studied knowing their potential involvement in rhizogenesis (Blakesley et al., 1994, Ashford and McCully, 1973, Jay-Allemand et al., 1995).

EXPERIMENTAL

Plant Material

Series of cotyledon fragments were isolated from mature walnuts (*Juglans regia* L.) collected on the variety Lara in September 1995. These fragments were cultured *in vitro* on a hormone-free gelified medium in order to induce ARF (Jay-Allemand et al., 1991). Most of the analyses were conducted on the root initiation zone (RIZ) defined as a zone of 2 mm below the petiole tip.

Histological Studies

Cotyledon fragments were sampled before *in vitro* introduction and at 0, 96 and 144 h of culture. They were fixed for 2 h in a 2,5% glutaraldehyde solution in 0,1M phosphate buffer pH 7,5. After several washes in buffer solution, samples were dehydrated in a graded ethanol series (50, 70 and 96%) and embedded in glycol methacrylate. Semithin sections of 2 μm were stained with toluidine blue O.

Triiodo-Benzoic Acid (TIBA) Treatments

For dose-response experiments a series of cotyledon fragments was cultured in medium containing 0, 5.10^{-8}, 10^{-7}, 10^{-6} or 5.10^{-6} M TIBA. After 24 h of culture, samples were transferred into a new culture medium without TIBA. Rooting rates were expressed as the percent of rooted fragments after 15 days of culture. To evaluate TIBA effects on endogenous IAA and IAAsp contents, cotyledon fragments were cultured in medium without or with 5.10^{-6} M TIBA and collected to be analysed after 0, 6, 12 and 18 of culture.

Indole-3-Acetic Acid (IAA) and Indole-3-Acetylaspartic Acid (IAAsp) Quantitation

Indolic compounds were extracted with cold 80% methanol from freeze-dried samples. After 30 min of homogeneization by sonication, samples were stirred overnight at 4°C in darkness and centrifuged at 15,000 x g for 5 min. The supernatants were purified on Sep-pak-C18 cartridges. IAA and IAAsp contained within the eluates were separated on reverse phase HPLC and quantified by ELISA techniques with polyclonal antibodies (Label et al., 1994).

Protein and β-Glucosidase (BG) Analysis

Soluble proteins were extracted in 50 mM phosphate buffer pH 7,5. Crude extracts were centrifuged at 15,000 x g for 10 min. Supernatants were quantified at 595 nm according to the method of Bradford (1976) with a plate detector using BSA as standard. Soluble protein extracts were analyzed by isoelectrofocalisation (IEF) in 7,5% polyacrylamide gels using the pH range 3,5 to 10 carrier ampholytes and silver-stained (Morrissey, 1981). Total BG activity was quantified at 405 nm using a plate detector with p-nitrophenol-β-D-glucoside as substrate in 0,2 M acetate buffer pH 5,5. Assays were incubated in microtiters plates at 40°C for 15 min.

Juglone (Jug) and Hydrojuglone-Glucoside (HJG) Analyses

Phenolic compounds were extracted with cold 80% acetone by sonication for 30 min. The extract was centrifuged at 15,000 x g for 10 min and the supernatant was vacuum-dried. The residue was then diluted in pure methanol, centrifuged at 15,000 x g for 2 min and the supernatant injected for HPLC analysis.

BG Inhibitor Treaments

Two compounds, D-glucono-δ-lactone (GδL) and castanospermine (CAS), known for their inhibitory effect on BG activity, were tested for their effects on root production. Assays were performed by applying for 24 h soaked paper strips on the RIZ at 0, 48 and 72 h of culture. Paper strips were either soaked with 50 nmol of GδL or 5 nmol CAS. Rooting rates were determined after 15 days of culture.

RESULTS AND DISCUSSION

Results

Histology. The initial stage of cotyledon is shown Fig. 1a. Breakage of embryonic axis reveals the RIZ which is composed of 3 main cell types: epidermal cells, parenchymatous cells and vessels initials (Fig. 1b). After removal of the embryonic axis, one can note a clear break through parenchyma cells and vessel initials (Fig. 1b). Rapid cell division and differentiation occurred within 24 h to 48 h of culture. Division and elongation of the parenchymatous cells led to the formation of a typical petiole-like structure. Whereas these divisions ceased after 3 days of culture in the petiole they continued in RIZ (Fig. 1c) giving rise to a pre-meristematic zone. After 4 days of culture meristematic rings surrounding each vascular bundles were clearly visible (Fig. 1d). Finally, the emerging root tip appeared to be in continuity with the vascular bundle below (Fig. 1e). In order to better understand the physiological mechanisms involved in ARF, we looked for potential physiological markers and studied the effects of different inhibitors.

Effects of TIBA. TIBA strongly inhibited ARF since 50% and 100% inhibition with 50 nM and 5 μM, were obtained respectively (Fig. 2). IAA and IAAsp endogenous levels in the RIZ were quantified upon 5.10^{-6} M TIBA treatment. As shown in Fig. 2, TIBA provokes both IAA and IAAsp accumulation between 6 and 12 h of culture. Indeed, IAA content is increased 10 times at 12 h (Fig. 3A) whereas IAAsp is increased 2 times at 6 h of culture (Fig. 3B).

Quantitative and Qualitative Evolution of Proteins. Quantitative analysis of soluble proteins was performed in RIZ during ARF. Soluble proteins content peaked at 36 h of culture (Fig. 4) reaching 40 mg g^{-1} DW. Qualitative analysis of soluble proteins was conducted by IEF (Fig. 5). Three main bands (a1, a2 and a3) and a cluster of bands (b) ap-

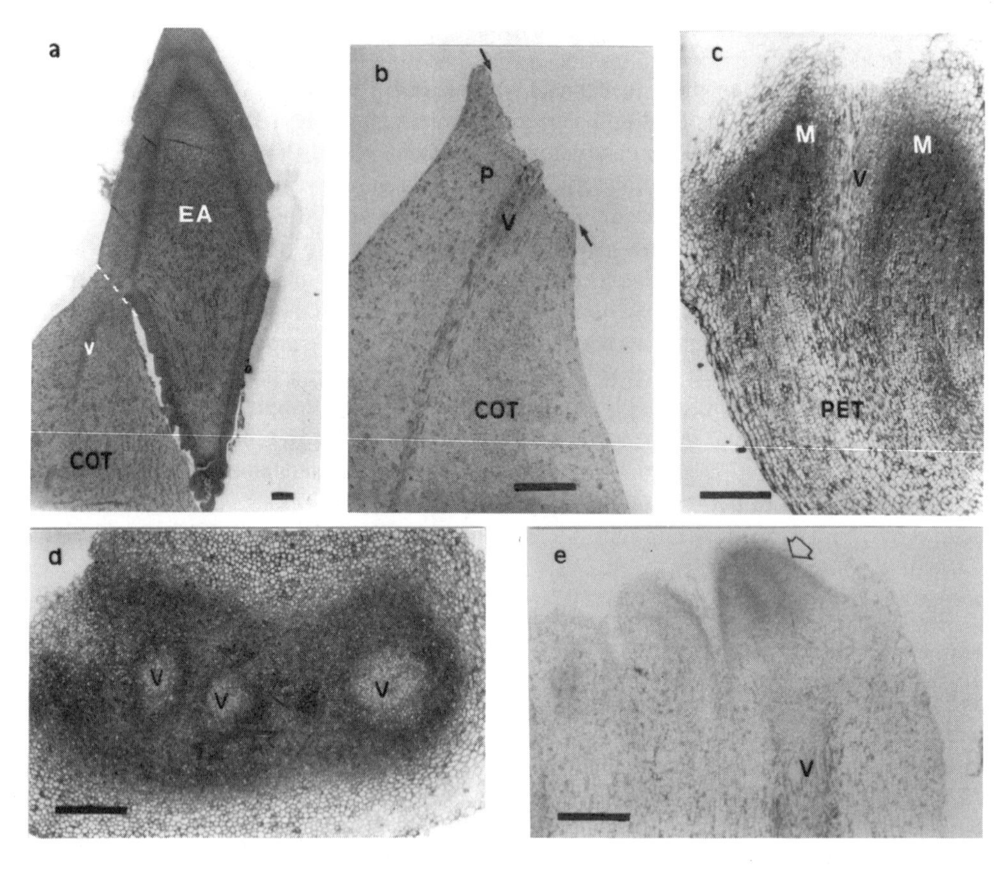

Figure 1. Different stages of cotyledon fragment development, toluidine blue O staining. (a) longitudinal section of cotyledon fragment at 0 h, still attached to the embryonic axis; EA, embryonic axis, V, vascular bundle, COT, cotyledon, dashed line indicates the future breakage zone. (b) longitudinal section of cotyledon fragment at 0 h after embryonic axis removal; P, parenchyma, arrows indicate the breakage zone. (c) longitudinal section of the petiole tip at 96 h of culture; M, meristematic zone, PET, petiole. (d) transversal section of the petiole tip at 96 h of culture; V, vascular bundle, arrows delimit the meristematic ring surrounding the vascular bundle. (e) longitudinal section of the petiole tip at 144 h of culture; an arrow indicates the emerging root. Bar = 200 μm (a,b), 50 μm (c–e).

peared after 36 h (Fig. 5) and were observed throughout root development. The expression of several isozymes was tested and revealed that bands a1, a2 and a3 had a potential BG activity (data not shown). Based on these results, we focused our work on these enzymes.

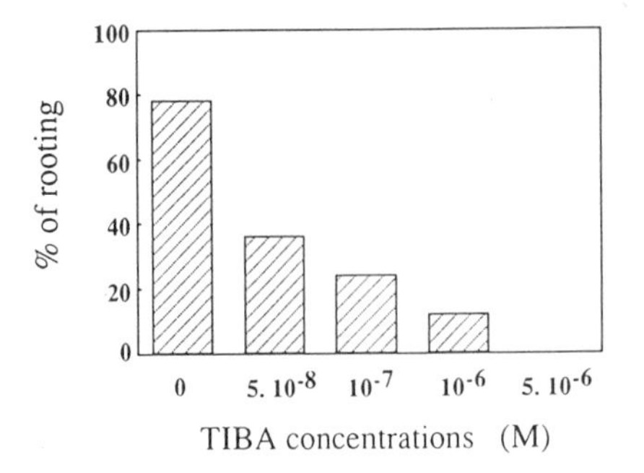

Figure 2. Dose-response effect of TIBA on rooting. For each value, 2 series of 10 cotyledon fragments were cultured in a medium containing increased concentrations of TIBA. Rooting percentages were determined as the number of explants forming at least one root.

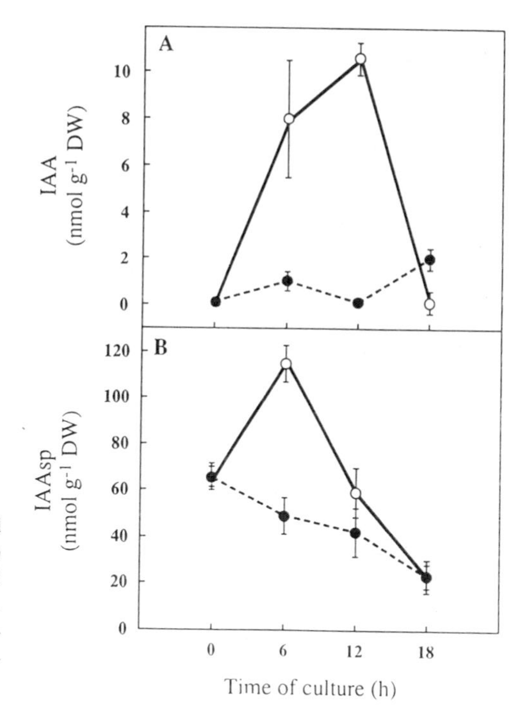

Figure 3. Effect of TIBA on endogenous IAA and IAAsp contents. Endogenous IAA (A) and IAAsp (B) contents were measured in RIZ in presence of 5.10^{-6}M TIBA (continuous line) or without TIBA (dotted line). Each value represents the mean of 3 replicates of 3 different samples composed by 12 fragments. Bars indicate SD.

BG Activity. Quantitative assays performed on RIZ and cotyledon crude extracts showed no activity within the first 24 h of culture (Tab. 1). A slight BG activity appeared at 48 h and was strongly enhanced at 96 h (Tab. 1). At this time, the average BG activity in RIZ was 15-fold higher than the cotyledon. Preliminary results showed that HJG, a typical naphtoquinone of *Juglans* species, was hydrolyzed by commercially available BG. Thus, we determined the time-course of naphtoquinones during ARF in our system (Tab. 1). No naphtoquinones (JUG and HJG) were detected at 0 h. JUG content increased between 24 h

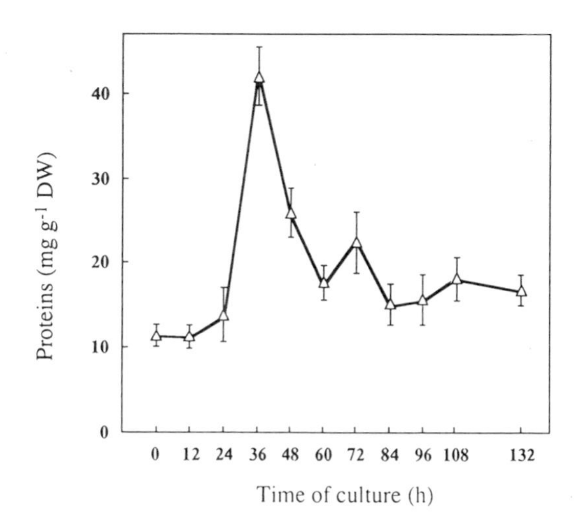

Figure 4. Evolution of soluble proteins in RIZ. Cotyledon fragments were collected every 12 h and the apical part was analysed. Each value represents the mean of 3 replicates of 3 different samples composed by 12 fragments. Bars indicate SD.

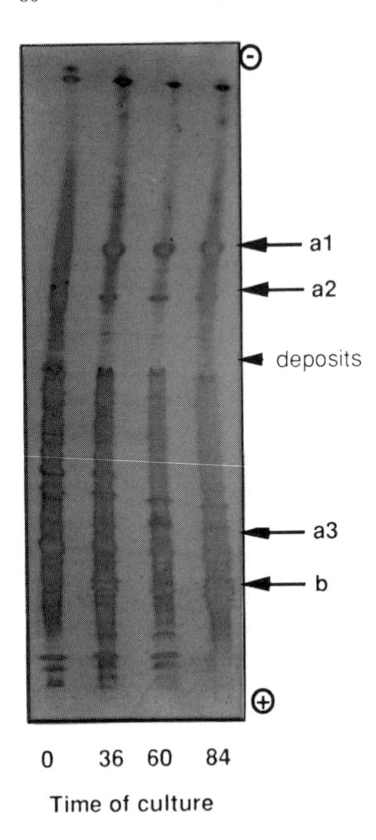

Figure 5. Qualitative analysis of soluble proteins from RIZ by IEF. Crude protein extracts were deposited on 7.5% acrylamide gel-slab containing carrier ampholytes in pH 3.5–10 range. Proteins were revealed by $AgNO_3$ staining. Arrows indicates proteins with BG activity.

to 96 h of culture while HJG content increased markedly (14 times) between 24 h and 48 h of culture (Tab. 1).

Effects of BG Inhibitors. Exogenous application of GδL or CAS led to a strong inhibition (88%) of ARF when applied at 72 h of culture (Tab. 2). Time-course study showed that CAS inhibited rooting with the same efficiency whenever the time of application. Interestingly, GδL inhibition efficiency was null when applied at 0 h but increased when applied later (38% at 48 h of culture).

Table 1. Evolution of soluble BG activity and RIZ naphtoquinones contents during rooting process[a]

Time of culture (h)	BG activity ($U.g^{-1}$ DW)		Naphtoquinones in RIZ ($mmol.g^{-1}$ DW)	
	RIZ	Cotyledon	Hydrojuglone-glucoside	Juglone
0	nd	nd	nd	nd
24	nd	nd	24 ± 8	2 ± 0.3
48	0.19 ± 0.22	0.06 ± 0.04	330 ± 54	18 ± 6
96	6.38 ± 0.36	0.42 ± 0.14	521 ± 53	53 ± 10

[a]Cotyledon fragments were collected every 12 h and BG activity measured both in RIZ or cotyledon whereas JUG and HJG contents were analysed in RIZ. One unit of enzyme activity is defined as the amount of enzyme hydrolyzing 1 μmole of pNPG per min. Each value represents the mean of 3 replicates from 3 different samples composed by 12 fragments. SD are indicated. nd = not detected.

Table 2. Effect of D-glucono-δ-lactone (GδL) and castanospermine (CAS) on rooting.[a]

Time of treatment	Treatment	% of rooting
	control	83
T 0h	water	100
	GδL	100
	CAS	17
T 48h	water	100
	GδL	62
	CAS	12
T 72h	water	100
	GδL	12
	CAS	12

[a]Inhibitors (5μl) were directly applied by paper strip on the root initiation zone during 24 h and then removed. Solution concentrations of GδL and CAS were 10 mM and 1mM respectively. Rooting percentages were determined as the number of explants forming at least one root. Each value represents the mean of 8 cotyledon fragments.

Discussion

The development of walnut cotyledon fragment leading to ARF is characterized by two distinct and presumably independent morphogenetic processes: 1) a limited growth of the cotyledon petiole corresponding to determined number of cell divisions occuring in the first 72 h of culture, and 2) the formation of ARM in close connection to the vascular strands which begins after 24 h of culture. The results reported here are relative to the latter point. Histological studies show that ARM is likely to be formed from a ring of intensively dividing cells surrounding the vascular bundle (Figs. 1c–e). Divisions affected the perivascular parenchymatous cells as well as the most outer vascular cell layers, in the RIZ. Thus, it seems that cells of different nature could take part in the formation of the root meristem. These results are in good accordance with the studies conducted in other species in which competent cells are often associated with peri- or interfascicular parenchyma (Chriqui, 1985).

The crucial role of the auxin IAA in ARF has already been reviewed by Blakesley et al. (1991). Indeed, several authors have shown that the first cell divisions leading to the formation of ARM were generally preceeded by a transient increase in endogenous IAA levels (reviewed by Blakesley, 1994). In our plant model, we also raised the question whether endogenous IAA could be considered as a possible inducer of the first cell divisions occurring within 24 h of culture (Gutmann et al., 1996). Our results show no transient accumulation of IAA within the first 18 h of culture. However, its involvement cannot be fully discarded since a complete inhibition of ARF was obtained with TIBA, known to inhibit cellular efflux of auxin (Lomax et al., 1995). Hence, TIBA treatment lead to a transient peak of IAA within the first 12 h of culture (Fig. 3A) which could be explained by IAA accumulation in the cells where it is synthesized. Three main hypotheses concerning the effects of IAA accumulation upon TIBA treatment can be seen: 1) cells in which IAA accumulates are submitted to abnormal fate thus preventing root meristem formation and, 2) competent cells in perivascular parenchyma are not supplied with IAA and consequently not induced to form the root meristem or, 3) in a same manner, provascular cells are not induced to differentiate thus preventing the efficient transport of metabolites from the cotyledon to the RIZ. We also showed that IAAsp content decreased within the first 18 h of culture (fig. 3B). This amide-conjugate form of IAA could be considered as a potential source of free IAA (reviewed by Sembdner et al., 1994) for root induction, although its significance in ARF is still unclear (Blakesley, 1994). TIBA treatment enhanced the accumulation of IAAsp (fig. 3B) thus revealing the potency of walnut cotyledon to also down-regulate IAA levels via conjugation.

One of the role of root-inducing growth regulators such as IAA is the induction of gene expression (see Van der Kop et al., 1996, and references herein). In our system, endogenous IAA could play such a role. Alternatively, it has been shown that other indolic compouds such as 5-hydroxytryptamine could be involved in this process (P. Doumas, unpublished results). Consequently to the early histological events started within the first 24 h of culture (Gutmann et al., 1996), we show that soluble protein content in RIZ peaked markedly at 36 h whereas no peak was recorded in the cotyledon (data not shown). This transient accumulation could either reflect the solubilization of proteins stored in protein bodies (Gutmann et al., 1996) or the expression of novel induced-proteins. Here, we have shown that the soluble protein content of the RIZ changes qualitatively throughout the rooting process and that BG activities are induced after 36 h of culture. Moreover, total BG activity was shown to increase during root meristem formation and more specifically in the RIZ. Ashford and MacCully (1973) showed that BG activity was localized in epidermis and cortex of the lateral root primordium, during lateral root formation in maize. For these authors, BG activity could be involved in cell-wall loosening or release of phenolics from their glucosides.

It is noteworthy that in our system, a strong accumulation of HJG occurred during root initiation reaching 180 $mg.g^{-1}$ DW after 96 h of culture. Besides, it was shown that the aglucone JUG is strongly correlated to the rooting capability of the walnut microshoot (Jay-Allemand et al., 1995) and can promote ARF in blackberry shoot cultures (Compton and Preece, 1988). Thus, BG could play an important role in releasing JUG from its glucoside, which may have a growth regulator-like activity (Ranade and David, 1985). Although no BG activity was detected in protein extracts at 24 h of culture (tab. 1), histochemical studies revealed that localized BG activities in RIZ were within cells peripheral to vascular bundle where HJG accumulates (C. Andary, pers. comm.). This activity could account for the JUG detected at 24 h of culture (tab. 1). BG activity increased between 48 h and 96 h was not accompanied with a consequent decrease in HJG content (tab. 1) revealing probably a different cellular compartmentation. In addition, ARF was strongly inhibited by GδL, a BG inhibitor, when applied at 72 h. This result could reveal an active involvement of BG in the organization phase of pre-meristematic cells since inhibition was null at 0 h and only of 38% at 48 h of culture. Compared to CAS, GδL inhibitory effect on ARF through BG activity inactivation could be more specific. Indeed, although CAS has previously been shown to inhibit rooting, this compound is also known to inhibit a wide range of hydrolases (Nagahashi et al., 1990) as well as glycoprotein processing (Elbein, 1988). This could explain why 83% of root inhibition was obtained with CAS even when applied at 0 h of culture.

CONCLUSION

This work conducted on walnut cotyledon fragments aimed, in a first attempt, to describe major events linked to the early stages of ARF. Several new elements were reported: 1) in this plant model, ARM seems to be formed from a ring of actively dividing cells surrounding a vascular bundle, 2) ARF is realized if endogenous IAA movements from cell to cell are possible and 3) ARM organization is partially depending on BG activity in possible conjunction with JUG production. These results define future lines of investigation in order to propose an integrated model of ARF. More detailed cyto-histological descriptions of ARF are yet needed to identify the pre-meristematic cells and to clarify their organization into a root meristem. Purification of different BGs is undertaken to both characterize their biochemical properties and their partial amino-acid sequence. This characterization will allow us to define their potential endogenous substrates (such as hormone-glucosides or phenol-glucosides) as well as their expression pattern *in situ* using molecular probes. The role of IAA as a potential inducer of these enzymes should also be investigated combining TIBA treatments and the above-mentioned approaches.

ACKNOWLEDGMENTS

The authors would like to thank the European Community for funding the research reported here through AIR3 CT92–0142 contract and the Region-Centre for L. Duroux's PhD fellowship.

REFERENCES

Aeschbacher, R.A., Schiefelbein, J.W., and Benfey, P.N., 1994, The genetic and molecular basis of root development. *Annu. Rev. Plant Physiol. Plant Mol. Biol.* 45:25.

Ashford, A.E., and McCully, M.E., 1973, Histochemical localization of β-glycosidases in roots of *Zea mays*. III. β-glucosidase activity in the meristems of lateral roots, *Protoplasma* 77:411.

Blakesley, D., 1994, Auxin metabolism and adventitious root initiation, *in*: "Biology of adventitious root formation" T.D. Davis and B.E. Haissig, eds., Plenum Press, New York.

Blakesley, D., Weston, G.D., and Hall, J.F., 1991, The role of endogenous auxin in root initiation. Part I: Evidence from studies on auxin application, and analysis of endogenous levels, *Plant Growth Reg.* 10:341.

Bradford, M.M., 1976, A rapid and sensitive method for the quantitation of microgram quantities of protein utilizing the principle of protein-dye binding, *Anal. Biochem.* 72:249.

Celenza Jr., J.L., Grisafi, P.L., and Fink, G.R., A pathway for lateral root formation in *Arabidopsis thaliana*, *Genes &Dev.* 9:2131.

Chriqui, D., 1985, Induction de prolifération des cellules prérhizogènes : auxines et polyamines. *Bull. Soc. bot. Fr.* 132:127.

Compton, M.E., and Preece, J.E., 1988, Effects of phenolic compounds on tobacco callus and blackberry shoot cultures, *J. Am. Soc. Hort. Sci.* 113:160.

Elbein, A.D., 1988, Glycoprotein processing and glycoprotein processing inhibitors, *Plant Physiol.* 87:291.

Gutmann, M., Charpentier, J.P., Doumas, P., and Jay-Allemand, C., 1996, Histological investigation of walnut cotyledon fragments for a better understanding of *in vitro* adventitious root initiation, *Plant Cell Rep.* 15:345.

Jay-Allemand, C., DePons, V., Doumas, P., Capelli, P., Sossountzov, S., and Cornu, D., 1991, *In vitro* root development from walnut cotyledons: a new model to study the rhizogenesis processes in woody plants. *C.R. Acad. Sci. Paris* 312:369.

Jay-Allemand, C., Doumas, P., Sotta, B., and Miginiac, E., 1995, Juvenility and physiology of rhizogenesis in two woody species (*Sequoia sempervirens* and *Juglans nigra* x *Juglans regia*), *in:* "Eurosilva - Contribution to forest tree physiology" H. Sandermann and M. Bonnet-Masimbert, eds., INRA ed., Paris.

Label, P., and Lelu, M. A., 1994, Influence of exogenous abscisic acid on germination and plantlet conversion frequencies of hybrid larch somatic embryos (*Larix leptoeuropaea*). *Plant Growth Reg.* 15:175.

Lomax, T.L., Muday, G.K., and Rubery, P.H., 1995, Auxin transport, *in*: "Plant hormones. Physiology, biochemistry and molecular biology" P.J. Davies, ed., Kluwer Academic Publishers, Dordrecht.

Mohnen, D., 1994, Novel experimental systems for determining cellular competence and determination, *in*: "Biology of adventitious root formation" T.D. Davis and B.E. Haissig, eds., Plenum Press, New York.

Morrissey, J.H., 1981, Silver stain for proteins in polyacrylamide gels: a modified procedure with enhanced uniform sensitivity. *Anal. Biochem.* 117:307.

Nagahashi, G., Tu, S.I., Fleet, G., and Namgoong, S.K., 1990, Inhibition of cell-wall associated enzymes *in vitro* and *in vivo* with sugar analogs, *Plant Physiol.* 92:413.

Nougarede, A., and Rondet, P., 1983, Cytophysiological basis of rhizogenezis induction in response to an auxinic treatment in the dwarf pea epicotyl, *Ann. Sci. Nat., Bot.* 5:121.

Ranade, S., and David, S.B., 1985, Quinones as plant growth regulators, *Plant Growth Reg.* 3:3.

Sembdner, G., Atzorn, R., and Schneider, G., 1994, Plant hormone conjugation, *Plant Mol. Biol.* 26:1459.

Van der Kop, A.M., Droog, F.N.J., Van der Zaal, B.J., and Hooykaas, P.J.J., 1996, Expression of an auxin-inducible promoter of tobacco in *Arabidopsis thaliana*, *Plant Growth Reg.* 18:1.

10

ADVENTITIOUS ROOT FORMATION IN WALNUT HYBRID ROOTSTOCKS (*Juglans hindsii* X *J. regia*)

James McKenna and Ellen G. Sutter

Department of Pomology
University of California
Davis, California 95616

ABSTRACT

The walnut hybrid "Paradox" is used extensively as a rootstock for English walnut cultivars grown in California orchards. "Paradox" rootstock produces more vigorous trees and increases yield compared to both *J. hindsii* (California black) or *J. nigra* (black). A major obstacle in the development of a clonal "Paradox" rootstock with superior disease resistance is that it has not been able to be rooted with much success. In addition, rooted cuttings usually died when transferred to the field. Recently we have been successful in rooting several genotypes of "Paradox" as well as other individual trees selected as potential rootstocks in the Walnut Improvement Program. Rooting percentages varied considerably ranging from 2 to 92%. Genotype was a major factor accounting for the variable results. Reliable, high (over 70%) rooting percentages, however, were obtained in several genotypes. The method of application was also an important factor in obtaining high rooting percentages. Application of IBA using a novel toothpick method was superior in most cases to that of a quick dip. The highest rooting percentages were first obtained with semi-hardwood cuttings. Careful attention to the amount of water applied to the basal sections of hardwood cuttings was essential for rooting of hardwood cuttings. Other factors that were critical for obtaining high rooting percentages included hedging of the stock plant, water status of the stock plant, and application of several compounds including abscisic acid, ferulic acid or spermine to hardwood cuttings. The addition of these organic compounds increased the rooting percentages significantly. Successful transfer in the field required the use of fiber plugs that kept the roots from being damaged when transferred.

1. INTRODUCTION

Production of walnut trees in California is commonly carried out by grafting selected varieties of English (Persian) walnut scion material onto seedling rootstocks. The

Biology of Root Formation and Development, edited by Altman and Waisel.
Plenum Press, New York, 1997.

great majority of these rootstocks are either *Juglans hindsii* (California black) or "Paradox", a hybrid resulting from crossing *J. hindsii* (female) with *J. regia*. "Paradox" is preferred as a rootstock in many parts of the state because it produces trees with greater vigor and yield than does *J. hindsii*. In addition "Paradox" is tolerant of most unfavorable soil conditions as well as being relatively resistant to *Phytophthora* and nematodes.

The desirability of "Paradox" as a rootstock has been further enhanced by selections from the Walnut Improvement Program (WIP) in the Department of Pomology at the University of California at Davis. Specific trees have been found to lack or have a greatly reduced hypersensitive reaction to cherry leaf roll virus (CLRV) which causes blackline, a devastating disease of walnut. In addition, selections of specific "Paradox" trees and trees resulting from a cross of "Paradox" trees with the parental walnut cultivar have yielded individual trees that exhibit great vigor and with it possibly the ability to better withstand *Phytophthora* spp. infection.

The production of "Paradox", however, is itself fraught with several problems including 1) being seedling material, it is non-uniform and heterozygous; 2) the yield of "Paradox" from *J. hindsii* trees is variable depending on the female parent and perhaps cultural or environmental conditions; 3) nurseries cannot determine in advance of germination which seeds will be "Paradox" and which will be California black, thereby using precious nursery space for germination of less desirable genotypes and 4) "Paradox" exhibits a hypersensitive reaction to cherry leafroll virus resulting in walnut blackline disease. The drawback of using seedling rootstocks is that they are heterozygous leading to variability in vigor and disease resistance. Production of clonal rootstocks would be a great improvement since it would provide uniform, genetically identical individuals available both for research as well as commercial use.

A major impediment to the production of clonal walnut rootstocks is the inherent difficulty of producing own-rooted walnut trees, whether by layering or cuttings (Lynn and Hartmann 1957, Serr 1954). Walnut cuttings and layers root poorly, if at all, and those cuttings that do root have low survival percentages when planted in the field. The objective of this research was to determine techniques and conditions that could be used in the production of clonal, own-rooted "Paradox" rootstocks from cuttings.

2. METHODS AND MATERIALS

Two main sources of rootstock material were used, "Paradox Bowman Kuhn", ("Bowman-Kuhn") a selection considered to be rootable, and selected WIP stock plants. "Bowman-Kuhn" trees had been grafted three years earlier using wood collected from mature trees. WIP stock plants were either grafted two to three years earlier or were three year old seedlings. Experiments were started in 1993. All trees were grown in the Department of Pomology orchard and were pruned back to their scaffold limbs and a few primary branches in February 1995.

2.1. Handling of Wood

Wood was collected throughout the growing season as well as during the winter when the trees were dormant. Preliminary experiments determined that late summer (semihardwood) and winter (hardwood) were the best time for harvesting wood. Branches were cut in the early morning and placed in buckets of water until they were brought back to the greenhouse for treatment. Twenty centimeter long cuttings with diameters ranging from 8 to 18 mm were prepared by cutting with clippers. The basal cuts of all cuttings were dipped in 1500 ppm citric acid prior to auxin treament. Leaves on semihardwood cuttings were trimmed to 2–3 per cutting, each with one pair of leaflets.

2.2. Hormone Treatments

The potassium salt of indolebutyric acid (KIBA) was prepared as aqueous solutions. Cuttings were treated either with a quick dip of auxin by dipping the basal end of the cutting into an 8000 ppm KIBA solution for 10 seconds or with a toothpick previously soaked in a 4000 ppm KIBA solution overnight. The toothpick was inserted into a hole drilled transversely through the stem 1 cm above the base of the cutting. It was important that the hole be a bit smaller than the toothpick so that the toothpick had to be tapped in gently and no air space surrounded the toothpick after it had been inserted.

Additional hormones and other organic compounds were also applied, either as basal dips or using a toothpick. These compounds included spermine (400 and 2,000 ppm), abscisic acid (ABA) (20 ppm), coumaric acid (12 μM) and ferulic acid (12 μM).

2.3. Field Survival

Both rooted and unrooted cuttings were planted in the field at various times, depending on the experiment. Semihardwood cuttings were planted into the field either in late autumn or early spring. In one treatment, semihardwood cuttings were planted in early spring when the ground was sufficiently dry. These cuttings had been potted up in late autumn/early winter (t) and had been kept in a lathhouse during the winter months. Other semihardwood cuttings were planted directly into the field approximately 50 days after hormone treatment and overwintered in the field. Hardwood cuttings were planted out in the spring when the soil could be worked.

3. RESULTS

3.1. Effect of Genotype on Rooting

Genotype had a strong effect on the percentage of rooting of the different selections. Rooting percentages varied from 0 to 91% depending on the genotype. Semihardwood cuttings of "Bowman-Kuhn" rooted as much as 83% whereas two WIP selections never rooted at all. The variation in rooting of other selections ranged from 4% to 91% depending on the genotype and method of application of auxin (Table 1; not all genotypes are

Table 1. The effect of genotype and method of auxin application on rooting percentages of semihardwood walnut cuttings

Genotype	Hybrid generation	Toothpick 4000 mg/l KIBA	Quick dip 8000 mg/l KIBA	Mean rooting %
84-121	F1 'Paradox'	91	52	71
84-128	F1 'Paradox'	79	52	65
Hind 3	F1 'Paradox'	50	44	47
87-026-4	F1 'Paradox' X J. regia	13	0	6
87-050-1	F1 'Paradox' X J. regia	6	22	14
87-117-2	F1 'Paradox' X J. regia	50	4	27
87-117-21	F1 'Paradox' X J. regia	38	29	33
87-117-12	F1 'Paradox' X J. regia	50	4	27

shown). It is notable that there was a tendency for the percentage of rooting to be highest in "Paradox" selections with a decrease in selections in which "Paradox" had been backcrossed to *J. regia*. The response of the different selections to the method of auxin application also differed markedly.

3.2. Effect of Type of Wood on Rooting of Cuttings

3.2.1 Hardwood Cuttings. The highest percentages of rooting of "Bowman-Kuhn" hardwood cuttings was obtained in December in both 1994 and 1995. The highest mean rooting percentage obtained was 14% on December 14 with December 3 being the next best time for obtaining rooted cuttings (7%). Rooting decreased considerably beginning in January with only 3% of the cuttings rooting. No rooting was obtained in February.

3.2.2 Semihardwood Cuttings. As with hardwood cuttings, the time of collection of semihardwood cuttings affected the rooting percentages significantly. Wood collected between early shoot elongation in April and early August was too soft, rotting easily and rarely producing roots. This wood was not used after the first year of experimentation. The greatest rooting percentages were obtained from August 15 to October 1 over a period of three years. For two years, wood of "Bowman-Kuhn" collected on September 15 produced cuttings with the highest rooting percentages. There was no single best date for collecting semihardwood cuttings for all the WIP genotypes. Although the dates varied, most genotypes produced cuttings with the highest rooting percentages between September 15 and October 1 (Figure 1).

3.2.3 Method of Auxin Application. The rooting of "Bowman-Kuhn" hardwood cuttings was not affected by the method of application and the greatest rooting percentage obtained was approximately 15% in both treatments in mid-December (data collected 1994). Hardwood cuttings of WIP selections varied considerably in their response to the method of auxin application, with most genotypes rooting poorly, if at all. Using toothpicks to

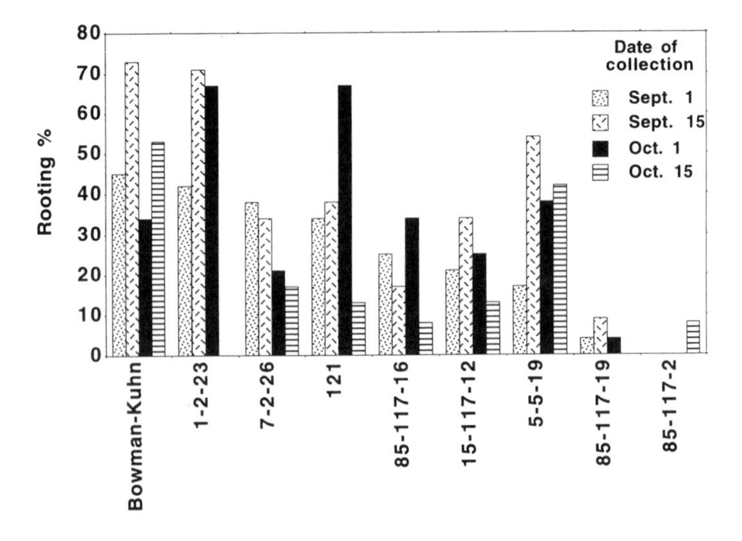

Figure 1.

Table 2. Effect of method of auxin application on rooting percentages of semihardwood cuttings of several genotypes of walnut

Genotype	Toothpick application[a]	Quick dip application[b]
85-117-12	46 ± 4	15 ± 11
128	79[c]	52[c]
121	79 ± 12	30 ± 22
87-026-4	28 ± 15	13 ± 12
87-050-1	32 ± 26	36 ± 14
87-032-1	48 ± 2	92[c]
85-117-19	2 ± 2	10 ± 7

[a]Toothpicks soaked in 4000 ppm K-IBA overnight and then placed into a hole drilled 1 cm above the base of the cutting.
[b]Base of cuttings dipped for 30 sec into 8000 ppm K-IBA solution.
[c]Genotype tested only one year.

apply auxin was effective in rooting hardwood cuttings of only one genotype, increasing the rooting percentage from 3% to 35%.

A pronounced promotive effect of using toothpicks to apply auxin was seen in selected WIP semihardwood cuttings, although there was no effect in the rooting of "Bowman-Kuhn" semihardwood cuttings. Rooting percentages increased from 50% to 200% depending on the genotype (Table 2). There was an equal number of genotypes, however, that had greater rooting percentages using a quick dip compared to the toothpick method (Table 2).

3.2.4. Use of Organic Supplements. Organic supplements were tested only on "Bowman-Kuhn" variety and only using the toothpick method of application. All organic compounds when used together with 4000 ppm K-IBA increased the rooting percentages of cuttings compared to K-IBA alone, but only in hardwood cuttings. Spermine (400 ppm) increased the rooting percentage from 20% to 80%. The other compounds also increased the rooting percentages compared to the use of K-IBA alone (Table 3).

3.2.5 Survival of Rooted Cuttings in the Field. Rooted cuttings of both hardwood and softwood cuttings had low survival percentages when transferred to the field. Most of the semihardwood cuttings that had been potted up and overwintered in the lathhouse died. When the plants were examined it was found that the roots had rotted. Initial survival was better in cuttings that had been rooted in plugs compared to those that were rooted directly in flats. The highest percentage of rooting for both hardwood and semihardwood cuttings was about 25%.

Table 3. Effect of a variety of organic compounds used in conjunction with 4000 ppm K-IBA on rooting percentages of "Bowman-Kuhn Paradox" walnut cuttings

	Compound	Concentration	% Rooting
K-IBA 4000 ppm	—	4000 ppm	15
K-IBA 4000 ppm	Spermine	2000 ppm	38
K-IBA 4000 ppm	Spermine	400 ppm	85
K-IBA 4000 ppm	Abscisic acid	20 ppm	58
K-IBA 4000 ppm	Coumaric acid	12 μM	64
K-IBA 4000 ppm	Ferulic acid	12 μM	78

4. DISCUSSION

Two areas of concern in the rooting of cuttings of walnut are the root induction and transfer to the field. Several factors appear to be most important in rooting walnut cuttings. Genotype, method of applying auxin, time at which the cuttings were taken, type of wood, and the use of additional organic compounds were all shown to be significant factors. Although as much as 90% of cuttings of a given genotype could be rooted subsequent survival and growth in the field was disappointly low. More work is required to determine the reasons for the low survival of rooted walnut cuttings in the field. Only with higher percentages of survival will clonal production of walnut rootstocks by cuttings be economically feasible.

REFERENCES

Lynn, C. And H.T. Hartmann. 1957. Rooting cuttings under mist. Calif. Agric. 11:15.
Serr, E.F. 1954. Rooting "Paradox" walnut hybrids. Calif. Agric. 8:7.

ROOTING INDUCTION AFTER *IN VITRO* COLD STORAGE OF CORK OAK CULTURES

A. Romano[1] and M.A. Martins-Loução[2]

[1]Unidade de Ciências e Tecnologias Agrárias, Universidade do Algarve, Campus de Gambelas, 8000 Faro, Portugal. [2]Departamento de Biologia Vegetal, Faculdade de Ciências de Lisboa, Campo Grande, Bloco C2, Piso 4, 1700 Lisboa, Portugal.

INTRODUCTION

Plant tissue culture techniques are the basis for *in vitro* conservation. The frequent transfer of *in vitro* cultures onto fresh medium is the simplest preservation method. However, it is costly in terms of laboratory management and can lead to serious problems, e.g., an increased risk of contamination, equipment failure and risks of genetic variation. Those increase with the culture duration in axenic conditions and can lead to the production of plants which are not true to type. These problems can be overcome by reducing the growth rate of cultures, under suboptimal conditions. Reduction of temperature is the most commonly used method for storing cultures *in vitro*. In this investigation we studied the rooting capacity of cork oak shoots after storage at 5°C for 1 to 12 months.

EXPERIMENTAL

The procedures and conditions for establishment and multiplication of shoot cultures, started from buds of cork oak adult trees, have been described previously (Romano and Martins-Loução, 1992; Romano et al., 1992; Romano et al., 1995). Cultures were stored *in vitro* on multiplication medium at 5°C in the dark without an intervening subculture for one to twelve months. For rooting, the basal ends of 3 cm long individual shoots, were harvested at the end of the 3rd multiplication cycle after storage.Those were dipped for 2 min in 2.5 mM indole-3-butyric acid followed by culture on GD (Gresshoff and Doy, 1972) growth regulator-free medium. Shoots were grown in the dark for 1 week and then placed under normal light conditions.

RESULTS

The survival rate of the shoots after twelve months of storage was 62%. This rate decreased with the increase of the storage duration. Multiplication rate of stored shoots and shoot elongation, after transfer to standard conditions, was not significantly different from unstored controls. The rooting capability of shoots derived from stored cultures for 12 months was similar to that of unstored controls, in terms of rooting percentage (92.5%), number of roots (N = 3) and length of the longest root (L = 6.5 cm). After 1 month of rooting induction the plantlets were successfully acclimatized to *ex vitro* conditions, and were similar to plantlets derived from unstored shoots.

CONCLUSIONS

The results show that cold storage of cork oak shoot cultures for one year is feasible. The storage of cultures at 5°C in the dark, without an intervening subculture, appears to offer considerable promise as a technique for medium-term conservation of cork oak germplasm.

REFERENCES

Gresshoff, P., and Doy, C., 1972, Development and differentiation of haploid *Lycopersicon esculentum* (Tomato), *Planta*, 170:161.
Romano, A., and Martins-Loução, M.A., 1992, Micropropagation of mature *Quercus suber* L.: Establishment problems, *Scientia gerundensis*, 18:17.
Romano, A., Noronha, C., and Martins-Loução, M.A., 1992, Influence of growth regulators on shoot proliferation in *Quercus suber* L., *Ann. Bot.*, 70:531.
Romano, A., Noronha, C. and Martins-Loução, M.A., 1995, Role of carbohydrates in micropropagation of cork oak. *Plant Cell Tiss. Org. Cult.*, 40:159.

NOVEL LIGNO-SILICON PRODUCTS PROMOTING ROOT SYSTEM DEVELOPMENT

Galina Telysheva,[1] Galina Lebedeva,[1] Natalya Zaimenko,[2] Datse Grivinya,[3] Tatiana Dizhbite,[1] and Olga Virzina[3]

[1]Latvian State Institute of Wood Chemistry, 27 Dzerbenes str., Riga, LV-1006, Latvia. [2]Botanic Garden of Academy of Science, Kiev, Ukraine. [3]National Botanic Garden of Academy of Science, Riga, Latvia.

Environmental friendly bio-active multipurpose silicon-containing lignocellulose products (LSP) have been synthesized using wastes of the wood processing industry or agricultural wastes. Combination of silicon with lignocellulose ensures a regulated transfer of silicon into the soil, the simplicity of its introduction and dosage. LSP application improves the quality of plants, increases the plants' resistance against diseases and other unfavourable environmental conditions, and positively influences the soil microbiota. The present investigation was designed to assess the role of LSP activity in the processes of the root system formation.

LSP activity was tested following LSP introduction into the soil and after dusting seeds and shoots. The LSP enhancing effect on the development of the root system was exemplified on different agricultural and ornamental plants.

LSP application to corn (dusting of seeds) promoted the development of the main and of the prop roots. Both, underground and air-roots, were more long, thicker and more branched. The length of lateral roots increased by 30–40%, as compared to the control corn seedlings. In micro field assays, the root system development, measured by 10-point scale, was 1 for control and 7 for LSP. The prop roots of LSP plants were continuing to develop during the period of corn grain formation. Such roots of control assays stopped their growth, and 8–12 days later died. For more than 20 ornamental plants such characteristics as root length, root volume, root dry mass increased significantly, under LSP influence, by 50–75% in comparison with those of the controls. Simultaneously with the enhanced development of the root system, flowering began earlier and was more prolonged. Rooting of the shoots was notably higher.

The mechanism of LSP promoting of root development is yet unknown and for this purpose the dynamics of the changes in the peroxidase and auxin activities was examined during the earliest stage of the root development in winter wheat seedlings treated with LSP. It was shown that following the application of LSP a rapid increase in peroxidase activity, followed by a rapid decrease, occurred during the 2nd day, as compared with the activity during the 3rd and 4th day, and was more pronounced as compared with the controls. The auxin activity, in seedlings at the earliest stage (1st-3rd days), was low both for the treatments and control samples. During this period gibberellins were not detected. However, later, the level of gibberellins, and of substances with auxin activity, increased in the LSP treatments more markedly than in control samples. Such a difference reached a peak on the 7th day of the seedlings' development.

REGULATION OF ROOTING ABILITY OF *Solanum tuberosum* L. EXPLANTS BY GASEOUS ENVIRONMENT, *IN VITRO*

G. Arente[1] and D. Gertnere[2]

[1]Department of Plant Physiology, Institute of Biology, Latvian Academy of Sciences, 3 Miera Str., LV-2169, Salaspils, Latvia. [2]Faculty of Biology, University of Latvia, 4 Kronvalda Blv., LV-1842 Riga, Latvia.

INTRODUCTION

It has been reported by many investigators that morphogenesis of explant tissue can be regulated by the composition of the nutrient medium as well as by the gaseous environment (Buddendorf-Joosten and Woltering, 1994). The aim of the present study was to investigate different effects on rooting ability of potato explants *in vitro*, caused by limited gas exchange and IBA and BAP treatment.

EXPERIMENTAL

Rooting ability of potato (*Solanum tuberosum* L. cvs. Agrie dzeltenie and Zile) stem explants and excised leaves were investigated. Explants (10–20 per flask) were cultivated in 50 ml glass flasks on 20 ml solidified agar, hormone free MS medium (Murashige and Skoog, 1962). In feeding experiments the explants and excised leaves were submerged for 5–30 min in indole-3-butyric acid (IBA) or benzylaminopurine (BAP) (10 mg/L) solution prior the micropropagation. Three different types of closure were used: cotton plugs, silicone stoppers and silicon stoppers with a cotton plug inserted in a 5 mm aperture of the stopper. Ethylene content inside the flask was analyzed on GC Chrom 5 (Czech Republic) equipped with a flame ionization detector and alumina filled glass column. Pigment content was estimated as described by Moran and Porath (1980) and catalase activity was measured according to Lück (1965).

RESULTS AND DISCUSSION

The content of ethylene, as well as that of ethane and propylene increased significantly in the flasks that were closed by silicone stoppers. At the same time the flasks that were closed by silicone stoppers with aperture, allowed a limited exchange of gases between the internal and external environment. In the case of hydrocarbon accumulation into internal space of *in vitro* flasks, inhibitory effect on root elongation was detected. When the flasks were closed with silicone stoppers with an aperture, the rooting density of *Solanum tuberosum* L. explants increased, with the subsequent stimulation of shoot and root growth. Cultivation in the presence of IBA in the MS medium effectively promoted root elongation as well as rooting density (200%) of stem explants. By submerging the explants and the excised leaves in IBA solutions, prior to micropropagation, the rooting density and the root elongation increased. In excised leaves, pretreated with IBA and cultivated under tight closing of the flasks, chlorophyll content decreased. This was less obvious in the case of BAP treatment. An increased production of ethylene was seen in explants, cultivated under a limited gas exchange. An enhanced activity of catalase (160%) and rooting, along the stem of potato shoots, was observed in flasks with limited gas exchange.

CONCLUSIONS

The results show the possibility to regulate the rooting of explants, *in vitro*, both by using active growth substances as well as by varying the gaseous environment by different means of closure.

REFERENCES

Buddendorf-Joosten, J.M.C., and Woltering, E.J., 1994, Components of the gaseous environment and their effects on plant growth and development *in vitro*. *Plant Growth Regul.* 15:1.

Lück, H., 1965, Catalase, *in*:"Methods of Enzymatic Analysis." H.U. Bergmayer, ed., Academic Press, New York, p. 885.

Murashige, T. and Skoog, F., 1969, A revised medium for rapid growth and bioassays with tobacco tissue culture. *Physiol. Plant.* 15:473.

Moran, R., and Porath, D., 1980, Chlorophyll determination in intact tissues using N,N-dimethylformamide. *Plant Physiol.* 65:478.

11

INCREASED INDUCTION OF ADVENTITIOUS ROOTING BY SLOW RELEASE AUXINS AND ELICITORS

W. M. Van der Krieken,[1] J. Kodde,[1] M. H. M. Visser,[1] D. Tsardakas,[2] A. Blaakmeer,[1] K. de Groot,[1] and L. Leegstra[1]

[1]AB-DLO, Department of Plant Physiology
PO Box 14, 6700 AA Wageningen
The Netherlands
[2]Technical Education Institute
PO Box 140, 71110 Heraklion (Crete)
Greece

SUMMARY

Indolebutyric acid (IBA) is the auxin that is commonly used for induction of adventitious roots. Our experiments show that IBA is not active itself but exerts its action by conversion (β-oxidation) into indoleacetic acid (IAA). This implies that IBA is a naturally occurring slow release source for IAA. We mimicked this slow release principle by synthesizing auxin slow release sources, and so decreasing the rate of auxin metabolic inactivation after uptake in plant tissue. The synthetic auxin slow release sources proved to be more effective than the standard auxins used in adventitious root induction. Also, a method was developed to increase auxin action: breakdown products of cell structures that are known to be involved in plants defense processes (elicitors) proved to enhance tissue sensitivity to auxin. The combined application of slow release auxins and elicitors resulted in high rooting percentages on very recalcitrant to root woody plants.

INTRODUCTION

The auxin most commonly used to induce root formation is IBA. However, the metabolism of IBA has received only little attention in contrast to IAA. IBA or IAA action in general depends on (A) the concentration of active auxin at the site of action and (B) tissue sensitivity. The active auxin concentration at the site of action depends on: 1) uptake by the plant tissue, 2) transport to the site of action, 3) metabolic inactivation.

Biology of Root Formation and Development, edited by Altman and Waisel.
Plenum Press, New York, 1997.

IBA or IAA Concentration at the Site of Action. After uptake, IBA can be conjugated with amino acids and sugar alcohols (Wiesman et al. 1989, Nordström et al. 1991). In general, comparable conjugation products of IBA and IAA are found (Andreae and Good 1957, Pythoud and Buchala 1989), however, the rate of the conjugation of IBA and IAA might be different. Like the IAA conjugates, the IBA conjugates are not physiologically active themselves but form a potential source of free (active) auxin (Hangarter and Good 1981, Wiesman et al. 1989, Cohen and Bandurski 1982). IBA can be converted into IAA (Epstein and Lavee 1984, Van der Krieken et al. 1992a, b). This conversion of IBA into IAA shows that IBA is a slow release source for IAA. Conversion of IAA into IBA was found in maize seedlings (Ludwig-Müller and Epstein 1991) but could not be detected in apple (Van der Krieken et al. 1993). For the better rooting effect of IBA than of IAA exist several possible options (for review see Epstein and Ludwig-Müller 1993). Firstly, IBA and IAA are both physiologically active in root induction, but the rate of conjugation of IBA might be slower than that of IAA. Secondly, the free acid of IBA might be more active than the free acid of IAA. Thirdly, IBA might not be active itself but exerts its action via conversion into IAA. Synthesis of new slow release sources of IAA and IBA may prove whether or not IBA is active via conversion into IAA: we need to synthesize a slow release source of IAA that after uptake leads to the same concentration of free IAA in the tissue (at the location where the roots are formed) as application of IBA does, both types of auxins then need to induce the same amount of roots.

Modified Auxins. Using chemical modification new auxin compounds were developed that are optimized with respect to activity: slow release auxins. Slow release auxins are coupled (conjugated) to a carrier molecule. The auxin can either be released enzymatically from the carrier (by e.g. esterases) or the bond between auxin and carrier is chemically unstable in the plant tissue. An example of a carrier molecule is bovine serum albumin (BSA). The BSA molecule is a relatively small globular carrier protein to which about 35 auxin molecules can be coupled. The BSA-auxin conjugate can enter the plant (apoplast) where it continuously releases (catalyzed by enzymes present in the apoplastic space) active auxin.

Tissue Sensitivity. When roots are induced on stem cuttings wounding may be involved in the rooting reaction. Wounding results in the destruction of cell compartments (vacuoles, vesicles, peroxisomes, plastids) which results in release of the catabolic enzymes (glucanases, peroxidases, phospholipases, lipoxygenases) that are present in these cell organelles. These enzymes lead to the further breakdown of certain cell structures (cell wall, cell membranes). Some of the breakdown products of the cell structures, the so called elicitors, are known to be involved in plants defense processes (for a review see Lyon et al. 1995). To assess the importance of wounding in adventitious rooting we tested if elicitors enhance tissue sensitivity for auxin.

So far, research dealing with auxin action related to root formation has been conducted with relatively large amounts of tissue (stem cuttings, plantlets). Since only a very small proportion of the cells in those tissues is (structurally) involved in rooting, the content of the active auxin components (e.g. the free IAA and/or IBA) in the total tissue need not be specifically related to the rooting process that takes place in the basal part of the tissue. Therefore, we studied uptake and metabolism of IBA and IAA in a test system of thin stem disks of in vitro cultured apple shootlets. In this system, a high number of roots (8) is formed synchronously on a very small amount of tissue (1.5 mg) without the interference of compounds originating from other parts of the plant (Van der Krieken et al.

1993). Using the thin stem disk system, it is possible to study auxin metabolism related to root induction at the location where the induction process is going on.

In this report we show, the biochemical and physiological levels the mode of action of auxins and elicitors on root regeneration. First, it is shown, by TLC analysis, that activity of IBA taken up from the medium is through slow release of IAA. Secondly, synthetic slow release sources can further enhance the auxin activity for adventitious rooting. Thirdly, elicitors formed via degradation of cell structures enhance auxin action.

MATERIALS AND METHODS

Rooting of Stem Disks

The experimental material of choice was the apple rootstock M."Jork". This clone has been micropropagated for more than seven years. Shoots were multiplicated on MS medium (Murashige and Skoog 1962) with 165 mM sorbitol, 4.4 μM benzyladenine and 0.6% (w/v) agar (BBL) at 20°C with a 16 hour light period (PAR 60 $\mu mol\ m^{-2}s^{-1}$). Rooting experiments were conducted with stem disks, 0.7 mm thick and a diameter of ca 1.5 mm, that were cut with an apparatus consisting of a block in which 10 razor blades are mounted at a distance of 0.7 mm from each other. For rooting experiments, groups of 20 or 30 slices were incubated in 9 cm petri dishes with the apical side on a nylon mesh on a medium described by Quoirin et al. (1977) with 88 mM sucrose and 0.6% (w/v) agar. Various IAA, IBA, auxin slow release sources and elicitor concentrations were used as indicated in the text. The dishes were incubated upside down in the dark at 25°C. After several different time intervals, the mesh with the slices attached was placed for a few seconds on sterile moist filter paper to minimize carry-over of hormones or elicitors. Hereafter, root outgrowth was in the light on medium without hormones and elicitors (PAR 60 $\mu mol\ m^{-2}s^{-1}$) at 25°C. The number of roots that emerged was counted 14 days after cutting of the disks.

Rooting of Scions in Vivo

Scions of apple (cv elstar) were rooted with rooting powder consisting of talc mixed with the slow release auxin indolehexanoic acid and the elicitor nonanoic acid (see expriments).

^{14}C-IAA and ^{3}H-IBA Labeling Experiments

Groups of twenty disks were incubated (in duplicate) on nylon mesh in petri dishes with a diameter of 5 cm on different concentrations of ^{14}C-IAA or ^{3}H-IBA as indicated in the experiments. Radioactivity was added to the petri dishes two days before the start of the experiment. In all incubations, the specific activity of ^{14}C-IAA was 0.2 TBq per mol and the specific activity of 3H-IBA was 3.0 TBq per mol. Radioactive IBA and IAA and their metabolites were extracted with methanol and butylhydroxytoluene (Van der Krieken 1992 a). The auxins and their metabolites were chromatographed on silica gel 60 PF_{254} plates (Merck) in two different solvents. The first solvent used to develop the plate was $CHCl_3$:HAc (95:5, v/v) and after drying, the plate was developed in the same direction in $CHCl_3$:MeOH:HAc; (75:20:5, v/v/v). To quantify radioactivity on the plates, the UV-absorbing spots and the sections between these spots were scraped off, mixed with 5 ml Ultima Gold scintillation liquid (Beckman) and counted in a liquid scintillation counter.

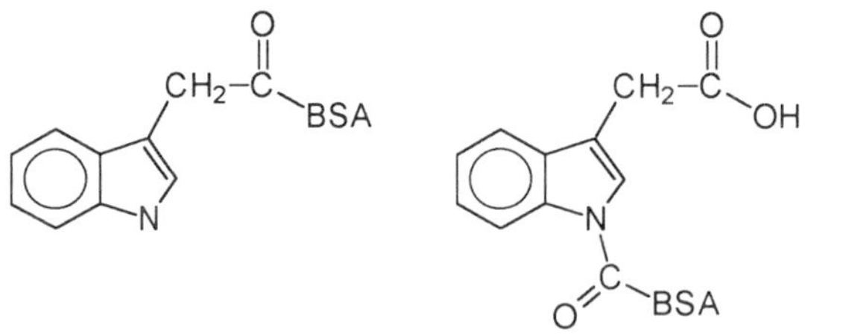

Figure 1. Structure of IAA coupled to BSA via the carboxylic group (IAA-C-BSA) and IAA coupled to BSA via the amide group (IAA-N-BSA). About 35 auxins can be coupled to one BSA molecule.

Finally, the authenticity of the free IBA and IAA acid was verified with 2D-TLC and reversed-phase HPLC (Van der Krieken et al. 1992 a).

Synthesis of Auxin Slow Release Sources

Auxins were coupled with an amide or ester bound to a carrier molecule (Fig. 1). The methods used were adaptations from methods described in literature (coupling of auxin to BSA via an ester bond: Weiler 1981, or via an amide bond: Pengelly 1977).

RESULTS

Uptake, Transport, and Metabolism of Auxin in Thin Stem Disks of Apple

As stated above auxin action depends on uptake, transport and inactivation by metabolism. Incubation of the shootlets on medium with radioactive IBA or IAA showed that auxin transport is not a problem for root regeneration (Van der Krieken et al. 1992B). Auxin accumulation is caused by the polar auxin transport system (directed from the apex towards the basis of the plant). The auxin accumulation in the stem base (the location where roots are formed) indicates that transport does not limit root regeneration. In the thin stem disks, used in the other experiments described in this paper, auxin levels were comparable to those in the stem base of shootlets after incubation on media with the same auxin concentrations.

The uptake and metabolism of IBA and IAA in thin stem disks of apple is presented in Table 1. Uptake and metabolism of IBA and IAA is linearly related to the concentration

Table 1. Uptake and metabolism of ^{14}C-IAA and ^{3}H-IBA in thin stem disks of apple[a]

Auxin	Uptake	Free IAA	Free IBA	Roots per disk
1 μM IBA	32	0.6	1.2	1.2
3.2 μM IBA	80	1.1	2.1	2.8
10 μM IBA	230	3.3	4.2	6.2
1 μM IAA	21	0.08	nd	0
3.2 μM IAA	51	0.12	nd	0.4
10 μM IAA	162	0.82	nd	1.6

[a]Groups of 20 disks were incubated for 1 day on media with different auxin concentrations. After incubation IBA or IAA and their metabolites were extracted and analyzed on TLC as described in materials and methods. Values are expressed in μmol kgFW^{-1}. The values represented are averages of experiments conducted in triplicate. All SE were less than 11% of the average value. nd = not detectable.

of the auxin in the medium. The uptake of IBA is higher than that of IAA: after 24 h the uptake (expressed as µmol kg FW^{-1}) of IBA and IAA was about 25 and 15 fold higher respectively than the medium concentration. This shows that the rate of auxin uptake is high (Table 1). So uptake is no limitation for auxin-induced root regeneration.

Auxin metabolism on the other hand is very fast. Of the IBA that was taken up in 24 hours about 2.5% was extracted as the free IBA acid and about 1.5 to 2% as free IAA acid. From IAA taken up from the medium only 0.5% was extracted as free IAA acid. So IBA application leads to a seven fold higher free (active) IAA concentration in the tissue than application of IAA does. Also, the concentration of IAA in the tissue derived from IBA uptake is present for a longer period of time (due to continuous conversion of IBA into IAA) than free IAA derived from IAA uptake. The results show that IBA acts as a slow release source for IAA. IAA was not converted into IBA. The rate of auxin metabolism and the conversion of IBA into IAA may be a key factor in auxin activity.

Release Kinetics of IAA Bound with the Amide-Moiety to BSA

The rate of auxin inactivation was found to decrease when slow-release conjugates of IBA and IAA were used. ^{14}C-IAA was coupled with an amide bond to BSA (IAA-N-BSA; Fig. 1). The release rate of IAA from the BSA was fast: after 24 hours after application 76% of the auxins were released in the tissue. This leads to a high concentration of free IAA in the tissue (see Table 2). The auxin release takes place at the same location as where the auxins are supposed to be active, i.e. the apoplast between the cells at the exterior of the plasma membranes because the relative large BSA-auxin conjugates can not pass the plasma membrane of the cells in the disks.

Rooting with BSA-auxin slow release sources. IAA induced 3.5 roots per disk at 32 µM in the medium (higher concentrations: 100 to 300 µM resulted in a rooting response of an average of 7 roots per disk, result not shown), while IAA-N-BSA induces 7 roots per disk at a concentration of 1 µM (based on the IAA concentration) in the medium (Fig. 2). So, the activity of the slow release formulation has dramatically increased the auxin activity.

Shoots of summer oak and cork oak, multiplied in tissue culture for a limited period of time (tissue not rejuvenated), root for only 20% with standard hormones, whereas the rooting percentage is 80–100 % with IBA-N-BSA conjugate.

Design of Other Auxin-Slow Release Sources

Besides the BSA slow release sources also other slow release sources were synthesized: indolehexanoic acid (IHA), IBA-anhydride, IBA-aminoacids, IBA-polyamine-IBA, IAA-polyamine-IAA. Most of these slow release compounds are stable and can be auto-

Table 2. Release of IAA from the IAA-N-BSA conjugate.[a]

Incubation	Uptake (µmol $kgFW^{-1}$)	Free IAA (µmol $kgFW^{-1}$)
BSA	—	—
IAA-N-BSA	180	3.7
IAA	480	nd

[a]Stem disks were incubated for 3 days on medium with 1 µM BSA, 32 µM IAA or 32 µM IAA coupled to BSA. The experiment was conducted in duplicate. All SE values were less than 13% of the average value. nd = not detectable, — = not measured.

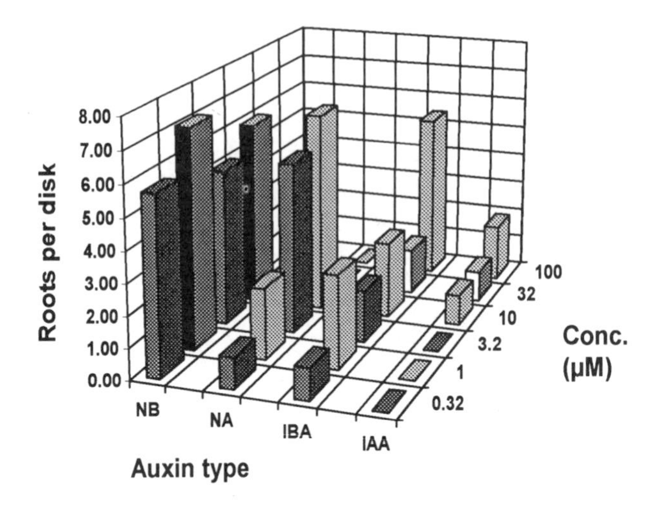

Figure 2. Root formation on apple stem disks with IBA-N-BSA (NB), IAA-N-BSA (NA), IBA and IAA. Disks were incubated for 6 hours on auxin medium and than transferred to auxin free medium as described by Van der Krieken et al. (1993). After 11 days root numbers were counted. All data are averages of 60 observations. The SE values were less than 6% of the average values.

claved. These slow release compounds proved to be very active in *in vivo* root induction especially in woody plants like rose, apple and oak and also in herbaceous plants. As an example the effect of IHA on rooting *in vivo* of recalcitrant to root apple cv elstar is shown in Fig. 3. IHA induced rooting in a maximum of 70% of the scions, whereas IBA did it in 30% of the scions at maximum.

Elicitors and Auxin Action

Different types of elicitors were used: breakdown products of the cell membrane (nonanoic acid and jasmonate) and of the cell wall (lignosulfonate) and elicitors of fungal origin (pythium extract). We tested the effect of different concentrations of the elicitors on IBA-induced rooting. The IBA concentration used was suboptimal for rooting, in this way a synergistic or antagonistic effect of the elicitor on auxin-induced rooting can easily be observed. In Table 3 the elicitor concentration that leads to a maximum increase in IBA-induced rooting is presented. These tested elicitors led to an increase (more than 100%) in the number of roots formed at suboptimal auxin concentration.

Not all elicitors are active: salicylic acid (a systemic elicitor) leads to a reduction in auxin induced rooting (results not shown), coumaric acid and nicotinamide (no breakdown

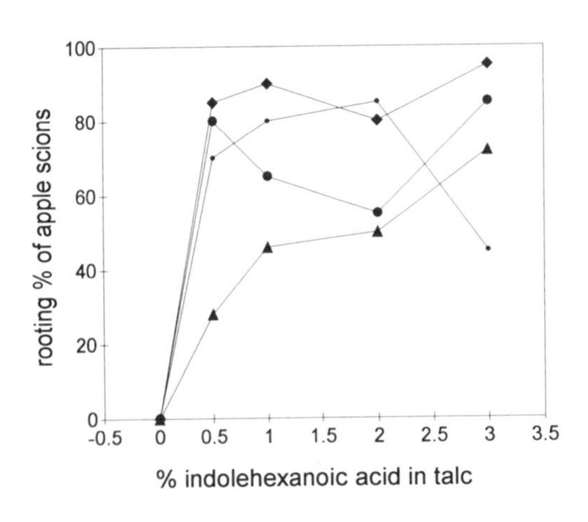

Figure 3. Effect of different concentrations of nonanoic acid (%, w/w in talc) and different concentrations of IHA (%, w/w in talc) on root formation on scions of apple cv elstar. The nonanoic acid concentrations used were: 0%, triangle; 0.25%, large circle; 0.5%, diamond and 1%, small circle. Root formation was scored 20 days after start of the experiment. The experiment was performed in thirty fold. All SE are less than 11% of the average value.

Table 3. Effect of different elicitors on auxin-induced rooting[a]

Elicitor: (optimal concentration)	# roots: 1 μM IBA, without elicitor	# roots: 1 μM IBA + elicitor	% increase in # roots	# roots without IBA+ elicitor
Lignosulfonate 1.4 μM	0.4	1.65	312	0
Pythium extract 3% (v/v)*	1.2	3.2	166	0
Methyljasmonate 0.1 μM	1.25	2.7	116	0
Nonanoic acid 1 μM	1.55	3.5	126	0

[a]Stem disks were incubated for one day on medium with elicitor (concentration see table) and 1 μM IBA. Roots were counted 14 days after start of the experiment. Values are averages of incubations of 20 disks performed in triplicate. All SE values are less than 9% of the average value.

*pythium extract: a sterilized filtrate was used of a pythium culture that had grown for 14 days in an erlenmeyer flask without shaking.

= average number of roots.

products of the cell wall or cell membrane) did not enhance auxin-induced rooting on thin apple stem disks *in vitro*.

The Effect of Combined Application of Slow Release Auxins and Elicitors on Rooting of Recalcitrant to Root Apple Scions *in Vivo*

Wounding or elicitor-induced tissue sensitivity for auxins precedes the time period after which the auxins can be active. So, combined application of slow release auxin with elicitor offers the following advantages for rooting: the elicitor and the wound reaction induce tissue sensitivity whereafter auxin that is released from the carrier molecule can exert its action.

Combination of the elicitor nonanoic acid and the slow release auxin indolehexanoic acid (IHA) results in rooting of very recalcitrant to root apple (cv elstar) scions (fig 3). IBA only induced 30% rooting, IHA without elicitor can induce about 70% rooting but IHA combined with nonanoic acid leads to 95% rooting. The rooting process induced with nonanoic acid and IHA appears to be much less depending upon the auxin concentration in the rooting powder: in the presence of 0.5% nonanoic acid the rooting percentage was high at all IHA-concentrations tested.

DISCUSSION

Studies describing the mode of action of auxins are often performed with systems consisting of relatively large amounts of tissue e. g. whole seedlings, shoots or large stem segments. Only few studies investigate auxin action in special test systems designed to unravel the mechanism of hormone action (Hand 1994). In this paper the mode of action of IBA, IAA, slow release auxins and elicitors on root formation was studied using a test system in which a relatively large percentage of cells was structurally related to the rooting process without interference of compounds originating from plant parts that are not near the location of root formation (Van der Krieken et al. 1993). Experiments were performed the first or second day after cutting the stem disks, during this period the tissue is most sensitive to auxin (the induction phase for root regeneration, Meins and Binns 1979, Christianson and Warnick 1983, 1988, De Klerk et al. 1993). In this way the mechanism of auxin and elicitor action can be specifically related to induction of adventitious rooting.

Analysis of the metabolism of IAA on TLC revealed that after IAA application in stem disks of apple the hormone was rapidly metabolized: within 24 hours only about

0.5% of the auxin taken up was extracted as free active IAA (Table 1). Generally, in most plant systems in which IAA metabolism was studied, IAA conjugation and decarboxylation were fast and only a small fraction (generally a few percent) of IAA taken up from the medium remained in the free IAA form (Cohen and Bandurski 1982). Uptake of IBA resulted in a higher fraction of active free IAA (1.5 to 2%) than uptake of IAA. IBA uptake is about twofold higher than IAA uptake so IBA application leads to a sevenfold higher free IAA concentration in the tissue than application of IAA. Also, the free IAA concentration derived from IBA uptake is present in the tissue for a longer period of time since conversion of IBA into IAA continues in time. The conversion of IBA into IAA is generally considered to play a certain, however still unclear, role in rooting (Epstein and Ludwig Müller 1993). To establish whether it is the free IAA or the free IBA fraction resulting from IBA uptake that is related to root induction we used IAA slow release sources to check if the slow release mechanism of IAA from IBA might account for the high rooting response of IBA. This slow release source was made of IAA that was coupled via the nitrogen atom to a BSA molecule (Pengelly 1977, Weiler 1981). BSA is well suited as an IAA carrier since it is a globular (transport) protein that can pass the apoplastic barrier (Venis et al 1990). Our results show that IAA-N-BSA was taken up in the tissue and after uptake the IAA was hydrolyzed from the conjugate (Table 2). This resulted in a relative high proportion of IAA in the tissue which was paralleled by an increased number of roots formed. An equal concentration of free IAA derived from either IBA uptake or IAA-N-BSA uptake resulted in an equal rooting response, proving that IAA produced from IBA is a key factor in the rooting response.

Many possibilities to synthesize other slow release molecules exist. The ones that we have synthesized and tested are IBA-, NAA- and IAA-anhydride, indolehexanoic acid, auxin-polyamine-auxin and auxin molecules coupled to sorbitol. Up to 6 auxin molecules can be coupled to one sorbitol molecule (Ellis and Honeyman, 1955). Many of these slow release sources are stable and can be autoclaved.

Tissue wounding is in many plants essential for the capacity for root regeneration. Wounding includes the destruction of cell compartments (vacuoles, vesicles, peroxisomes, plastids) which results in the release of the catabolic enzymes (glucanases, peroxidases, phospholipases, lipoxygenases) that are present in these cell organelles. These enzymes cause the breakdown of specific cell structures (cell walls, cell membranes). Some of the breakdown products of the cell structures are known to be involved in plants defense processes and are therefore elicitors. Our results show that combined application of elicitor and auxin lead to a synergistic effect on auxin-induced rooting (Table 2), elicitor alone did not induce rooting (Fig. 3). The mode of action of these elicitors is currently under investigation in our lab. Possible explanations are: (1) effect on auxin uptake and auxin transport. This is unlikely since our results indicate that these factors are not limiting auxin action. Moreover, preliminary results indicate that elicitors do not affect auxin uptake and transport. (2) effect of elicitors on auxin metabolism. Elicitors might decelerate the rate of metabolic inactivation of auxins. In our metabolism experiments, however, we could not find any indication for this. (3) the elicitors might speed up the endogenous synthesis of auxins. This is unlikely since combined application of elicitor plus slow release auxin lead to a high rooting percentage that can not be reached by any concentration of auxin alone (see next paragraph); also, elicitors alone (without auxin) can not induce roots. (4) elicitors might enhance the sensitivity of the tissue for plant hormones. The fact that we found that subsequent application of elicitors and auxin leads to an even higher rooting response then their simultaneous application (data not shown) makes this explanation even more plausible. This is consistent with the situation in nature: in the plant elicitors are produced after

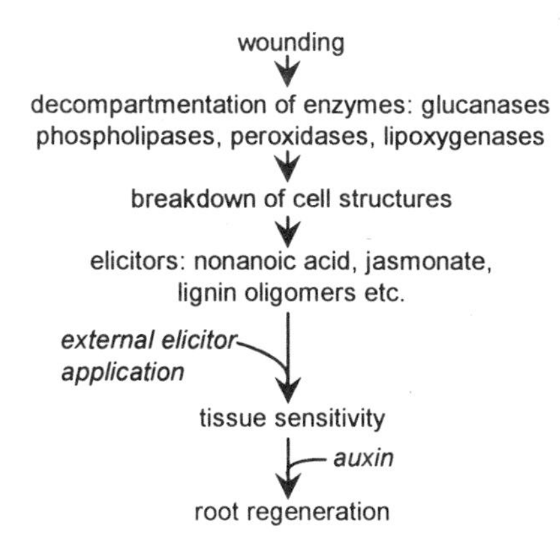

Figure 4. Proposed scheme for the mode of action of elicitors on auxin-induced rooting.

wounding for defense purposes; its concomitant induction of tissue sensitivity results subsequently in the repair of the damaged tissue. A scheme for the proposed elicitor activity is depicted in Fig. 4.

Combination of these elicitors with slow release auxins also results in a strong increase in the rooting response (Fig. 3). From literature it is known (De Klerk et al. 1995) that it takes about 1 to 1.5 days in apple stem disks before the tissue responds to the auxin applied in the culture medium. We have found (results not shown) that elicitors enhance auxin-induced rooting maximally when applied one day before the auxin. Combined application of slow release auxins and elicitors leads to a high auxin concentration in the tissue after a certain period of time (the time needed for release) during this period the elicitor can enhance the tissue sensitivity for auxin (Table 2).

It is obvious that the auxin-conjugates in combination with elicitors as described here can be used in commercially interesting markets, especially in the field of vegetative propagation (the slow release-elicitor concept is patent pending). Many millions of plants are propagated annually, *in vitro* or via cuttings. Also, the induction of meristems in callus tissue, whether or not preceded by genetic manipulation is of great interest when put in an economic perspective, for example concerning the production of "artificial seeds". There are also many other interesting applications: in principle it is possible to use auxin conjugates and elicitors in all situations where standard hormones are presently applied. The principle of increased activity of plant hormones by chemical modification to improve uptake, transport or metabolism can also be applied to other plant growth regulator groups (cytokinins, (anti)-ethylene, abscisic acid, (anti)-gibberellins, polyamines etc.) where a slow release mechanism to inhibit metabolic inactivation or to extend the period of activity is desirable.

ACKNOWLEDGMENTS

We thank Dr. Bob Veen (AB-DLO) and Désirée Engelen for critically reading of the manuscript.

REFERENCES

Andreae W.A. and Good E.G. 1957. Studies on 3-indoleacetic acid metabolism. IV. conjugation with aspartic acid and ammonia as processes in the metabolism of carboxylic acids. Plant Physiol. 32: 566–572.

Cohen, J.D. and Bandurski, R.S. 1982. Chemistry and physiology of the bound auxins. - Annu. Rev. Plant Physiol. 33: 403–430.
Christianson, M.L. and Warnick, D.A. 1983. Competence and determination in the process of in vitro shoot organogenesis. Dev. Biol. 95: 288–293.
Christianson M.L. and Warnick D.A. (1988) Organogenesis in vitro as a developmental process. HortSc 23: 515–519.
De Klerk, G.J., Ter Brugge, J. and Keppel, M. 1993. Successive phases during rooting of microcuttings of Malus. Acta Hort. 336: 249–256.
De Klerk, G.J. Keppel M., Terbrugge J. and Meekes H. 1995. Timing of the phases in adventitious root formation in apple microcuttings. J. of Exp. Bot. 46: 965–972.
Ellis G.P. and Honeyman J. (1955) Glycosylamines. Adv in Carboh Chem 10: 95–168.
Epstein, E. and Lavee, S. 1984. Conversion of indole-3-butyric acid to indole-3-acetic acid by cuttings of grape vine (Vitis vinifera) and olive (Olea europea). Plant Cell. Physiol. 25: 679–703.
Epstein, E. and Ludwig-Muller, J. 1993. Indole-3-butyric acid in plants: occurrence, synthesis, metabolism and transport. Physiol. Plant. 88: 382–389.
Hand P. 1994. Biochemical and molecular markers of cellular competence for adventitious rooting. In: biology of adventitious root formation Ed. T.D. Davis and B.E. Haissig. Plenum Press New York, London. pp 111–123.
Hangarter, R. P. and Good, N. E. 1981. Evidence that IAA conjugates are slow release sources of free IAA in plant tissues. Plant Physiol. 68: 1424–1427.
Ludwig Muller J. and Epstein E. 1991. Occurrence and in vivo biosynthesis of indole-3-butyric acid in corn (Zea mays L.). Plant Phys. 97: 765–770.
Lyon G.D., Reglinski T. and Newton C. 1995. Novel disease control compounds: the potential to "immunize" plants against infection. Plant Path. 44: 407–427.
Meins F. and Binns A.N. 1979. Cell determination in plant development. Bio. Sci. 29: 221–225.
Murashige, T. and Skoog, F. 1962. A revised medium for rapid growth and bioassays with tobacco tissue cultures. Physiol. Plant. 15: 473–497.
Nordström, A.C., Alvarado Jacobs, F. and Eliasson, L. 1991. Effect of exogenous indole-3-acetic acid and indole-3-butyric acid on internal levels of the respective auxins and their conjugation with aspartic acid during adventitious root formation in pea cuttings. Plant Physiol. 96: 856–861.
Pengelly W. 1977. A Specific Radioimmunoassay for nanogram quantities of the auxin, indole-3-acetic acid. Planta 136: 173–180.
Pythoud, F. and Buchala, A.J. 1989. The fate of vitamin D3 and indolebutyric acid applied to cuttings of populus tremula L. during adventitious root formation. Plant, Cell and Environment 12: 489–494.
Quoirin, M., Lepoivre, P. and Boxus, P. 1977. Un premier bilan de 10 années de recherches sur les cultures de méristemès et la multiplication in vitro de fruitiers ligneux. In C.R. Rech. 1976–1977 et Rapports de synthèse Stat. Fruit. et Maraîch. Gembloux. Belgium: 93–117.
Van der Krieken, W.M., Breteler, H. and Visser, M.H.M. 1992A. The effect of the conversion of indolebutyric acid into indoleacetic acid on root formation on microcuttings of Malus. Plant Cell Physiol. 33: 709–713.
Van der Krieken, W.M., Breteler, H., Visser, M.H.M. and Jordi, W. 1992B. Effect of light and riboflavin on indolebutyric acid-induced root formation on apple in vitro. Physiol. Plant. 85: 589–594.
Van der Krieken, W.M., Breteler, H., Visser, M.H.M. and Mavridou, D. 1993. The role of the conversion of IBA into IAA on root regeneration in apple: introduction of a test system. Plant Cell Rep. 12: 203–206.
Venis, M.A, Thomas, E.M., Barbier-Brygoo, H., Ephritikhine, G and Guern, J. 1990. Impermeant auxin analogues have auxin activity. Planta 182: 232–235.
Weiler E.W. 1981. Radioimmunoassay for pmol-quantities of indole-3-acetic acid for use with highly stable [125I]- and [3H]IAA derivatives as radiotracers. Planta 153: 319–325.
Wiesman, Z., Riov, J. and Epstein E. 1989. Characterization and rooting ability of indole-3-butyric acid conjugates formed during rooting of mung bean cuttings. Plant Physiol. 91: 1080–1084.

12

JASMONIC ACID AFFECTS CELL DIVISION IN MERISTEMS OF CULTURED POTATO ROOTS

Barbara Vilhar,[1] Maja Ravnikar,[1] and Dennis Francis[2]

[1]National Institute of Biology
Karlovška 19, Ljubljana
Slovenia
[2]School of Pure and Applied Biology
University of Wales, Cardiff
United Kingdom

INTRODUCTION

Recently, jasmonic acid (JA) and other jasmonates have been the focus of much attention as endogenous signals that stimulate expression of wound-inducible and defence-related genes, and as regulators of various developmental responses (Farmer, 1994).

Jasmonates have marked effects on the development of potato plants. For example, tuberonic acid, a member of the jasmonate family, is an endogenous inducer of tuber formation in potato stolons (Koda, 1992); changes in the microtubular system of the competent potato cells were suggested to be the underlying molecular mechanism (Matsuki et al., 1992).

In the potato plants grown *in vitro* (cv. Vesna), increasing concentrations of JA resulted in a root system with shorter and thicker roots, with many lateral branches (Ravnikar et al., 1992). The enhanced growth response in potato plantlets (cv. Ulster Sceptre) grown on a medium with 1 μM JA included expanded root systems, extended leaf areas and enlarged stem diameters (Dermastia et al., 1994). The biochemical changes in JA-treated plants comprised an increase in the endogenous levels of active cytokinin derivatives (Dermastia et al., 1994), alterations in protein composition in leaves (Dermastia et al., 1994; Vilhar et al., 1995), altered mineral and water uptake (Ravnikar et al., 1995), and altered root permeability and/or changed reduction potential within the roots (Vilhar et al., 1991).

One of the most striking effects of jasmonic acid in potato stem node culture is the altered morphology of roots. Our hypothesis was that JA inhibited root elongation but caused the root to thicken through cell division and not cell expansion. To test this hypothesis, we analysed various aspects of development in detached roots treated with JA, lateral root formation and changes in the root tip.

Biology of Root Formation and Development, edited by Altman and Waisel.
Plenum Press, New York, 1997.

EXPERIMENTAL

Root Culture

Single-node stem cuttings of potato plants (*Solanum tuberosum* L. cv. Ulster Sceptre) were grown *in vitro* on modified MS medium (Murashige and Skoog, 1962) without growth regulators. The concentration of sucrose in the medium was 90 mM. Plants were kept at 20±1°C in 16h day/8h night cycles (Osram L18W 20 lamps, 50 $\mu Mm^{-2}s^{-1}$). After 10 days of culture, adventitious roots (length 1.0–2.5 cm) that developed on the stem cuttings were cut at the base and immersed in liquid MS medium of the same composition, but without agar. In some cases, the medium was supplemented with 1 μM JA (Apex Organics, UK). The roots were kept for 5 days at 20±1°C in 16h light / 8h dark cycles. At regular intervals, roots were fixed in 3:1(v/v) absolute ethanol : glacial acetic acid, and different morphological parameters were measured. The position of the most distal (closest to the tip) protoxylem tracheary element was detected in cleared roots on the basis of typical secondary wall formation.

In our experiments, we analysed development of adventitious roots. However, in this paper the following terminology is used to describe root structure. The adventitious root formed on the stem cutting in stem node culture is termed the primary root. Consequently, the lateral roots formed on the primary root are the secondary roots.

Root Tip Sections

The fixed root tips were dehydrated in an alcohol series prior to wax-embedding using standard procedures (Taylor and Francis, 1989). Sections were cut (10 μm thickness) through the root meristem, stained with Feulgen reagent and counter-stained with Fast Green.

Root-Tip Squash Preparations

Roots were fixed and stained with the Feulgen reagent. The meristems were then dissected with fine needles in a droplet of 45% (v/v) acetic acid and a coverslip overlaid. The cells were gently tapped out to form a monolayer, without squashing (Armstrong and Francis, 1985). The coverslip was then removed by the dry-ice method (Conger and Fairchild, 1953) and the cells counter-stained with 0.1% (w/v) Fast Green in ethanol for 60 sec (Armstrong and Francis, 1985) and made permanent with DPX.

Feulgen Staining

The fixed root tips or rehydrated sections of root tips were hydrolyzed in 5 M hydrochloric acid (optimum hydrolysis time 30 min, 25°C), washed twice in ice-cold deionized water and stained with Feulgen reagent (pararosaniline chloride, BDH, UK) for 2½ h at 25°C.

Statistics

All the experiments were repeated twice. Statistical analysis was performed using the Mann–Whitney test, the 5% level being regarded as significant.

RESULTS

The absolute increase in primary root length was significantly lower in the JA-treated roots compared to the control treatment (Tab. 1).

JA also markedly affected the branching pattern of the root. The total number of secondary roots and secondary root primordia was higher in the JA-treated roots in all repeats of the experiment, although the differences were not statistically significant (Tab. 1). Besides, the lateral root zone (the region between the root base and the youngest secondary root primordium) was longer in the JA-treated roots, despite the fact that the roots were shorter than those in the control medium. Likewise, the distance between the youngest (the most distal) secondary root primordium and the primary root tip was significantly shorter in the JA treatment compared to the control (Tab. 1). However, there was no significant difference in the distance between successive primordia in either treatment (results not shown).

The development of the primordia was enhanced in the JA-treated roots. Thus, after 5 days of culture, 42% of all primordia emerged as secondary roots in the JA-medium, whereas no secondary roots emerged in the control medium (results not shown). At this stage, 52% of all primordia that formed in the control medium and 42% of those that formed in the JA-medium were still at an early stage of development (hardly visible when Feulgen-stained) (results not shown). Notably, we did observe a few emerged roots after 3 days of culture in the control treatment, but again the number was lower than in the JA-treatment.

Root tip meristems were larger in the JA-treated roots compared to the control (Fig. 1, Tab. 2). Thus, the JA-treated meristems were significantly longer and wider after 24 h of treatment. Analysis of root tip sections showed that the number of cell files in the stele increased in the JA-treated roots, whereas the number of cell files in the cortex did not change (Tab. 2). Nevertheless, analysis of the mitotic index in the root tip squash preparations revealed no significant difference between the control and the JA-treated plants (results not shown).

In addition, in the JA-treated roots the distance increased between the root tip (without root cap) and the first observed xylem element with typically modified cell walls (Tab. 1).

Table 1. The influence of JA on root growth, root branching, and xylogenesis[a]

Time	Treatment	Increase in primary root length (mm)	Distance youngest primordium – root tip (mm)	Total number of secondary roots and secondary primordia	Distance root tip – xylem = (μm)
Day 0		—	10.5 ± 1.3	0.54 ± 0.27	1417 ± 74
Day 1	C	2.0 ± 0.2	13.5 ± 2.0	0.46 ± 0.21	1055 ± 52
	JA	1.2 ± 0.2 *	8.8 ± 1.3 ns	1.00 ± 0.47 ns	934 ± 37 ns
Day 3	C	6.8 ± 0.4	13.8 ± 1.2	1.61 ± 0.43	1024 ± 35
	JA	3.5 ± 0.2 ***	4.6 ± 0.2 ***	3.85 ± 0.98 ns	1163 ± 39 *
Day 5	C	10.2 ± 0.6	15.9 ± 1.5	1.61 ± 0.56	710 ± 34
	JA	5.8 ± 0.5 ***	7.2 ± 0.7 **	3.25 ± 0.99 ns	1163 ± 69 ***

[a]C – roots grown in the control medium, JA – roots grown in the medium with 1 μM JA. The mean value ± standard error is shown. N = 13. Differences were tested between the control and the JA-treated samples at the respective day of the experiment (ns – not significant, * p≤0.05, ** p≤0.01, *** p≤0.001).

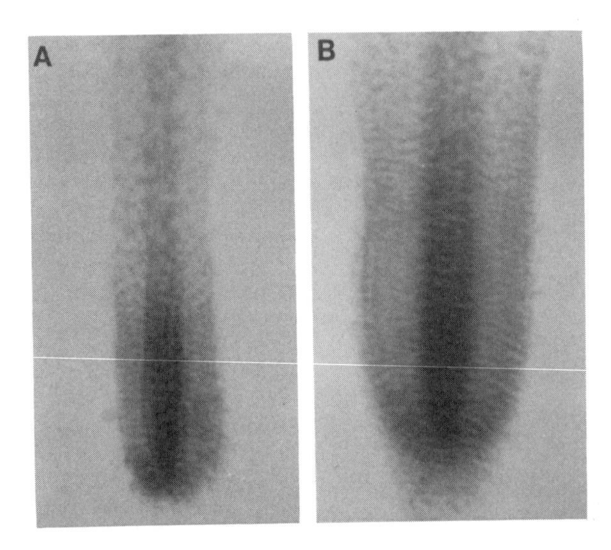

Figure 1. Distal part of a root grown in the control medium (A) and in the medium with 1 μM JA (B) after 5 days of culture.

DISCUSSION

Various correlations exist between root growth rate, meristem size and development of vascular elements (Rost and Baum, 1988). For example, an almost linear relationship between meristem length and the position of the most apical protoxylem tracheary element has been observed in pea seedlings (Rost and Baum, 1988). Throughout the JA-treatment, both the length and the width of the primary root meristem were significantly larger compared with the control treatment (Tab. 2). Although the growth rate of the control roots was faster than that in the JA roots, protoxylem was observed closer to the tip in the controls (Tab. 1). Hence, presumably JA maintains more cells in a dividing state, and as a consequence protoxylem differentiation is located at a greater distance from the proximal edge of the meristem compared with the controls. This is consistent with the well-established positive relationship between the position of the most proximal protoxylem cells and meristem length (Rost and Baum, 1988).

Table 2. The influence of JA on the size of the primary root meristem and the number of cell files in cortex and stele of roots[a]

Time	Treatment	Meristem length (μm)	Meristem width (μm)	Number of cell files in cortex	Number of cell files in stele
Day 0		701 ± 20	233 ± 5	3.2 ± 0.2	6.0 ± 1.0
Day 1	C	596 ± 21	230 ± 6	4.7 ± 0.2	9.0 ± 0.6
	JA	698 ± 25 *	261 ± 6 **	4.2 ± 0.2 ns	10.0 ± 0.6 ns
Day 3	C	537 ± 15	246 ± 6	—	—
	JA	702 ± 14 ***	325 ± 10 ***	—	—
Day 5	C	422 ± 16	235 ± 4	5.0 ± 0.0	8.3 ± 0.7
	JA	713 ± 28 ***	378 ± 10 ***	4.7 ± 0.2 ns	14.7 ± 1.2 ns

[a]The mean value ± standard error is shown. N = 13 for meristem size, N = 3 for number of cell files. Differences were tested between the control and the JA-treated samples at the respective day of the experiment (ns – not significant, * $p \leq 0.05$, ** $p \leq 0.01$, *** $p \leq 0.001$).

Moreover, this effect of JA is also reflected in the increased number of secondary root primordia in the secondary root zone. In the JA-treatment, the most distal secondary root primordium was positioned closer to the root tip than in the control roots (Tab. 1). In both treatments, the majority of primordia developed in the section of the root which grew in the agar medium without JA before excision. Notably, the spacing between succesive primordia was not changed significantly during the JA-treatment. Our data are consistent with the idea that while roots were grown in the stem-node culture a regular pattern of pericycle cells, competent to form primordia, was established along the root axis (Barlow and Adam, 1988). After excision of roots and transfer to the liquid medium, in the JA-treatment a larger number of competent pericycle cells were induced to divide and subsequently form secondary root primordia.

The primordium formation comprises several stages. Initiation of primordia occurs when a pair of elongated and highly vacuolated pericycle cells in the same column divides transversely and asymmetrically (Casero et al., 1995). Both asymmetrical transverse divisions are polarized, so that the smaller daughter cells are side-by-side (Casero et al., 1995). The pericycle cells which divide periclinally previously undergo radial expansion; only very short and radially expanded pericycle cells divide periclinally (Casero et al., 1995). Succesive cell divisions and cell expansion result in growth along the new axis. Instances where the polarity of growth shifts by 90 degrees often involve preceding shifts in microtubule alignment (Shibaoka, 1991). JA disrupts cortical microtubules in cultured tobacco cells (Abe et al., 1990) and in cultured potato cells (Matsuki et al., 1992). JA also induces expansion of cells in potato tubers (Takahashi et al., 1995). Besides, JA has been shown to stimulate cell division in early stages of protoplast culture of potato (Ravnikar et al., 1992) and in a fern, *Platycerium bifurcatum* (Camloh et al., 1996). Hence, apart from enhancing cell divisions JA might also be involved in the shift of the cell division plane and cell expansion through its interaction with cytoskeletal elements.

Various treatments have been reported to cause swelling of the root tip, such as cotyledon removal, growing roots in narrow glass tubes, trifluralin (Rost and Baum, 1988 and refs. therein), digitonin, monensin, cytochalasin B (Baskin and Bivens, 1995). However, the radial root expansion caused by JA is at least partly a consequence of formation of additional cell files in the stele, and not merely cell expansion. The roots that were grown in stem-node culture in agar medium with JA also had wider and longer root meristems and an increased number of cell files in the stele than the control roots (data not shown), but the root tips were not swollen; they were wider than the controls throughout root development. The additional cell files in the stele arise through longitudinal cell divisions. Interestingly, in *Platycerium* JA enhances longitudinal cell divisions and thus promotes transition of gametophytes from filamentous to planar growth (Camloh et al., 1996).

The two marked effects of JA on roots are inhibition of primary root growth and stimulation of root branching. These effects of JA on root development were first observed in stem-node culture (Ravnikar et al., 1992). Similar developmental changes have been associated with the effects of auxin (Hinchee and Rost, 1992). Therefore, some of the alterations in JA-treated roots might be a consequence of a signal, such as auxin, that might have been synthesized in the shoot and transported to the root system. However, similar morphological changes arose when the roots were first detached from plants and then treated with JA (Tabs. 1 and 2). This excluded the shoot-synthesized auxin as the regulator of morphological changes in our experiments. However, one still intriguing explanation is that JA acts as a secondary mediator downstream of auxin in the auxin-signalling cascade, hence triggering auxin-like effects when applied to plant cells.

ACKNOWLEDGMENTS

We thank the British Council and the Ministry of Science and Technology of Slovenia for financial support which enabled interchange between the laboratories.

REFERENCES

Abe, M., Shibaoka, H., Yamane, H., Takahashi, N., 1990. Cell cycle-dependent disruption of microtubules by methyl-jasmonate in tobacco BY-2 cells. Protoplasma 156: 1.

Armstrong, S.W., Francis, D., 1985. Differences in cell cycle duration of sister cells in secondary meristems of Cocos nucifera L. Ann. Bot. 56: 803.

Barlow, P.W., Adam, J.S., 1988. The position and growth of lateral roots on cultured root axes of tomato, *Lycopersicon esculentum (Solanaceae)*. Plant Syst. Evol. 158: 141.

Baskin, T.I., Bivens, N.J., 1995. Stimulation of radial expansion in arabidopsis roots by inhibitors of actomyosin and vesicle secretion but not by various inhibitors of metabolism. Planta 197: 514.

Camloh, M., Ravnikar, M., •el, J., 1996. Jasmonic acid promotes division of fern protoplasts, elongation of rhizoids and early development of gametophytes. Physiol. Plantarum, in press

Casero, P.J., Casimiro, I., Lloret, P.G., 1995. Lateral root initiation by asymmetrical transverse divisions of pericycle cells in four plant species: *Raphanus sativus, Helianthus annuus, Zea mays* and *Daucus carota*. Protoplasma 188: 49.

Conger, A.D., Fairchild, L.M., 1953. A quick-freeze method for making smear slides permanent. Stain Technol. 28: 281.

Dermastia, M., Ravnikar, M., Vilhar, B., Kovaè, M., 1994. Increased level of cytokinin ribosides in jasmonic acid-treated potato (*Solanum tuberosum*) stem node cultures. Physiol. Plantarum 92: 241.

Farmer, E.E., 1994. Fatty acid signalling in plants and their associated microorganisms. Plant Mol. Biol. 26: 1423.

Hinchee, M.A.W., Rost, T.L., 1992. The control of lateral root development in cultured pea seedlings. III. Spacing intervals. Bot. Acta 105:127.

Koda, Y., 1992. The role of jasmonic acid and related compounds in the regulation of plant development. Int. Rev. Cytol. 135: 155.

Matsuki, T., Tazaki, H., Fujimori, T., Hogetsu, T., 1992. The influences of jasmonic acid methyl ester on microtubules in potato cells and formation of potato tubers. Biosci. Biotech. Bioch. 56 (8): 1329.

Murashige, T., Skoog, F., 1962. A revised medium for rapid growth and bio assays with tobacco tissue cultures. Physiol. Plantarum 15: 473.

Ravnikar, M., Vilhar, B., Gogala, N., 1992. Stimulatory effects of jasmonic acid on potato stem node and protoplast culture. Journal of Plant Growth Regul. 11: 20.

Ravnikar, M., Bevc, L., Gogala, N., 1995. The influence of jasmonic acid on water, Ca and some other ion uptake of the potato (*Solanum tuberosum* L.) in vitro. Acta Pharmaceu. 45: 241.

Rost, T.L., Baum, S., 1988. On the correlation of primary root length, meristem size and protoxylem tracheary element position in pea seedlings. Am. J. Bot. 75(3): 414.

Shibaoka, H., 1991. Microtubules and the regulation of cell morphogenesis by plant hormones. In: The cytoskeletal basis of plant growth and form. Academic Press LTD. ISBN 0–12–453770–7.

Takahashi, K., Fujino, K., Kikuta, Y., Koda, Y., 1995. Involvement of the accumulation of sucrose and the synthesis of cell wall polysaccharides in the expansion of potato cells in response to jasmonic acid. Plant Sci. 111: 11.

Taylor, M., Francis, D., 1989. Cell cycle changes in the shoot apex of *Silene coeli-rosa* during the second and third days of floral induction. Ann. Bot. 64: 625.

Vilhar, B., Ravnikar, M., Schara, M., Nemec, M., Gogala, N., 1991. The influence of jasmonic acid on biophysical properties of potato leaf protoplasts and roots. Plant Cell Rep. 10: 541.

Vilhar, B., Dermastia, M., Ravnikar, M., 1995. Jasmonic acid and sucrose affect patterns of soluble leaf proteins in potato. Acta Chim. Slovenica 42: 479.

13

CHOICE OF AUXIN FOR *IN VITRO* ROOTING OF APPLE MICROCUTTINGS

Geert-Jan De Klerk,[1] Jolanda Ter Brugge,[1] Jan Jasik,[1,2] and Svetla Marinova[1]

[1]Centre for Plant Tissue Culture Research
PO Box 85
2160 AB Lisse, The Netherlands
[2]Department of Plant Physiology
Comenius University
Mlynska dolina B2
842 15, Bratislava, Slovakia

ABSTRACT

We studied rooting of microcuttings of *Malus* cv "Jork 9" exposed *in vitro* to a range of concentrations of various auxins. Indoleacetic acid (IAA) gave the best results by far as compared to indolebutyric acid (IBA), naphthaleneacetic acid (NAA) and 2,4-dichlorophenoxyacetic acid (2,4-D). This result is likely due to the high instability of IAA in the light: Because the cultures were initially in the dark and after that transferred to the light, IAA was present at a high concentration only during the first period of the rooting treatment. IAA and IBA were very efficient in inducing root meristems but inhibited the outgrowth of the meristems only weakly. NAA, and in particular 2,4-D, were less efficient in inducing root meristems, but strongly blocked their outgrowth. The different efficacies of the various auxins during formation and outgrowth of root meristems, may be explained by the occurrence of two auxin-receptors.

INTRODUCTION

Cuttings are mostly rooted by dipping them in a rooting powder or in a concentrated solution of auxin (Hartmann et al., 1990). In this way, the cuttings are exposed to auxin for a short period. Microcuttings may be rooted in the same way (*ex vitro* rooting), but often it is preferable to root them *in vitro* (George, 1996). In *in vitro* rooting, the microcuttings are exposed to auxin for an extended period of time. For *ex vitro* rooting, IBA is most frequently used. However, because of the long duration of application, for *in vitro* rooting another auxin may be optimal. In this article, we examine the performance of the auxins commonly used for rooting, *viz.*, IAA, IBA and NAA, and of 2,4-D.

Biology of Root Formation and Development, edited by Altman and Waisel.
Plenum Press, New York, 1997.

MATERIALS AND METHODS

In vitro shoot production by *Malus sylvestris* Mill., cv "Jork 9" microcuttings was maintained as described previously with subculture cycles of 5 weeks at 25°C (De Klerk et al., 1995). For rooting, microcuttings of 1.5 cm in length were cultured in high Petri dishes (2.5 cm high, ∅ 9 cm; 10 microcuttings per dish) on 40 ml of a modified Murashige-Skoog medium (De Klerk et al., 1995). Auxins were added after autoclaving. After 5 days in the dark, the cultures were transferred to the light (16h per day; 70 μmol m^{-2} s^{-1}). Roots were counted after 21 days. For each determination, three Petri dishes with 10 microcuttings each were used.

Rooting of 1-mm thick stem slices, cut from 2-cm long defoliated microcuttings was also examined. Slices cut from the middle 1 cm of stems have an equal capability to root (De Klerk and Caillat, 1994). The slices were cultured with the apical side down on a nylon mesh (4 x 4 cm) on 25 ml of the same rooting medium as used for microcuttings in a 9-cm Petri dish (Van der Krieken et al., 1993). The Petri dish was incubated upside down in the dark in a culture room at 25°C. The concentrations of auxins and the durations of the dark period before the transfer to the light are indicated for each experiment. In each Petri dish, 30 slices, originating from six microcuttings, were cultured. The five slices taken from each shoot were non-adjacent, because there is no correlation between the capability to root of non-adjacent slices (De Klerk and Caillat, 1994). Roots were counted after 21 days under a dissecting microscope. For each determination, 90 slices were used.

Histological techniques have been described previously (Jasik and De Klerk, 1997).

RESULTS

Rooting with Various Auxins

Microcuttings were cultured with increasing concentrations of IAA, IBA, NAA or 2,4-D (Fig. 1a). For the initial 5 days, the cultures were in the dark and after that in the

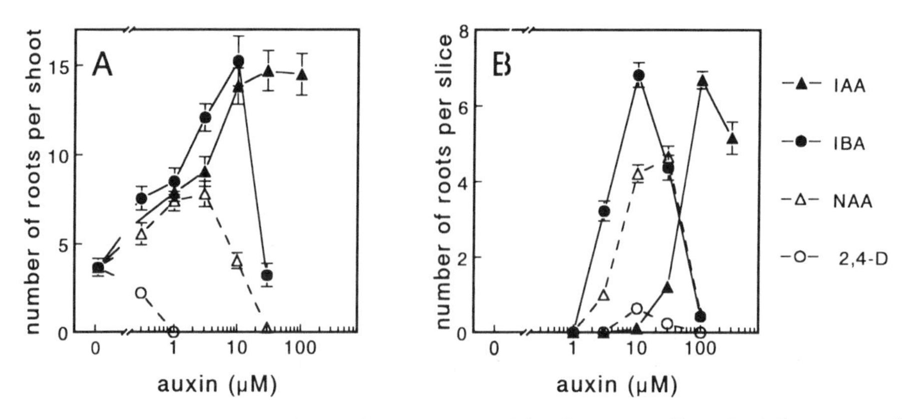

Figure 1. Rooting of microcuttings after continuous exposure (a) or 1-mm stem-slices after 1 day exposure (b) to a range of IAA, IBA, NAA or 2,4-D concentrations. The cultures were 5 days (a) or 1 day (b) in the dark and after that transferred to hormone-free medium and to the light.

light. IBA and IAA treatments resulted in the same maximal root number. However, only with IAA the maximal root number was reached at a broad range of concentrations (10–100 μM). NAA induced a significantly lower root number. 2,4-D inhibited rooting at all concentrations tested. At the optimal concentration, the appearance of the microcuttings, root length and the absence of callus were much better with IAA than with one of the other auxins. With 2,4-D, even the low concentrations resulted in poor shoot quality (data not shown).

To examine auxin effects specifically during the first period of rooting, we applied them during the initial 24h only. This experiment was carried out with 1-mm stem slices (Fig. 1b). Slices had an absolute requirement for auxin: they never regenerated roots if incubated without auxin. The required concentrations of IAA were 10 to 30 times higher than those of IBA. In slices continuously exposed to auxin, there was -just as in shoots- only a three-fold difference in the required IBA and IAA concentrations (data not shown). The relative performance of NAA had improved in the short exposure in comparison to the continuous exposure. However, major differences remained among 2,4-D on one hand, and IAA and IBA on the other. The backgrounds of this were examined in more detail.

Effect of 2,4-D

With 2,4-D added for the initial 24h, the maximal root number was low (Fig. 1b). A possible explanation for this is that at 8 μM 2,4-D many meristems had been formed, but that their outgrowth was blocked because of the persistence of 2,4-D after the 24h pulse. Figure 2 shows that this hypothesis is not valid since at 8 μM 2,4-D rooting was increased to the maximal value by simultaneous addition of IBA. Thus at the optimal 2,4-D concentration only few meristems are formed.

It is well known that auxins become inhibitory after root meristems have been formed (De Klerk et al., 1990). In slices, meristems have been formed at 72h. Therefore, we added 2,4-D and IBA at day 4 (72–96h). Almost complete inhibition of outgrowth of root meristems occurred when 8 μM 2,4-D was added (Fig. 3). In contrast, IBA did not inhibit rooting when added at day 4 (highest concentration tested 25 μM). The inhibition by 2,4-D may be caused by action on the root meristems directly, or by callus formation from surrounding cells that overgrow the meristems. Histological analysis showed that after the 2,4-D pulse callus developed from the meristems (Fig. 4). Thus, 2,4-D acted on the meristems directly.

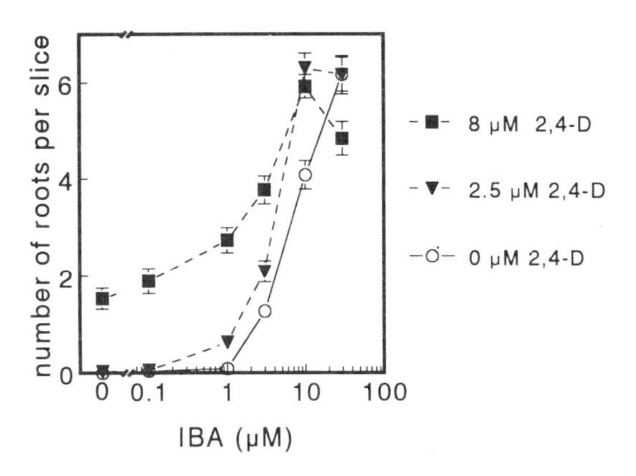

Figure 2. Rooting of 1-mm stem stem-slices after a 24h exposure to 0, 2.5 or 8 μM 2,4-D together with a range of IBA concentrations. During the auxin-treatment the slices were in the dark, and after that transferred to auxin-free medium and to the light.

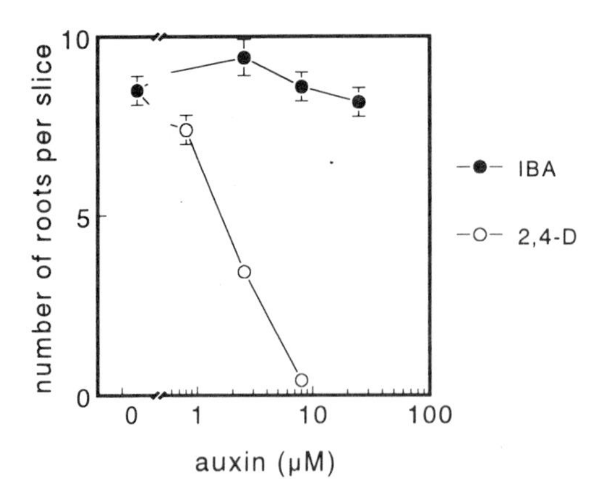

Figure 3. Rooting of 1-mm stem-slices after exposure from 0 to 24h to 25 µM IBA and from 72 to 96h to a range of IBA or 2,4-D concentrations. The cultures were up to 96h in the dark and after that transferred to the light.

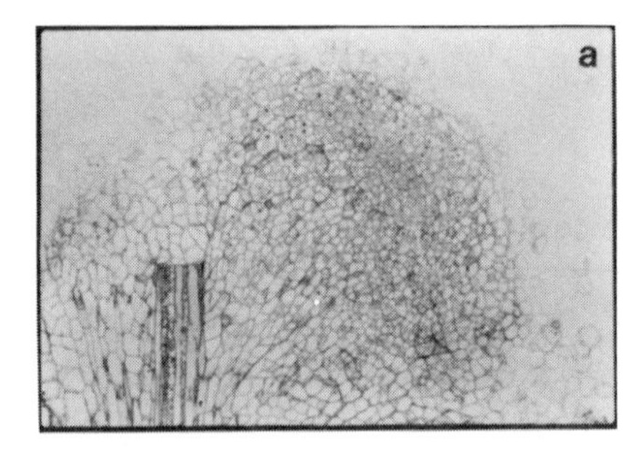

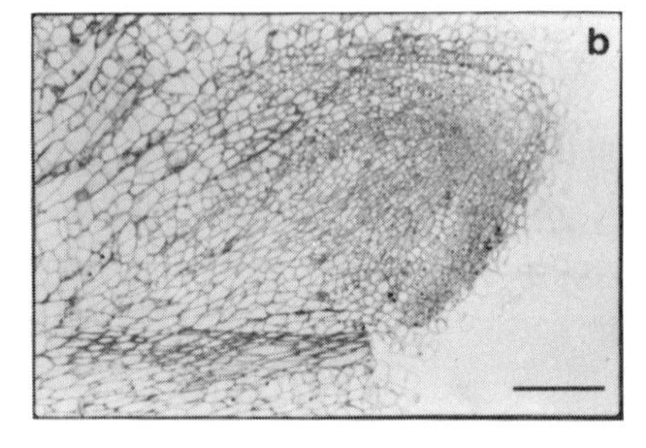

Figure 4. Micrograph of sections of apple stem slices at 144h. The slices had been treated with 25 µM IBA from 0 to 24 h. After that, they were (a) or were not (b) treated with 10 µM 2,4-D from 72 to 96h. bar = 50 µm.

DISCUSSION

Preferable Auxin for *in Vitro* Rooting of "Jork 9"

IAA was optimal for *in vitro* rooting of "Jork 9" microcuttings: a high number of roots were induced over a broad range of concentrations. The roots were long, the stems showed little callus formation and the microcuttings (leaves, apex and stem) were well developed. All these characteristics are essential for rapid reassumption of growth after transfer to soil (Mohammed and Vidaver, 1990; Van Telgen et al., 1992; Wisniewski et al., 1986). When auxin was applied for a short period only, IBA was optimal: Under such conditions IAA has to be applied in a very high concentration (Fig. 1b). From the four auxins tested, NAA and in particular 2,4-D showed a poor performance in both the long- and short-time exposures.

Causes of the Differences in Efficacy

In the light, IAA is rapidly decomposed by photo-oxidation (Nissen and Sutter, 1990). Thus, when the cultures were transferred at day 5 from the dark to the light, IAA was present at high concentration during the auxin-requiring phase (De Klerk et al., 1995). After that, IAA was disintegrated and did not interfere with the outgrowth of the meristems. The other auxins are either slowly (IBA) or hardly photo-oxidized (NAA and 2,4-

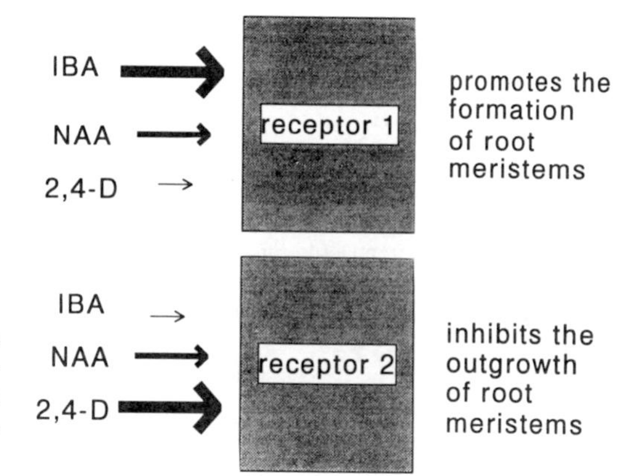

Figure 5. Model to explain the different efficacies of the various auxins during the formation of root-meristems and their subsequent outgrowth. The thickness of the arrow indicates the putative affinity to the receptor.

D). Therefore, IAA has the best performance in *in vitro* rooting. On the other hand, when auxin is applied for a short period, IBA is preferable because IAA is present for a too short only due to its rapid breakdown within the tissue (Van Der Krieken et al., 1993). IBA is more stable and therefore remains present during a larger portion of the auxin-sensitive phase than IAA.

When the auxins were supplied for one day only, we still observed major differences among them. Under such conditions, differential photo-oxidation did not play a role because the cultures were in the dark during auxin application. So there is a second reason for the differences in efficacy. This was studied for IBA and 2,4-D. 2,4-D induced only few roots (Fig. 1b). This may be caused by the persistence of 2,4-D after uptake resulting in a high inhibitory level during the outgrowth of the meristems. Thus, at the optimal 2,4-D concentration (8 μM) a high number of meristems might have been formed which could not develop because of the persistence of the 2,4-D. However, this explanation does not hold since 8 μM 2,4-D added along with IBA did not inhibit rooting (Fig. 2). Thus, 2,4-D is a weak auxin with respect to root induction. This is in sharp contrast with the very strong inhibitory action of 2,4-D during outgrowth of root meristems (Fig. 3; 8 mM almost completely abolishes rooting). IBA showed the opposite behaviour: strong promotion during meristem induction (Fig. 1b) and little inhibition during meristem outgrowth (Fig. 3). It is hard to believe that these results can be explained by differences in uptake, transport and metabolism between the two auxins. We assume that the formation of root-meristems and the inhibition of their outgrowth are regulated by two different auxin-receptors. The first one is related to root meristem formation and has a high affinity for IAA and IBA. The other one is related to the inhibition of root meristems and has a high affinity for 2,4-D (Fig. 5). The occurrence of various auxin receptors, and the possible relationship between an isolated auxin-receptor and adventitious rooting has been suggested for tobacco callus cultures (Libbenga and Mennes, 1995).

REFERENCES

De Klerk, G.J., Ter Brugge, J., Smulders, R, and Benschop, M., 1990, Basic peroxidases and rooting in microcuttings of *Malus*. *Acta Hortic.* 280:29.

De Klerk, G.J., and Caillat, E., 1994, Rooting responses of stem-disks excised from the same "M9 Jork" microcutting, *Adv. Hort. Sci.* 8:15.

De Klerk, G.J., Keppel, M., Ter Brugge, J., and Meekes, H ., 1995, Timing of the phases in adventitious root formation in apple microcuttings, *J. Exp. Bot.* 46:965.

George, E.F., 1996, "Plant Propagation by Tissue Culture. Part 2. In Practice", Exegetics, Eddington.

Hartmann, H.T., Kester, D.E., and Davies, F.T., 1990, Plant Propagation: Principles and Practices, Prentice Hall, Englwood Cliffs, NJ.

Jasik, J., and De Klerk, G.J. 1997, Anatomical and ultrastructural examination of adventitious root formation in stem slices of apple, *Biol. Plant.* 39: 79.

Libbenga, K.R., and Mennes, A.M., 1995, Hormone binding and signal transduction, *in*: "Plant Hormones", P.J. Davies, ed., Kluwer, Dordrecht.

Mohammed, G.H., and Vidaver, W.E., 1990, The influence of acclimatization treatment and plantlet morphology on early greenhouse performance of tissue-cultured Douglas fir [*Pseudotsuga menziesii* (Mirb.) Franco], *Plant Cell Tiss. Org. Cult.* 21:111.

Nissen, S.J., and Sutter, E.G., 1990, Stability of IAA and IBA in nutrient medium of several tissue culture procedures, *HortScience* 25:800.

Van Der Krieken, W. M., Breteler, H., Visser, M.H.M., and Mavridou, D., 1993, The role of conversion of IBA into IAA on root regeneration in apple: introduction of a test system, *Plant Cell Rep.* 12:203.

Van Telgen, H.J., Van Mil, A., and Kunneman, B., 1992, Effect of propagation and rooting conditions on acclimatization of micropropagated plants, *Acta Bot. Neerl.* 41:453.

Wisniewski, L.A., Frampton, L.J., and McKeand, S.E., 1986, Early shoot and root quality effects on nursery and field development of tissue-cultured loblolly pine, *HortScience* 21:1185.

14

BIOCHEMICAL AND MOLECULAR FACTORS AFFECTING *IN VITRO* ROOTING ABILITY IN ALMOND

E. Caboni, P. Lauri, M. G. Tonelli, P. Iacovacci, and C. Damiano

Fruit Trees Research Institute
00040 Ciampino Aeroporto
Rome, Italy

INTRODUCTION

Rooting initiation involves cell divisions of induced cells followed by the formation of a root meristem. *De novo* formation of the root meristem involves complex changes in the metabolism and it is evident that numerous endogenous factors interact in the developmental shift leading to adventitious root formation, both at biochemical and molecular level.

There is substantial evidence that auxins play a crucial role in the formation of adventitious roots (Davis *et al.*, 1988). Exogenous auxins induce adventitious root formation more than any other chemical both *in vitro* and *in vivo* (Haissig, 1986) and a direct correlation between the high rooting ability and the endogenous IAA levels in cuttings has been already reported in herbaceous and woody species (Alvarez *et al.*, 1989; Blakesley *et al.*, 1991; Kracke *et al.*, 1981; Weigel *et al.*, 1984)

However, although auxins play an indisputable role in root initiation, peroxidases have been also proposed as biochemical factors involved in the rooting process. However, despite years of studies, the supporting results are often contradictory: adventitious root formation of some species was shown to occur after the cuttings had reached and passed a peak of activity (Gaspar *et al.*, 1992). Changes of these enzymes have been found to correspond to parallel changes in IAA-oxidase activity (Mato and Vieitez, 1986), basic peroxidase have been shown to be effective in IAA oxidation, and changes of these isoenzymes related to the rooting process have been also reported (Berthon *et al.*, 1989; Gaspar *et al.*, 1985). On the other hand, in some studies no correlation was found between peroxidase and rooting differentiation (Pythoud and Buchala, 1986; De Klerk *et al.*, 1990). The stimulating action of some phenolic compounds, supplied to the rooting medium, on rhizogenesis has been described (Jones and Hatfield, 1976; Zimmermann and Broome, 1981). It can be speculated that phenols would not act directly in rooting induction but by inhibiting IAA-oxidase activity and, consequently, maintaining high endogenous level of IAA neces-

Biology of Root Formation and Development, edited by Altman and Waisel.
Plenum Press, New York, 1997.

sary for root induction. In fact, an inverse variation of total peroxidase activity and phenolic compound has also been described during the rooting process (Moncousin, 1986; Gaspar *et al.*, 1992).

In recent years, advances in molecular biology allowed to detect synthesis of new mRNA populations leading to interacting changes involved in cell division and adventitious root formation (Hackett *et al.*, 1990; Van der Krieken *et al.*, 1991). The above mentioned studies, like most of the other comparative analysis of mRNAs involved in other developmental processes, have been relied largely on subtractive cDNA library or differential hybridizations (Duguid and Dinauer, 1992). In 1992, easier and faster methods were developed using random primed cycles of PCR, after the reverse transcription reaction (Liang and Pardee, 1992; Welsh *et al.*, 1992). This method was then studied and improved to obtain reproducible results both in animal cells (Liang *et al.*, 1993; Bauer *et al.*, 1993; Ackland-Berglund and Leib, 1995) and in tomato cells. Most of the mentioned changes during the rooting process have been observed in several species, including woody species, but only few data are available on almond (Caboni, 1994).

In this work we studied relationships between *in vitro* rooting ability and some biochemical and molecular aspects (peroxidase activity, total phenol content, IAA endogenous levels and change in mRNA populations) in several almond rootstocks having different rooting ability.

EXPERIMENTAL

Plant Material and Shoot Proliferation Conditions

In vitro cultures of two seedlings of almond (*P. dulcis* Mill), Sel. M50 and M51, very difficult and easy-to-root, respectively, have been used. The axillary shoots of all the genotypes were maintained for 2 years on a multiplication medium, previously described (Caboni and Damiano, 1994), consisting of MS salts and vitamins (Murashige and Skoog, 1962). Shoots were subcultured every 15 days. Cultures were maintained in a 16-h photoperiod under a cool white light, density of 30 $\mu E\ m^{-2}\ s^{-1}$, provided by Osram 140 white fluorescent tubes at 21°C.

Rooting Conditions

Microcuttings, 1.5 to 2.0 cm in length, were excised from 15 day old cultures and transferred into a rooting medium consisting of MS salts and vitamins (Murashige and Skoog, 1962), containing 30 $g\ l^{-1}$ sucrose, supplemented or not with indole-3-butyric acid (IBA) (5μM) and solidified with 7 $g\ l^{-1}$ Difco Bacto agar. The pH was adjusted to 5.6, prior to autoclaving. Microcuttings were maintained for 5 days in dark, then transferred to the light. Growth chamber conditions were the same as reported for the multiplication phase. The rooting percentages were recorded after 30 days in the rooting medium. Triplicate treatments consisting of 10 plants each were used. The experiment was repeated twice. Percentage scores of explants forming roots were transformed with the formula arcsin %.

Peroxidase Activity and Phenolic Compound Assays

Enzyme assays were performed on the crude extract of the microcuttings according to Moncousin and Gaspar (1983). Activity was expressed in units of enzyme (U = the enzyme amount catalyzing the oxidation of 1 μM of substrate · min^{-1} · mg^{-1} protein). Proteins were measured by the method of Bradford (1976). Total phenolic compounds were extracted by Legrand's method (1977) and assayed with the Folin-Ciocalteu reagent according to Bray and Thorpe (1954). Total phenol content was expressed as μg · mg^{-1} of fresh mass (f.m.). Triplicate treatments, consisting of 10 explant each, were used both in the peroxidase and phenol analysis. Spectrophotometrical determinations were repeated twice, with similar results. Samples were taken each day till the 4th day after the transplanting in the rooting medium in auxin-induced and non-induced microcuttings.

IAA Determination

Free indole-3-acetic acid (IAA) and IAA-aspartate were extracted on samples of basal part of shoot stems taken from 15-d-old cultures and purified by solid-liquid extraction according to Nordstrom and Eliasson (1991). Samples were analyzed on reversed-phase on HPLC (Waters 600) coupled to a Waters 470 fluorescence detector (excitation 280, emission 360) and to a Waters 746 integrator. The column was a Water Resolve C18, 15cm long and with 5μm particle size. The mobile phase was 8% acetonitrile, 1% acetic acid and water with a flow rate of 0.5 ml min^{-1}. The system was operated isocratically. Identification of free IAA and IAA-aspartate peaks was confirmed adding authentic IAA or IAA-aspartate (Sigma) to a part of the sample and the calculation of recovery was done on authentic IAA and IAAasp dissolved in extraction buffer and purified as above reported. Data on free-IAA were also confirmed by GC-MS.

Duplicate treatments, consisting of 30 explants each, were used. IAA quantitative determinations at HPLC were repeated twice, with similar results.

mRNA Population Analysis

Total RNA was extracted from the basal part (0.5–1.0 cm) of 100 microcuttings 48h after transferring to the inducing media. The cDNA fragments were reverse synthesized and amplified with a PCR reaction according to Liang and Pardee (1992) using 3 different random primers. PCR products were run on vertical, non denaturating, 6% polyacrylamide (Bio-Rad) gel in TBE buffer (Sigma), at 2 Volts cm^{-1} for 16h at room temperature. Gels were stained in TBE buffer containing 1 mg l^{-1} of ethidium bromide (Sigma). Random cDNA fragments were then cut and eluted. Then, fragments were reamplified by PCR and they have been testing with reverse Northen blot analysis (Cordewener *et al.*, 1996) using as probe total RNA extracted by auxin induced and non-induced material. Only fragments confirmed to be rooting-specific will be cloned.

RESULTS AND DISCUSSION

Rooting

Rooting in almond is strictly genotype-related. The two genotypes used in our study have very different rooting ability. In fact, it was evident that rooting of M51 is highly induced by the auxin treatment (75% of explants were induced to root by 5μM IBA treatment) while only 5% of the M50 microcuttings produced adventitious roots in presence of the same auxin concentration. Furthermore, some M51 explants (15.5%) rooted also when auxin was not supplemented in the medium (Table 1).

Table 1. Almond rooting percentages with 5 μM IBA and 5 days in dark[a]

	Rooting %	
Genotype	IBA (5 μM)	Control
M50	5 ± 0.3	0
M51	75 ± 3.0	15.5 ± 0.2

[a]Values are the means of two independent experiments (each performed with 3 replicates) ± SE.

Peroxidase Activity and Phenols

Peroxidase activity, measured daily during the first 4 days of the IBA treatment, slightly decreased after one day, steeply increased after 3 days, then decreased in the easy-to-root genotype (M51); only a small peak was detected in the control microcuttings (Fig. 1A). In M50, the difficult-to-root one, only a very slight increase of activity was detected after 3 days both in the auxin supplemented and control medium (Fig. 1A). Total phenol content increased after 1 day, then decreased till day 3 and afterwards again increased slightly at day 4 in M51 (Fig. 1B). In M50, it only poorly increased after one day (Fig. 1B). In the hormone free medium only a slight increase was detected in both the genotypes (Fig. 1B).

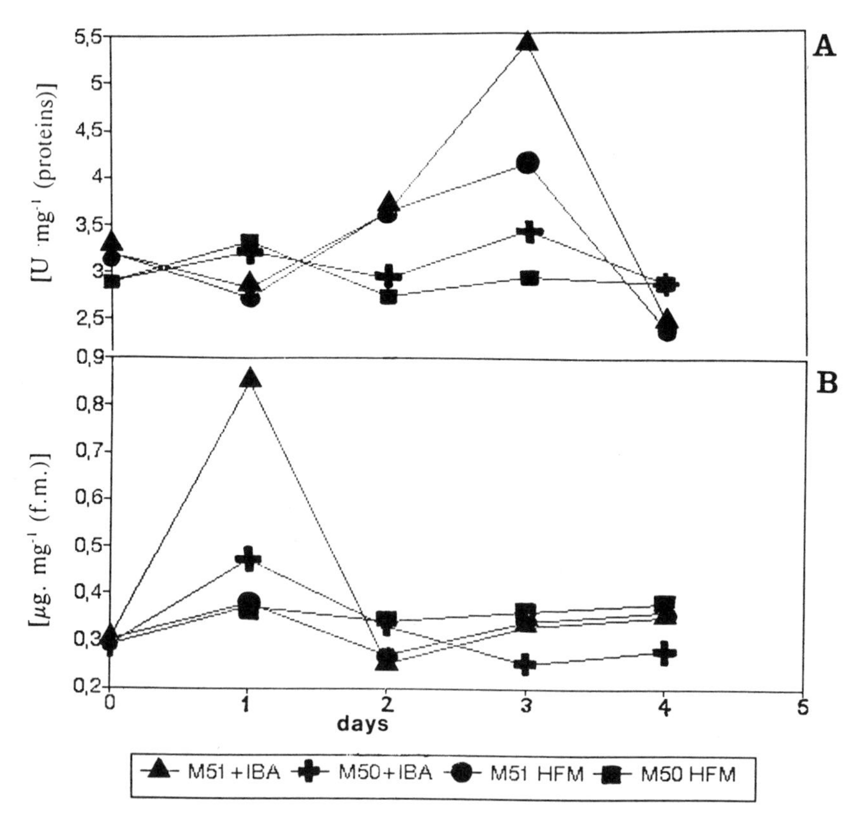

Figure 1. Polyacrylamide gel electrophoresis of random amplified cDNA fragments from almond. lane 1, M51 microcuttings treated with auxin and lane 2 control, lane 3, M50 microcuttings treated with auxin and lane 4 control.

Peroxidases have been shown to be involved in IAA catabolism and in the rooting process. In some cases they have been used as predictors of the rooting performance of cuttings for propagation (Quoirin *et al.*, 1974). Peroxidase activity of microcuttings of many species, when transferred to a rooting inducing medium, underwent a typical curve with an early transient minimum followed by an increase up to a peak and then a decline (Gaspar *et al.*, 1992). This variation normally corresponds to an inverse trend of the level of total phenol compounds which, presumably modulate peroxidase activity (Gaspar *et al.*, 1992). Measuring the enzyme activity during the first 4 days of inducing treatment, in the easiest-to-root genotype (M51) and the most difficult-to-root one (M50), showed that in M51 the peroxidase activity variations follow the curve mentioned above. It could be presumed that this variation corresponds to opposite changes in free IAA endogenous level and this aspect is now being investigated.

Furthermore, in this genotype it was also found that the total phenol content changed inversely to peroxidase activity, suggesting that phenols act in modulating activity of the enzyme also in the root induction phase of almond as reported for other species (Moncousin, 1986). Thus, our data show that interacting changes of total phenols and peroxidases during the first days of the rooting process are involved in the rooting process in almond, and that these variations are critical for obtaining a satisfactory rooting response.

IAA Endogenous Levels

Previous researchers have investigated the rooting performance of genotypes in some woody species in relation to constitutive IAA endogenous levels (Le, 1985; Foret *et al.*, 1986). Findings of Alvarez *et al.* (1989) suggested that differences in *vitro* rooting ability of 2 apple rootstocks are related to differences in free IAA levels in basal section: the higher the free endogenous level, the higher the rooting response. Also Blakesley *et al.* (1991) reported correlation between rooting percentage and the endogenous level of the species tested. An inverse relationship was found by the same authors for conjugated IAA endogenous levels of basal part of microcuttings. There is no available information on IAA levels in tissue cultured almond. We measured the free IAA and IAA aspartate endogenous level in the basal part of one difficult-root almond genotype (M50) and in a very easy-to-root one (M51). The level of free IAA was significantly higher in M51 than in M50 (Fig. 2A). These data support Alvarez *et al.* (1989) and Blakesley *et al.*(1991) findings on the direct relationship of free endogenous levels of basal part of microcuttings and rooting ability. However, we also found higher IAA-aspartate endogenous level in M51 then in M50 (Fig. 2B). Thus, we can hypothesize that in almond root formation is related to the level of the free IAA but that also constitutive higher level of the IAA-aspartate is present in the easy-to-root genotypes, perhaps due to a higher capacity of these genotypes to synthesize IAA.

Changes in mRNA Populations

Samples of induced and control microcuttings showed patterns of amplified cDNA species almost similar each other, but it was possible to identify some species unique for treated or untreated microcuttings in both the genotypes (Fig. 3). The use of different random primers during amplification reactions produced different patterns of cDNA species. From a total of 15 specific fragments of auxin induced M51 microcuttings, eluted and reamplified, only 2 hybridized with mRNA isolated from M51 microcuttings treated for 48 h with 5μM IBA and they are now being cloned. For the other genotype, elution, amplification and evaluation of unique fragments are now in progress.

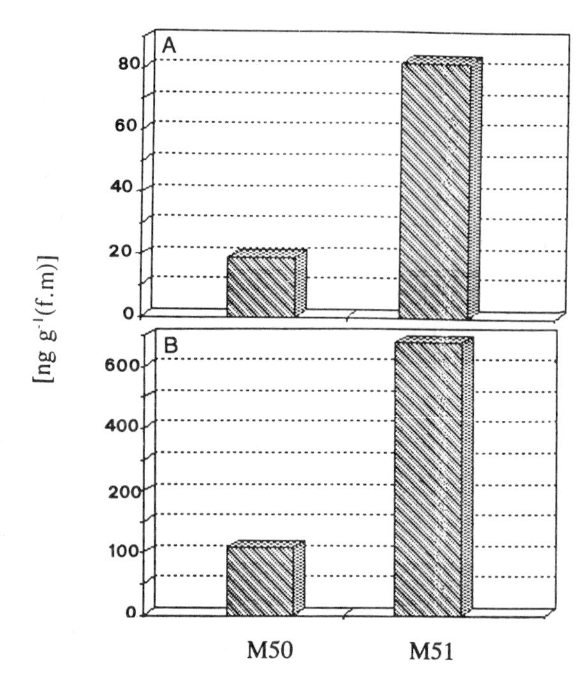

Figure 2. Content of endogenous free IAA (A) and IAA-aspartate (B) in basal part of shoots of M51 and M50 almond genotypes.

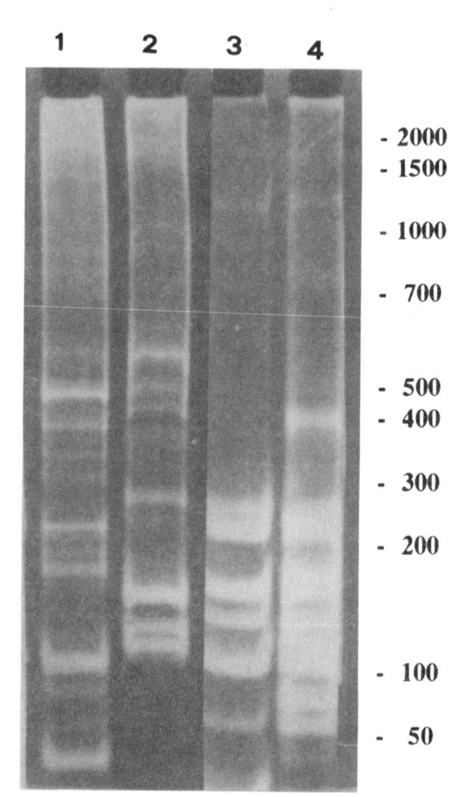

Figure 3. Peroxidase activity (A) and total phenol content (B) in microcuttings of M51 and M50 treated or not with 5μM IBA during the first 4 days in the rooting medium.

ACKNOWLEDGMENTS

This work was supported by the finalized research project "Fruitculture" of the Italian Ministry of Agriculture (MiRAAF) and by National Research Council of Italy, Special Project RAISA, Sub-project N.2.

The authors are grateful to Loreta Marinaccio for her work in subculturing the almond cultures *in vitro*.

REFERENCES

Ackland-Berglund, C.A., Leib, D.A., 1995, REN display, a rapid and efficient method for nonradioactive differential display and mRNA isolation, BioTechniques 18(2):200.

Alvarez, R., Nissen S.J., Sutter, E.G., 1989, Relationship between indole-3-acetic acid levels in apple (*Malus pumila* Mill) rootstocks cultured *in vitro* and adventitious root formation in the presence of indole-3-butyric acid, Plant Physiol. 89:439.

Bauer, D., Müller, H., Reich, J., Riedel, H., Ahrenkiel, V., Warthoe, P., Strauss, M., 1993, Identification of differentially expressed mRNA species by an improved display technique (DDRT-PCR), Nucl. Acids Res. 21(18):4272.

Berthon, J.Y., Maldiney, R., Sotta, B., Gaspar, T., and Boyer, N., 1989, Endogenous levels of plant hormones during the course of adventitious rooting in cuttings of *Sequoiadendron giganteum* (Lindl.) *in vitro*, Biochem. Physiol. Pfl. 184:405.

Blakesley, D., Weston, G.D., Elliot M.C., 1991, Endogenous levels of indole-3-acetic acid and abscisic acid during rooting of *Cotinus coggygria* cuttings taken at different times of the year, Plant Growth Reg. 10:1.

Bradford, M.M., 1976, A rapid and sensitive method for the quantization of microgram quantities of protein utilizing the principle of protein-dye binding, Anal. Biochem. 72:248.

Bray, M., and Thorpe, G., 1954, Analysis of phenolic compounds of interest in metabolism, In:"Methods of Biochemical Analysis" Intersection Publication p. 27.

Caboni, E., 1994, Peroxidase activity and *in vitro* rooting in an almond (*Prunus dulcis*, Mill) genotype, Acta Hort. 373:201.

Caboni, E., and Damiano, C., 1994, Rooting in two almond genotypes, Plant Sci. 96:163.

Cordewener, J., Jansen,H., and Van Lookeren Campagne, M., 1996, Markers of early embryogenesis isolation by differential display RT-PCR, in:"Mechanisms and markers of regeneration and genetic stability", Cost 822, Turku, 1–5 May

Davis, T.D., Haissig, B.E., Sankhla, N., 1988, Adventitious Root Formation in Cuttings Discorides Press, Portland, p.150.

De Klerk, G.J, ter Brugge, J., Smulders, R., Benshop, M., 1990, Basic peroxidase and rooting in microcuttings of *Malus*, Acta Hort. 280:29.

Duguid, J.R., Dinauer, M.C., 1992, Library subtraction of *in vitro* cDNA libraries to identify differentially expressed genes in scrapie infection, Nucl. Acids Res. 18:2789.

Foret, Y., Maldiney, R., Sotta, B., Miginiac, E., 1986, Rhizogenesis and auxin and abscissic acid levels in free clones of different aged *Sequoia sempervirens* (Endl.) trees, CR Acad Sci Paris 303:135.

Gaspar, Th., Penel, C., Castillo, F.J., and Greppin, H., 1985, A two-step control of basic and acidic peroxidases and its significance for growth and development, Physiol. Plant. 64:418.

Gaspar, Th., Kevers, C., Hausman, J.F., Berthon, J.Y., Ripetti, V., 1992, Practical uses of peroxidase activity as a predictive marker of rooting performance of micropropagated shoots, Agronomie 12:757.

Hackett, W.P., Murray, J.R., Woo, H.H., Stapfer.R.E., and Geneve, R., 1990, Cellular, biochemical and molecular characteristics related to maturation and rejuvenation in woody species, in:"Plant Ageing: Basic and Applied Approaches", R. Rodriguez, R. Sanchez Tames, and D.J.Durzan, eds., NATO ASI Ser. A, vol. 186, Plenum Press, New York.

Haissig, B.E, 1986 Metabolic process in adventitious rooting of cuttings, In:"New Root Formation in Plants and Cuttings" Jackson, M.B., ed., Martinus Nijhoff Publ., Dordrecht p.191.

Jones, O.P., Hatfield, G.S., 1976, Root initiation in apple shoot cultured *in vitro* with auxin and phenolic compounds, J. Hort. Sci. 51:495.

Kracke, H., Cristoferi, G., Marangoni, B., 1981, Hormonal changes during the rooting of hardwood cuttings of grapevine rootstocks, Americ. J. Enol. Vitic. 32:135.

Le, C.L., 1985, Influence of temperature on in vitro root initiation and development of apple rootstock M26, HortScience 20:451.

Legrand, B., 1977, Action de la lumière sur les peroxydases et sur la teneur en composés phénoliques de tissus de feuilles de *Cichorium intybus* L. cultivés *in vitro*, Biol. Plant. 19:27.

Liang, P., Pardee, A.B., 1992, Differential display of eukaryotic messenger RNA by means of the polymerase chain reaction, Science 257:967.

Liang, P., Averbourkh, L., Pardee, A.B., 1993, Distribution and cloning of eukaryotic mRNAs by means of differential display: refinements and optimization, Nucl. Acids Res. 21(14):3269.

Mato, M.C., and Vieitez, A.N., 1986, Changes in auxin protectors and IAA oxidases during the rooting of chestnut shoots *in vitro*, Physiol. Plant. 66:491.

Moncousin, C., 1986, Peroxidase as markers for rooting improvement of clones of *Vitis* cultured *in vitro*, in:"Molecular and Physiological Aspects of Plant Peroxidases" Greppin H., Penel C., Gaspar Th., eds., Univ. Geneve p. 379.

Moncousin, C., and Gaspar, Th., 1983, Peroxidase as marker for rooting improvement of *Cynara scolymus* L. cultured *in vitro*, Biochem. Physiol. Pflanzen. 178:263.

Murashige, T., and Skoog, F., 1962, A revised medium for rapid growth and bioassays with tobacco tissue cultures, Physiol. Plant. 15:437.

Nordstrom, A.C., Eliasson, L., 1992, Levels of endogenous indole-3-acetic acid and indole-3-acetylaspartic acid during adventitious root formation in pea cuttings, Physiol. Plant. 82:599.

Pythoud., F., and Buchala, A.J., 1985, Peroxidase activity and adventitious rooting in cuttings of *Populus tremula*, Plant Physiol. Biochem. 27:503.

Quoirin, G., Boxus, P., Gaspar, Th., 1974, Root initiation and isoperoxidases of stem tip cuttings from mature *Prunus* plants, Physiol. Vég. 12:165.

Van der Krieken, W.M., Breteler H., Visser M.H.M., 1991, Indolebutyric acid induced root formation in apple tissue culture. Design of an experimental system, auxin metabolism and isolation of cDNA clones related to root initiation, Acta Hortic. 289:343.

Weigel, U., Horn, W, Hock, B., 1984, Endogenous auxin levels in terminal stem cuttings of *Chrysanthemum morifolium* during adventitious rooting, Physiol. Plant. 61:422.

Zimmerman, R., Broome, O.C., 1981, Phloroglucinol and *in vitro* rooting of apple cultivar, J. Am. Soc. Hortic. Sci. 106:648.

15

ABSCISIC ACID RELATIONS IN STRESSED ROOTS

Wolfram Hartung[1] and Neil C. Turner[2]

[1]Julius von Sachs Institut für Biowissenschaften
Lehrstuhl Botanik I, Universität Würzburg
Mittlerer Dallenbergweg 64
D 97082 Würzburg, Germany
[2]CSIRO, Division of Plant Industry
Private Bag, P.O. Wembley
WA 6014, Australia

INTRODUCTION

It is now well established that abscisic acid (ABA) plays a significant role as a stress hormone in higher and lower plants. Abscisic acid is synthesised in plant tissues under stress conditions, released to transport paths such as xylem or phloem and transported to target cells, tissues or organs where it improves the water relations of the plants by acting on stomata (closure) or on roots (increasing hydraulic conductance). An important role of roots has been demonstrated by Blackman and Davies (1985). They performed split root experiments in which half of the roots were adequately watered and half were stressed. Despite of an unchanged leaf water potential, the stomata of those plants closed. This was explained by a root to shoot stress signal formed by the stressed roots that was transported to the guard cells and induced stomatal closure.

In the meantime a large body of evidence has been published indicating that abscsisic acid (ABA) is the stress signal (reviews by Davies and Zhang, 1991 Hartung and Davies, 1991, 1993), although there may be some exceptions where other substances such as ABA-precursors or ABA-conjugates may act as stress signals (Munns and King 1988, Bano et al. 1993). Additionally it seems that ABA may be transported also from the shoot to the roots where physiological processes of stress physiological significance may be initiated (Slovik et al. 1995, Zhong et al. 1996).

ROOTS AS A SOURCE OF STRESS-INDUCED ABA

When bean roots were allowed to lose up to 80% of their water and then kept for 1–2 h under these conditions, cortical and stelar tissues have exhibited a similar ABA accumula-

Biology of Root Formation and Development, edited by Altman and Waisel.
Plenum Press, New York, 1997.

tion (Hartung and Davies 1993). Dehydrated tips of maize roots showed the highest capacity for ABA accumulation, indicating biosynthesis in the non-vacuolated root tip cells.

When root tips of seven different lupin species lost 50% of their water, very different degrees of ABA accumulation and ABA degradation were observed. Roots of the drought tolerant *Lupinus digitatus*, a species which occurs on the edge of the Sahara desert and is considered to be very drought tolerant, showed the highest response to water deficits (+ 465%). While there were 3-fold differences in the inherent levels of ABA in the fully hydrated roots, drying the roots to 50% water content increased the ABA concentration in all species (Table 1). The use of 10^{-5}M tetcyclacis, a nornbornanodiacetine derivative known to inhibit the formation of phaseic acid and hence prevent ABA breakdown, demonstrated that the tested species half of the produced ABA was metabolised by the roots. *L. digitatus* was notable in having no significant breakdown of free ABA. The concentration of conjugated ABA in the roots (predominantly ABA glucose ester) of the seven species, was one tenth of the concentration of free ABA and did not increase with water stress (data not shown).

LATERAL TRANSPORT OF ABA FROM THE CORTEX TO THE XYLEM VESSELS

We lack information on the lateral transport pathway of ABA in roots. If ABA would be transported exclusively symplastically, as assumed by the computer simulation of Slovik et al. (1995), an increased water flow followed by an enhanced transpiration, would cause dilution of ABA in the xylem (Slovik et al. 1995; Else et al. 1994). However, Steudle (1995) has pointed out that water and some solutes can be transported apoplastically across the endodermis. We cannot exclude the possibility that ABA is translocated by solvent drag in the apoplast to the xylem elements which would counteract the dilution of ABA. In a series of preliminary experiments we have applied reduced pressure to the cut surface of a maize stem simulating an increased transpiration. *Zea* roots seem to have a

Table 1. Concentration of free abscisic acid (ABA) (nmol g dw^{-1}) with and without the inhibitor of ABA metabolism, 10^{-5}M tetcyclacis, in roots of 3–5 days old lupin seedlings at two water contents. Data are means ± one standard error of the mean of four replicates

	With inhibitor Root water content			Without inhibitor Root water content		
Species	100%	50%	Difference	100%	50%	Difference
Lupinus digitatus	3.16 ± 0.4	17.88 ± 4.24	+14.72 (+465%)	3.55 ± 1.59	22.12 ± 2.19	+18.57 (+503%)
Lupinus angustifolius	5.24 ± 1.35	19.58 ± 3.70	+14.34 (+273%)	2.69 ± 1.34	10.11 ± 1.78	+7.42 (+275.8)
Lupinus albus	6.13 ± 1.28	18.31 ± 1.50	+12.18 (+198%)	4.63 ± 1.34	9.70 ± 1.07	+5.07 (+109%)
Lupinus luteus	4.32 ± 1.26	10.57 ± 2.49	+6.25 (+144%)	2.43 ± 0.27	3.57 ± 0.56	+ 1.14 (+46.9%)
Lupinus pilosus	3.77 ± 1.10	8.11 ± 1.14	+4.34 (+133%)	1.55 ± 0.48	8.66 ± 0.75	+7.11 (+458.7%)
Lupinus cosentinii	10.25 ± 1.5	16.97 ± 4.7	+ 6.72 (+65%)	8.47 ± 1.44	14.91 ± 1.52	+6.44 (+76.0%)
Lupinus atlanticus	5.15 ± 1.21	7.97 ± 2.91	+2.82 (+54%)	2.63 ± 1.32	4.92 ± 1.19	+2.29 (+87%)

distinct apoplastic bypass for ABA. Large fluctuations of ABA concentration, in relation to rapid changes of lateral water flow, would be avoided.

Radin and Matthews (1989) have pointed out that under certain stress situations, such as phosphorus deficiency, an apoplastic bypass can play a significant role for lateral water transport. According to Steudle (pers. comm.) the lateral apoplastic bypass is particularly significant in root systems of several trees.

ABA-REDISTRIBUTION IN ROOTS UNDER STRESS

Intracellular redistribution of ABA follows the anion trap concept with the non membrane permeating ABA anion being trapped in alkaline compartments (Hartung and Slovik, 1991). The stress physiological significance of ABA redistribution in leaves has been reviewed in detail by Hartung and Slovik (1991). Drought- and salt stress-dependent redistribution of ABA in roots does not play that role because only small stress dependent effects on intracellular pH gradients can be observed (Daeter et al. 1993). On the other hand, pH gradients are induced by nutritional changes in the rhizosphere such as ammonium, iron and phosphate deficiency. pH gradients across the plasmalemma become steeper by acidification of the rhizosphere, which causes accumulation of ABA in the cytosol. In roots supplied with nitrate the apoplast of the cortex and stele is alkalised which flattens the pH gradient and increases the apoplastic accumulation of ABA (Daeter et al. 1993).

EFFECT OF ENVIRONMENTAL FACTORS OTHER THAN WATER DEFICITS ON ABA ACCUMULATION IN ROOTS AND XYLEM SAP

Salt Stress

Roots that grow in a drying soil are not only exposed to reduced water content, they have to cope also with high salt concentrations, alkaline conditions, nutrient deficiency and high soil impedance. Most of these effects have been studied with intact lupins or castor bean plants using the flow modelling technique as described by Wolf et al. (1990) and Peuke et al. (1994). In lupins and castor beans, that were salt stressed, ABA synthesis was significantly stimulated in roots which caused a 10-fold increase in ABA transport in the xylem. Roots received also significantly higher amounts of ABA synthesised in the leaves and transported in the phloem to the roots. Deposition of ABA was also strongly increased.

Soils of extreme habitats often contain salts different from NaCl. Hartung et al. (1990) found high NaCl- and $CaCl_2$-concentration of the soil solution close to the root system of the winter annual desert plant *Anastatica hiërochuntica*. When applied separately both salts had only small effects on ABA accumulation in *Anastatica* roots. However, when applied together, an ABA accumulation up to 25-fold could be detected.

Phosphorus Nutrition

Soils with high salt concentrations are often very alkaline. In alkaline soils, nutrients such as phosphorus, iron or manganese are very low. Similarly as in the case of salt stress, ABA synthesis and ABA deposition in roots, and long distance transport in the xylem, is drastically increased in phosphorus-deficient castor beans. Adult leaves of phosphorus deficient castor bean plants receive high amounts of root-derived ABA which reduces leaf conductance significantly. However contrary to salt stress, phloem transport of ABA to the roots is not stimulated. Deposition of ABA in leaves, however is very small because of a strong net ABA degradation (Jeschke et al. 1997).

N Nutrition

Nitrate supply has only negligible effects on ABA biosynthesis and deposition in *Ricinus* roots. However, if nitrate is replaced by ammonium, ABA synthesis, deposition and xylem transport is significantly increased. This is accompanied by characteristic changes of root morphology such as root thickening and an increase of number of lateral roots. Such changes can also be observed when roots are treated with high concentrations of ABA (Peuke et al. 1994).

Soil Compactness

Soils of extreme habitats very often show a high mechanical impedance. We have observed that mechanical impedance caused a transient 10-fold increase of ABA transport in the xylem of maize plants. This was correlated with reduced leaf growth and reduced leaf conductance. Roots penetrating compacted soils exhibited an increased number of root hairs, increased radial growth and an increased size of the root cap. The same changes were observed when root tips were treated with 1 μM ABA. Root hairs can act as an anchor when penetrating the compacted soil. We conclude that elevated ABA aids root penetration through compacted soil layers by inducing anatomical changes in the root tip. This involves secondary effects on shoot physiology (Hartung et al. 1994). An increased formation of root hairs after ABA treatment has also been observed by Eshel and Waisel (this volume).The involvement of increased ABA biosynthesis in roots that grow in compacted soil has been shown also by Mulholland et al. (1996) and by Hussain et al. (this symposium volume).

In Fig. 1 the relation between total hydraulic conductance (G) and xylem ABA concentration of almond trees grown in the Negev desert is shown. With decreasing hydraulic conductance, ABA concentration of the xylem is affected when a threshold G, caused by reduced soil moisture, is reached. Hydraulic conductance can be decreased by xylem cavitations and by a restricted water movement from soil particles to the roots. Wartinger (1991) and Heilmeier et al. (1993) have shown that it is the formation of soil aggregates in the drying loess soil that affects water conductance. This also shows that not only soil water content and the solutes in the soil solution but also soil structure influences ABA accumulation in roots and the ABA release to the xylem (Fig. 1 drawn, using data of Wartinger 1991 and Wartinger et al. 1990).

THE ORIGIN OF ROOT ABA

A significant part of ABA in stressed roots originates from biosynthesis in roots. Abscisic acid flow studies have also shown, that a significant portion of ABA that has been synthesised in leaves is transported in the phloem to the roots. Part of this ABA may be recirculated to the shoot. This portion can also be enhanced under stress conditions (Wolf et al., 1990). Mathematical models established by Slovik et al. (1995) also have predicted a recirculation of leaf synthesised ABA within intact plants.

Another source of root ABA was recently proposed by Daeter et al. (1993) and Slovik et al. (1995). By developing a mathematical model of stress-dependent ABA redis-

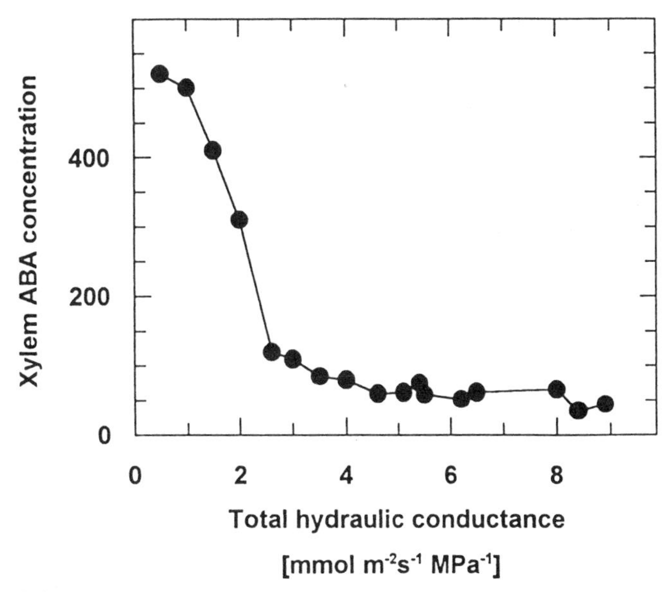

Figure 1. The relation between the total hydraulic conductance of an almond tree grown under the extreme conditions of the desert and the ABA concentration in the xylem (redrawn, using data of Wartinger 1991 and Wartinger et al., 1990).

tribution in roots and shoots of intact rosette plants a drastic release of ABA from the roots to the rhizosphere was predicted. This results in severely ABA deficient plants assuming that the soil solution was free of ABA. According to the computer simulation the dramatic loss of ABA would be prevented if approximately 1 nM ABA is present in the soil solution at a slightly acid pH. We have analysed soil solutions under different crops and under different conditions and found that ABA was present in the soil solution in the concentration range of 0.5–4.0 nM (Tab. 2). Highest concentrations were found in soils with low pH, low water and high salt contents, and lowest were observed in moist soils with a neutral or slightly alkaline pH. Such ABA can be taken up by the roots, hereby influencing root development and growth. It can also be transported in the xylem to the shoot. (Hartung et al. 1996, Müller et al. 1989).

Table 2. Abscisic acid concentration in the soil solution, taken from fields with different crops

Crops	ABA in soil solution [nM]
Wheat	0.48
Lupin	0.8
Sugar beet	1.03
Sunflower	1.07
Rye	1.2
Spring barley	1.22
Maize	1.46
Winter barley	1.49
Rape seed	2.07
Oat	2.69
Conifer trees	4.05

THE BENEFICIAL ROLE OF ABA IN STRESSED ROOTS

It has been shown repeatedly that ABA is a stimulator of root growth (Abou Mandour and Hartung 1980, the early literature is cited therein). Roots can also maintain their growth in a drying soil as along as they are supplied with high amounts of ABA (Saab et al. 1990).

The increased number of root hairs and the increased diameter of ABA-treated roots (Eshel Waisel, this volume; Biddington and Dearman, 1982; Hartung at al., 1994) facilitates the penetration of compacted soil layers by root tips. Behl and Jeschke (1979) observed an ABA-increased Na^+- and K^+-accumulation in vacuoles of cortical cells of barley roots. Thus, an increased osmotic potential of the cell sap of cortical cells of roots may have a positive role on water uptake under stress conditions.

ABA effects on hydraulic conductance have been investigated by several investigators in the past, unfortunately with contradictory results. We have reinvestigated this problem with sunflower root systems. In contrast to earlier investigations we applied reduced pressure to the cut surface of hypocotyl stumps, thus imitating the transpiration stream. It seems that ABA significantly increases hydraulic conductance of the root system under these conditions. The beneficial role of ABA can also be demonstrated with plantlets of *Ruta divaricata* that have been regenerated from callus. These plants do not survive transplanting to soil because they are very low in ABA and therefore very stress sensitive. When they are pretreated with ABA or tetcyclacis (an inhibitor of ABA breakdown that increases endogenous ABA), 100% of plants survive the transplant shock because of the beneficial ABA effects on the root system as described above (Fig. 2, Hartung and Abou-Mandour, 1996).

Figure 2. Plantlets of *Ruta divaricata* were regenerated from callus cultures and then transplanted into soil. Only plantlets that have been pretreated with 10 μM ABA or 10 μM tetcyclacis have survived the transplantation shock (1, control, 2, .pretreatment with 1 μM ABA, 3, pretreatment with 10 μM ABA, 4, pretreatment with 10 μM tetcyclacis; (after Hartung and Abou-Mandour, publication in preparation).

CONCLUSIONS

The plant stress hormone ABA is synthesised in stressed roots and released to the xylem vessels. In the shoot, ABA can improve plant water relations by reducing leaf conductance (stomatal closure) and the transpiring area (inhibition of leaf growth). Stress-dependent ABA accumulation may be caused by stimulated biosynthesis in roots and increased phloem transport of leaf synthesised ABA to the roots (shoot to root stress signal). Abscisic acid of the soil solution around the root seems to have an important function as well. In root systems that are nutrient deficient and ammonium stressed ABA redistribution between apoplast and symplast is strongly affected (Daeter et al., 1993; Peuke et al., 1994). The stimulating effects of ABA on elongation of drought stressed roots, lateral root formation and root hair formation accord with the beneficial role of this stress hormone in plant root systems.

ACKNOWLEDGMENTS

W.H. and N.C.T. are grateful to the Deutsche Forschungsgemeinschaft (SFB 251, TP A3) and the Alexander von Humboldt Foundation, respectively, for financial support and to Barbara Dierich for skilful technical help.

REFERENCES

Abou-Mandour, A.A., and Hartung, W., 1980, The effect of abscisic acid on growth and development of intact seedlings, root and callus cultures and stem and root segments of *Phaseolus coccineus*. *Z. Pflanzenphysiol.* 100: 25.

Bano, A, Dörffling, K, Bettin, D, and Hahn, H., 1993, Abscisic acid and cytokinins as possible root to shoot signals in xylem sap of rice plants in drying soil. *Austr J Plant Physiol* 20: 109.

Behl, R., and Jeschke, W.D., 1979, On the action of abscisic acid on transport, accumulation and uptake of K^+ and Na^+ in excised barley roots. Effects of the accompanying anions. *Z Pflanzenphysiol* 95: 335–353

Biddington, N.L., and Dearman A.S., 1982, The effect of abscisic acid on root and shoot growth of cauliflower plants. *Plant Growth Regulation* 1:15.

Blackman, P.,G., and Davies, W.,J., 1985, Root to shoot communication in maize plants and the effect of soil drying. *J exp Bot.* 36: 39.

Daeter, W., Slovik, S., and Hartung, W. 1993, The pH-gradients in the root system and the abscisic acid concentration in the xylem- and apoplastic saps. *Phil. Trans. Royal Society, London, Ser B*, 341: 49.

Davies, W.,J., and Zhang, J., 1991, Root signals and the development of plants growing in drying soil. *Ann Rev. Plant Physiol. Mol. Biol.* 42: 55.

Else, M.A., Davies, W.J., Whitford, P.N., Hall, K.C., and Jackson, M.B., 1994, Concentration of abscisic acid and other solutes in xylem sap from root systems of tomato and castor oil plants are distorted by wounding and variable sap flow rates. *J. exp. Bot* 45: 317.

Eshel, A., and Waisel Y., 1997, Aeroponics: An important tool for root research. *in*: "The Biology of Root formation and development." A. Altman, and Y. Waisel eds. Plenum Press. pp. 335–340.

Hartung, W., and Davies, W.J., 1991, Drought induced changes in physiology and ABA. *in:* "Abscisic Acid. Physiology and Biochemistry". W.J. Davies, and H.G. Jones. eds., Bios, Scientific Publ. Oxford, p. 63.

Hartung, W., and Slovik, S., 1991, Physico-chemical properties of plant growth regulators and plant tissues determine their distribution and redistribution. *New Phytologist* 119: 361.

Hartung, W., and Davies, W.J. 1993, Abscisic acid under drought and salt stresses. *in:* "Handbook of crop stress". Pessarakli, M. ed., Marcel Dekker New York, p. 401.

Hartung, W., and Abou-Mandour, A.A. 1996, The beneficial role of abscisic acid for regenerates of *Ruta divaricata* Tenore (Gams) suffering from transplant shock. *Angew. Botanik.* 70: 221.

Hartung, W., Heilmeier, H., Wartinger, A., Kettemann, I., and Schulze, E.D., 1990, Ionic and abscisic acid relationships of *Anastatica hiërochuntica* L. under arid conditions. *Israel J. Bot.* 39: 373.

Hartung, W., Zhang, J., and Davies, W.J., 1994, Does abscisic acid play a stress physiological role in maize plants growing in heavily compacted soil? *J. exp. Bot.* 45: 221.

Hartung, W., Sauter, A., Turner, N.C., Fillery, I., and Heilmeier, H., 1996, Abscisic Acid in Soils: What is its function and which factors influence its concentration?. *Plant and Soil.* 184: 105.

Heilmeier, H., Kaune A., Mahr, A., Türk, T., Horn, R., and Schulze, E.D., 1993, Der Wasserfluß im System Boden Pflanze. *Mittlg. Dtsch. Bodenk. Ges.* 72: 123.

Husain, A., Black, C.R., Roberts, J.A., and Taylor, I.B., 1997, Does ethylene have a role in controlling leaf growth in barley (*Hordeum vulgare*) growing on compacted soil? *in*: Proc. 2nd International Symposium on the Biology of Root Formation and Development, Jerusalem, 1996. (Abstract), p. 21.

Jeschke, W.D., Peuke, A.D., Pate, J.S., and Hartung, W., 1997, Transport, Synthesis, and Catabolism of Abscisic Acid (ABA) in intact plants of castor bean (*Ricinus communis* L.) under phosphate deficiency and moderate salinity. *J. Exp. Bot.* 48: 1737.

Mulholland, B.J., Black, C.R., Taylor, O.B., Roberts, J.A., and Lenton, J.R., 1996, Effect of soil compaction on barley (*Hordeum vulgare* L.) growth.I. *J exp Bot* 47:539.

Müller, M., Deigele, C., and Ziegler, H., 1989, Hormonal interactions in the rhizosphere of maize (*Zea mays* L.) and their effects on plant development. *Z. Pflanzenern Bodenk* 152: 247.

Munns, R., and King, R.,W., 1988, Abscisic acid is not the only stomatal inhibitor in the transpiration stream *Plant Physiol* 88:703.

Peuke, A.,D., Jeschke, W.,D., and Hartung, W., 1994, The uptake and flow of C, N and ions between roots and shoots in *Ricinus communis* L. III. Long distance transport of abscisic acid depending on nitrogen nutrition and salt stress. *J Exp Bot* 45: 741.

Radin, J.W., and Matthews, M.A., 1989, Water transport properties of cortical cells in roots of nitrogen and phosphorus deficient cotton seedlings. *Plant Physiol.* 89:264.

Saab I.N., Sharp R.E., Pritchard J., and Voetburg G.S., 1990, Increased endogenous abscisic acid maintains primary root growth and inhibits shoot growth of maize seedlings at low water potential. *Plant Physiol.* 93:1329.

Slovik, S., Daeter, W., and Hartung, W., 1995, Compartmental distribution and redistribution of abscisic acid (ABA) in roots as influenced by environmental changes. A biomathematical model. *J. exp. Bot.* 46:881.

Steudle, E., 1995, The regulation of plant water at the cell, tissues and organ level. Role of active processes and compartmentation. in: Flux control in biological systems, E.D. Schulze, ed., Academic Press p.237

Wartinger, A., 1991 Der Einfluß von Austrocknungszyklen auf Blattleitfähigkeit, CO_2-Assimilation, Wachstum und Wassernutzung von *Prunus dulcis* (Miller) D.A.Webb. Ph.D. Dissertation, University Bayreuth.

Wartinger, A., Heilmeier, H., Hartung, W., and Schulze, E.D., 1990, Daily and seasonal courses of leaf conductance and abscisic acid in the xylem sap of almond trees [*Prunus dulcis* (Miller) D.A.Webb] under desert conditions. *New Phytol.* 116: 581.

Wolf, O., Jeschke, W.D., and Hartung, W., 1990, Long distance transport of abscisic acid in salt stressed *Lupinus albus* plants. *J. exp. Bot.* 41:593.

Zhong, W., Hartung, W., Komor, E., and Schobert, Ch., 1996, Phloem transport of abscisic acid in *Ricinus communis* L. seedlings. *Plant Cell Env.* 19:471.

16

CONVERSION OF PUTRESCINE TO γ-AMINOBUTYRIC ACID, AN ESSENTIAL PATHWAY FOR ROOT FORMATION BY POPLAR SHOOTS *IN VITRO*

Jean-François Hausman,[1] Claire Kevers,[2] Danièle Evers,[1] and Thomas Gaspar[2]

[1]CRP-Centre Universitaire
CREBS, 162A, Av. de la Faïencerie
L-1511 Luxembourg, Luxembourg
[2]Institute of Botany B22
University of Liège
Sart-Tilman, B-4000 Liège, Belgium

1. INTRODUCTION

Polyamines have been involved in the control of the inductive phase of rooting in *Nicotiana tabacum* (Malfatti *et al.* 1983, Kaur-Sawhney *et al.* 1988, Altamura *et al.* 1991, Altamura 1994), *Vigna radiata* (Friedman *et al.* 1985), *Prunus avium* (Biondi *et al.* 1990) and *Beta vulgaris* (Biondi *et al.* 1993). We have come to the same conclusions for poplar shoots raised *in vitro* (Hausman *et al.* 1994, 1995a, b) and also for walnut shoots (Heloir *et al.* 1996). On the contrary, spermidine and spermine inhibited the induction of the rooting process in the same material. Such specific effects of the most prevalent polyamines in plants, i.e. putrescine, spermidine and spermine, were also found by other laboratories (Tiburcio *et al.* 1989, Altamura *et al.* 1991, Rey *et al.* 1994). In poplar shoots, putrescine was suggested to act through its catabolic pathway (Hausman *et al.* 1994).

Putrescine can be converted into Δ^1-pyrroline by diamine oxidase (Fig. 1). This compound is metabolized to γ-aminobutyrate (GABA) by pyrroline dehydrogenase (Flores and Filner 1985, Davies *et al.* 1990). The produced GABA is transformed to succinic semialdehyde by a GABA-transaminase and then to succinate via a succinic semialdehyde dehydrogenase. This succinate is available for the Krebs cycle (Bown and Shelp 1989, Shelp *et al.* 1995). It has been reported that an accumulation of GABA is observed after various stresses (Bown and Shelp 1989). However, the accumulation of GABA is often attributed to the so-called GABA shunt (Shelp *et al.* 1995). In the same pathway, spermidine can also be converted to 1,3-diaminopropane and Δ^1-pyrroline via polyamine oxidase; then Δ^1-

Biology of Root Formation and Development, edited by Altman and Waisel.
Plenum Press, New York, 1997.

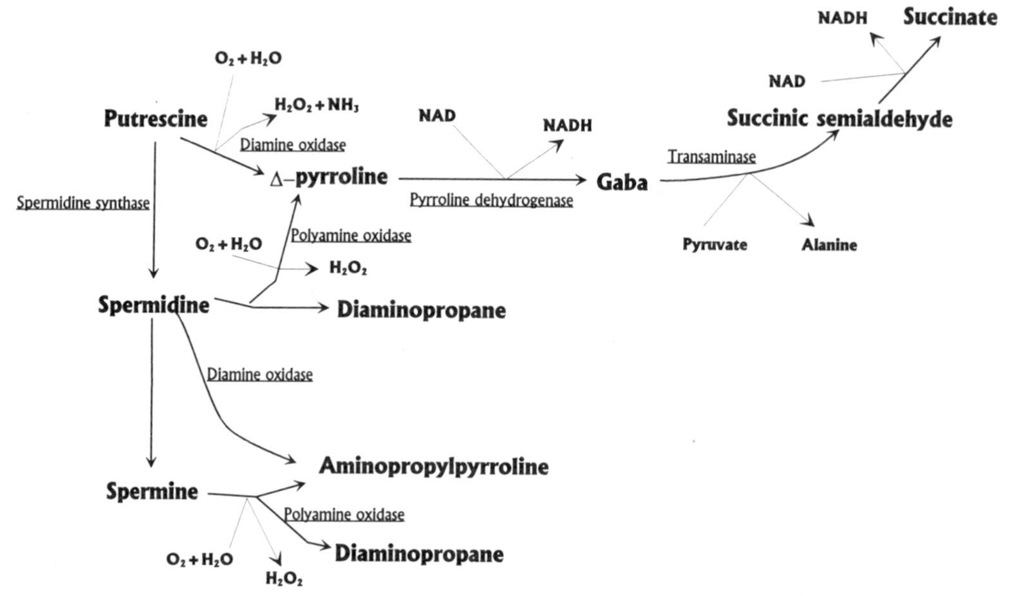

Figure 1. Some metabolic pathways associated with polyamine metabolism.

pyrroline is metabolized to GABA via pyrroline dehydrogenase (Fig. 1) (Flores and Filner 1985, Davies *et al.* 1990). Spermidine can also be converted to aminopropylpyrroline via diamine oxidase (Evans and Malmberg 1989). Polyamine oxidase oxidizes spermidine to give aminopropylpyrroline (Smith 1985). In the present study, we used micropropagated poplar shoots (*Populus tremula x P. tremuloides*, cv. Muhs 1) to focus on the role of putrescine and its γ-aminobutyrate metabolism in the induction of roots. First, we determined the activity of enzymatic pathways that convert putrescine or spermidine to γ-aminobutyrate. Second, we supplied exogenous compounds of the GABA pathway to induce rooting.

2. MATERIAL AND METHODS

2.1. Plant Material and Culture Conditions

Stock cultures of *Populus tremula* L. x *P. tremuloides* Michx. cv. Muhs 1 proliferated *in vitro* (obtained from Dr. G. Douglas of the Kinsealy Research Centre in Dublin, Ireland) were used. Cultures were subcultured every 3 weeks on Murashige and Skoog (MS, 1962) medium supplemented with 3.9 μM 2-isopentenyladenine. For rooting, isolated shoots (30 mm-long) were excised at their base and transferred to one-half strength MS medium supplemented with 1.7 μM NAA; the same medium without growth regulators was used as control. All media were supplemented with 30 g. l^{-1} sucrose and adjusted to pH 5.5–5.6 with KOH or HCl prior to the addition of "Roland" agar (8 g. l^{-1}) and subsequently autoclaved for 20 min at 121°C and 118 kPa. Shoot cultures were grown on 100 ml medium in 600 ml (10 cm diameter) "Le Parfait" glass jars, with glass lids held in place by transparent plastic Reynolds film. These containers (20 explants each) were maintained in a growth chamber at 24°C with a 16h photoperiod and an irradiance of 2.9 $W.m^{-2}$ (Sylvania Gro-Lux fluorescent lamps) at culture level inside the jar.

2.2. Chemical Treatments

Shoots sampled from the proliferating cultures at the end of the multiplication phase were cut at the base and transferred to an auxin–containing or control medium containing the filter-sterilised compound(s) to be tested. Since root induction was shown to terminate at hour 7 after the transfer on the auxin-containing medium (Hausman *et al.* 1994), all the shoots were transferred to fresh control medium at beginning of hour 8 from the auxin or control medium. The percentages of rooted shoots (% of rooting) were assessed after 13 days. Putrescine, spermidine, GABA, 1,3-diaminopropane, aminopropylpyrroline, cyclohexylamine and aminoguanidine were supplied at concentrations of 10 and 100 μM.

2.3. Determination of Enzymatic Activity Responsible for Catalysing the Conversion of Putrescine or Spermidine into GABA

GABA production was estimated by the formation of NADH, spectrophotometrically detected at 340 nm as described by Flores and Filner (1985). A similar process was used for the estimation of the activity responsible of the conversion of spermidine into GABA (Flores and Filner 1985).

2.4. Extraction and Determination of Polyamines

Shoots were harvested at intervals during rooting, and during or after exposure to various treatments. Their basal regions, where the roots appeared, were separated, weighed and stored at −40°C. Free polyamine extraction, separation, identification and measurement by direct dansylation and HPLC were done as described by Walter and Geuns (1987).

All results are the mean of at least three separate experiments.

3. RESULTS

3.1. Variation of Enzymatic Activity Responsible for the Conversion of Putrescine or Spermidine into GABA

Enzymatic conversion of putrescine to GABA increased until the 3rd hour in the auxin-containing medium (Fig. 2). A rapid decrease of this activity occurred between hours 6 and 8, followed by a stabilization until day 4; the activity then dropped and was undetectable from day 8 on.

On control medium, enzymatic conversion of putrescine to GABA did not show any specific variation during the 13 days of rooting. No enzymatic conversion of spermidine to GABA via Δ^1pyrroline was detected in shoots on either auxin or control media.

3.2. Variation of the Polyamine Levels during the Rooting Process

Putrescine in the basal part of the shoots on auxin medium increased dramatically at hour 6, followed by a decrease to a constant level between hours 8 and 24. From day 4, putrescine levels increased and remained high until day 13 (Fig. 3). Such variations were

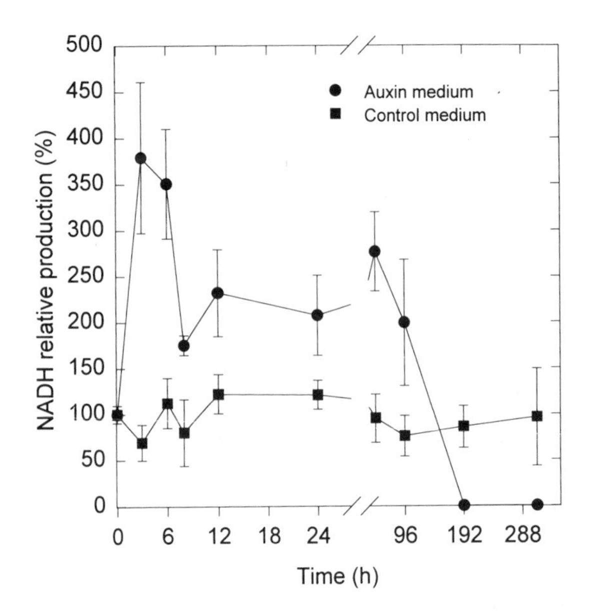

Figure 2. Relative variation of the enzymatic activity of putrescine conversion into GABA in the basal part of the shoots, during the rooting process with or without auxin medium (means±SD).

not observed in shoots on control medium. During the first 24 hours of the rooting process neither spermidine nor spermine showed any significant variation, neither on auxin nor on control medium. At day 2 spermidine levels peaked in the basal part of the shoots and then remained constant at a lower level. This variation was not observed in control medium. Spermine levels on either auxin or control media did not show any specific variation.

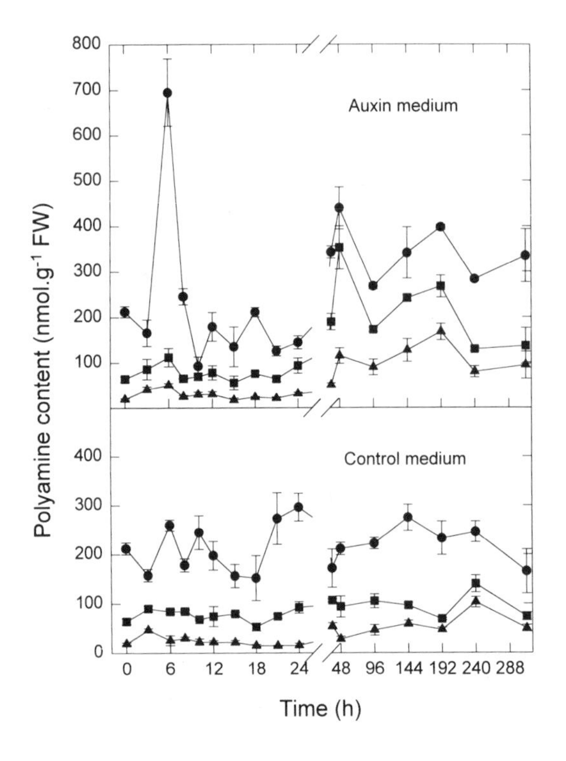

Figure 3. Concentration of free polyamines in the basal part of the shoots during the rooting process with or without auxin (means±SD). Putrescine (●), spermidine (■), spermine (Δ).

3.3. Effects of Exogenously Supplied Compounds and Inhibitors of Polyamine Biosynthesis on the Percentage of Rooted Shoots

The results in Table 1 show 100% rooting of poplar shoots that were kept for 7 h on the auxin medium followed by 13 days on control medium. Poplar shoots that were cultured in the control medium for 7 h and then transferred to a fresh control medium did not root at all. The addition of putrescine, GABA or cyclohexylamine (at a concentration of 100 μM) to the control medium during the 7 h inductive phase, had a stimulatory effect and resulted in respectively 40%, 23% and 24% of rooted shoots. These compounds did not have any effect when applied to an auxin medium. Spermidine, diaminopropane, aminopropylpyrroline and aminoguanidine at 100 μM had inhibitory effects on rooting in auxin medium, and the percentage of rooted shoots was 75%, 84%, 52% and 79%, respectively. These compounds had no significant effect when supplied to the shoots on the control medium.

4. DISCUSSION

Although previous work already demonstrated the importance of putrescine for root differentiation (Tiburcio *et al.*, 1989), no work has shown the essentiality of the conversion of putrescine to GABA for the rooting process. The present results show that the endogenous level of putrescine increased (Fig. 3) when the enzymatic conversion of putrescine into GABA decreased (Fig. 2) between hours 6 and 8 of rooting. This relationship is quite normal considering that putrescine is the major substrate of the enzymatic GABA-producing pathway. Spermidine synthase activity is insufficient to convert excess putrescine to spermidine. This is sustained by the fact that the peak of putrescine is not followed by an increase in spermidine or spermine content. The short-lived peak in putrescine content indicates that the accumulated putrescine is consumed rapidly, probably through production of GABA between hours 8 and 96.

Table 1. Effects of polyamines and compounds related to polyamine biosynthesis, applied during the induction phase (0–7h), on rooting with auxin or without auxin and then transferred to a medium without auxin (% of rooted shoots after 13 days, ±SD)

Added compounds during the 7 h induction phase	Medium	Concentration of the supplied compound (μM)		
		0	10	100
Putrescine	auxin	100	100	100
	no–auxin	0	24±9	40±5
Spermidine	auxin	100	86±11	75±11
	no–auxin	0	0	7±2
GABA	auxin	100	100	100
	no–auxin	0	15±4	23±7
1,3–diaminopropane	auxin	100	95±8	84±11
	no–auxin	0	0	0
Aminopropylpyrroline	auxin	100	74±16	52±9
	no–auxin	0	0	3±1
Aminoguanidine	auxin	100	91±7	79±13
	no–auxin	0	0	0
Cyclohexylamine	auxin	100	100	100
	no–auxin	0	9±5	24±8

The importance of putrescine in root induction was confirmed by the promotive effect of exogenous putrescine, when applied during the 7h induction period. Under similar conditions cyclohexylamine, which inhibits spermidine synthase, enhanced rooting while aminoguanidine, which inhibits diamine oxidase, had the opposite effect. The results indicate that the conversion of putrescine to GABA is essential for the induction of rooting in poplar shoots. This hypothesis is sustained by the promotive effect of exogenously applied GABA.

The inhibitory effect of exogenously applied spermidine may be due to a diminution of putrescine production in the arginine or ornithine pathway, caused by inhibition of the production of spermidine. Such an inhibition could have a negative effect on the availability of free putrescine for the GABA pathway. Another hypothesis would be to consider spermidine and spermine as precursors of 1,3-diaminopropane and aminopropylpyrroline, which inhibit rooting in poplar shoots (Table 1). This could be due to the conversion of diaminopropane into uncommon polyamines (Kuehn *et al.* 1990).

Classically, root induction has been associated with a transient increase in endogenous free indole-3-acetic acid (Blakesley 1994, Gaspar *et al.* 1994). The necessity of the conversion of putrescine to GABA for rooting and the peak of putrescine during root induction should be considered in relation to auxin metabolism.

5. ABSTRACT AND CONCLUSIONS

Poplar shoots cultured *in vitro* rooted to 100% in the presence of auxin, while in its absence they did not root. A transient increase and accumulation of free putrescine, but not of spermidine nor spermine, was measured in the basal rooting portion of poplar shoots at the 6th hour of culture in the presence of naphthalenacetic acid (NAA). Exogenously applied putrescine (100 μM) caused up to 40% rooting in the absence of NAA (0% rooting without putrescine). On the contrary, spermidine inhibited rooting in the presence of NAA. γ-Aminobutyrate (GABA) also induced rooting (without NAA) while 1,3-diaminopropane and aminopropylpyrroline inhibited it in the presence of NAA. A typical NAA-induced increase in the enzymatic conversion of putrescine to GABA was evident in the rooted shoots. The involvement of the conversion of putrescine to GABA in the rooting inductive phase of poplar was further demonstrated by the effects of aminoguanidine, which is an inhibitor of the diamine oxidase and of cyclohexylamine which is an inhibitor of spermidine synthase. These two compounds reduced the rooting process in the presence of NAA and promoted it in the absence of NAA, respectively. Such compounds were active even when applied only during the 7 hours duration of the inductive phase.

6. ABBREVIATIONS

GABA, γ-minobutyrate; NAA, naphthalene acetic acid.

REFERENCES

Altamura, M.M. 1994. Rhizogenesis and polyamines in tobacco thin cell layers. Adv. Hort. Sci. 8: 33–35.

Altamura, M.M., Torrigiani, P., Capitani, F., Scaramagli, S., & Bagni, N. 1991. *De novo* root formation in tobacco thin layers is affected by inhibition of polyamines biosynthesis. J. Exp. Bot. 42: 1575–1582.

Biondi, S., Diaz, T., Iglesias, I., Gamberini, G., & Bagni, N. 1990. Polyamines and ethylene in relation to adventitious root formation in *Prunus avium* shoot cultures. Physiol. Plant. 78: 474–483.

Biondi, S., Hagège, D., Rossini, P., & Bagni, N. 1993. Polyamine metabolism and ethylene biosynthesis in normal and habituated sugar beet callus. Physiol. Plant. 89: 699–706.

Blakesley, D. 1994. Auxin metabolism and adventitious root initiation. *In* Biology of Adventitious Root Formation (T.D. Davis and B.E. Haissig, eds), pp. 143–154. Plenum Press, New York.

Bown, A. & Shelp, B.J. 1989. The metabolism and physiological roles of 4-aminobutyric acid. Biochem. Life Sci. Adv. 8: 21–25.

Davies, P.J., Rastogi, R. & Law, D.M. 1990. Polyamines and their metabolism in ripening tomato fruit. *In* Polyamines and Ethylene: Biochemistry, Physiology and Interactions (H.E. Flores, R.N. Arteca, and J.C. Shannon, eds.), pp.112–125. American Society of Plant Physiologists,

Evans, P.T. & Malmberg, R.L. 1989. Do polyamines have roles in plant development? Annual Review of Plant Physiology. 40: 235–269.

Flores, H.E. & Filner, P. 1985. Polyamine catabolism in higher plants: characterization of pyrroline dehydrogenase. Plant Growth Regul. 3: 277–291.

Friedman, R., Altman, A. & Bachrach, U. 1985. Polyamines and root formation in mung bean hypocotyl cuttings. II. Incorporation of precursors into polyamines. Plant Physiol. 79: 80–83.

Gaspar, T., Kevers, C., Hausman, J.F. & Ripetti, V. 1994. Peroxidase activity and endogenous free auxin during adventitious root formation. *In* Physiology, Growth and Development of Plants in Culture (P.J. Lumsden, J.R. Nicholas, and W.J. Davies, eds.), pp.289–298. Kluwer Academic Publishers, The Netherlands.

Hausman, J.F., Kevers, C., & Gaspar, T. 1994. Involvement of putrescine in the inductive rooting phase of poplar shoots raised *in vitro*. Physiol. Plant. 92: 201–203.

Hausman, J.F., Kevers, C., & Gaspar, T. 1995a. Putrescine control of peroxidase activity in the inductive phase of rooting in poplar shoots *in vitro*, and the adversary effect of spermidine. J. Plant Physiol.146: 681–685.

Hausman, J.F., Kevers, C., & Gaspar, T. 1995b. Auxin-polyamine interaction in the control of the rooting inductive phase of poplar shoots *in vitro*. Plant Sci. 110: 63–71.

Heloir, M.C., Kevers, C., Hausman, J.F. & Gaspar, T. 1996. Changes in the levels of auxins and polyamines in the cause of rooting of walnut shoots *in vitro*. Tree Physiol. 16: 515–519.

Kaur-Sawhney, R., Tiburcio, A.F. & Galston, A.W. 1988. Spermidine and flower bud differentiation in thin-layer explants of tobacco. Planta. 173: 282–284.

Kuehn, G.D., Rodriguez-Garay, B., Bagga, S., & Phillipps, G.C. 1990. Novel occurrence of uncommon polyamines in higher plants. Plant Physiol. 94: 855–857.

Malfatti, H., Vallee, J.C., Perdizet, E., Carre, M. & Martin C. 1983. Acides aminés libres d'explants foliaires de *Nicotiana tabacum* cultivés *in vitro* sur des milieux induisant la rhizogenèse ou la caulogenèse. Physiol. Plant. 57: 492–498.

Murashige, T. & Skoog, F. 1962. A revised medium for rapid growth and bioassays with tobacco tissue culture. Physiol. Plant. 15: 473–497.

Rey, M., Diaz-Sala, C., & Rodriguez, R. 1994. Exogenous polyamines improve rooting of hazel microshoots. Plant Cell Tissue Org. Cult. 36: 303–308.

Shelp, B.J., Walton, C.S., Snedden, W.A., Tuin, L.G., Oresnik, I.J., & Layzell, D.B. 1995. GABA shunt in developing soybean seeds is associated with hypoxia. Physiol. Plant. 94: 219–228.

Smith, T.A. 1985. Polyamines. Annual Review of Plant Physiology. 36: 117–143.

Tiburcio, A.F., Gendy, C.A., & Tran Thanh Van, K. 1989. Morphogenesis in tobacco subepidermal cells: putrescine as marker of root differentiation. Plant Cell Tissue Org. Cult. 19: 43–54.

Walter, H.J. & Geuns, J.M. 1987. High speed HPLC analysis of polyamines in plant tissues. Plant Physiol. 83: 232–234.

17

ROOT DEVELOPMENT IN *IN VITRO* POTATO EXPLANTS AS AFFECTED BY JASMONIC ACID

A. M. Pelacho, J. Pérez-Catalan, and Ll. Martín-Closas

Department Horticulture, Botany and Gardening
ETSEA
University of Lleida
25198 Lleida, Spain

INTRODUCTION

Jasmonic acid is a natural plant substance, known for a long time in the cosmetic industry, which has been later re-discovered as a growth regulator. Its effects are not completely well-known. Yet there are already a number of experimental and review papers which include an increasing, and sometimes contradictory, number of effects (Koda 1992, Parthier et al. 1992, Sembdner and Parthier 1993, van den Berg and Ewing 1991). Initially, jasmonic acid was related to senescence promotion; more recently, it has been associated with a wide range of effects, including promoting growth activity. It has also been related with the synthesis of determined storage proteins and with stress and pathogen responses (Sembdner and Parthier 1993).

In some reports it has been demonstrated that jasmonic acid is a potent tuber-inducing substance (Pelacho and Mingo-Castel 1991, Koda et al. 1991). In addition, the variations in the endogenous levels of jasmonates, in the leaves of potato plants, suffer photoperiod related changes (Koda and Okazawa 1988). Both facts together, and in accordance with the hormonal theory of tuberization in potato, indicate that jasmonic acid might be the tuber inducing substance, formed in the leaves in response to short photoperiods, and transported to the stolons where it would induce tuberization. However, other reports suggest that it may increase potato vegetative growth (Ravnikar et al. 1992), which contrasts a tuber-inducing effect.

The *in vitro* systems, in which jasmonic acid has induced tuberization, were kept in continuous darkness, where vegetative development is hard to be evaluated. Thus, the aim of the present work is to analize the possible growth promoting activity of jasmonic acid, both in roots and in shoots, and to contrast it with the possible tuber promoting activity, using equivalent systems under varying photoperiods.

Biology of Root Formation and Development, edited by Altman and Waisel.
Plenum Press, New York, 1997.

EXPERIMENTAL

Potato plantlets (*Solanum tuberosum* cv. Kennebec) were micropropagated at 22°C and 16h photoperiod. Single-node sections were subcultured in an MS medium (Murashige and Skoog, 1962) containing 6% sucrose. The solutions contained 0.1, 1.0 or 10.0 $mg \cdot l^{-1}$ jasmonic acid (JA) when culturing under long photoperiod (16 h light) at 45 $\mu mol \cdot m^{-2} \cdot s^{-1}$, or 0.1, 1.0 or 2.5 $mg \cdot l^{-1}$ jasmonic acid when culturing under shorter photoperiods (8 and 12 h light) at 90 $\mu mol \cdot m^{-2} \cdot s^{-1}$. After adding the JA, pH was adjusted to 5.7–5.8 with 1N NaOH and the media were autoclaved. In both cases, explants cultured in equivalent media without JA, were used as controls. Cultures were kept for 45 days at 22 ± 1°C.

Each treatment consisted of twenty explants and was repeated twice. At the end of the experiment explant length, root initiation and length, tuber initation, number of branches, total fresh and dry weight, tuber fresh and dry weight, and root fresh and dry weight were recorded. For practical reasons a new shoot borne root was considered when it was over two millimeters long.

RESULTS AND DISCUSSION

Development under Long Photoperiod

The two lower JA concentrations significantly increased explant length (Table 1). They also caused a more than four fold increase in the number of roots (Table 1), although these were significantly shorter. The branching response for all treatments paralleled that of rooting (Table 1). Fresh and dry weight of the explants at the end of the experiment significantly decreased in the 10.0 and 1.0 $mg \cdot l^{-1}$ concentrations; while under the lowest JA dose, the slight decrease was not significant (Table 1).

JA was previously found to be the strongest tuber inducing substance in *in vitro* cultured stolons of the Kennebec cultivar, and under an equivalent system in the dark (Pelacho and Mingo-Castel 1991), in which it hardly allowed shoot or root development. Since there is, *in vivo*, a known relationship between tuber initiation and the appearance of some morphological changes, e.g., a decrease in rooting and in branching, and cesation of longitudinal growth (Steward et al. 1981); and since the highest JA concentration tested caused a significant decrease in explant length, a 90% inhibition of root longitudinal growth and no branching, some tuber promotion was expected. However, explants grown with 10 $mg \cdot l^{-1}$ JA showed an unhealthy appearance, with some very dark roots, and only 10% of the explants developed tubers. It was concluded that this JA concentration acts more at supraoptimal than at a physiological range.

Table 1. Effect of jasmonic acid (JA) treatments on explant elongation, number of branches, root number, length of the longest root, and total fresh and dry weight, after 45 days in culture under 16 h photoperiod

JA conc. ($mg \cdot l^{-1}$)	Elongation (mm)	Branch number	Root length (mm)	Root number	Total fresh weight (mg)	Total dry weight (mg)
0.0	71.3 b	1.0 a	32.2 c	3.4 a	176.1 c	16.6 c
0.1	156.5 d	2.4 b	18.7 b	14.3 b	156.4 bc	13.5 bc
1.0	105.6 c	1.9 b	16.7 b	14.8 b	111.2 b	11.0 b
10.0	16.2 a	1.1 a	3.5 a	4.9 a	22.1 a	3.6 a

Numbers within a column followed by the same letter are not significantly different (p = 0.05).

In practice, the common way to induce microtubers, using a wide range of growth regulating chemicals, is not under long photoperiods. Although no specific tuber inhibiting substance has been isolated to date from explants grown under long photoperiods, the possibility remains. In any case, long photoperiods lead to lower tuberization and higher vegetative growth than short ones. It seemed convenient to test if any of the JA effects found were photoperiod dependent.

Development under Shorter Photoperiods

Both under 8 and under 12h photoperiod JA effects were stronger than in the previous experiment and growth promotion was the general trend under any of the JA concentrations tested. As before, elongation and branching were highest for the 0.1 $mg{\cdot}l^{-1}$ JA concentration. However, in this case, results were also significantly over controls for all JA treatments, with minor differences among photoperiods (Table 2).

A 4h increase in photoperiod length caused explants to be heavier. Moreover, the effectivity of JA as enhancer of plant weight depended on photoperiod. Under short photoperiod (8h) the lowest JA concentration tested was as effective in increasing fresh and dry weight as the middle one (Table 2). However, under the 12h light photoperiod the capacity of the explants to increase their weight in response to JA applications changed and it rose considerably more than under 8h photoperiod; at the same time, and apparently contradictory, a 1 $mg{\cdot}l^{-1}$ JA application became more effective (Table 2).

Root formation in JA-treated explants was so abundant that quantifying the effect of JA on root development was not possible through counting the number of new adventitious roots or the root length. The fact that roots were thicker and usually shorter than in controls difficulted the task from very early stages. Roots of the explants were also heavier under the longer photoperiod (Table 3). JA-induced increases were also considerably higher under the 12h than under the 8h photoperiod, the rise under both photoperiods being significantly higher for the treatments with 1.0 $mg{\cdot}l^{-1}$ JA. Moreover, the root weight was the only parameter in which the highest JA application did not cause the lowest response.

Under the most favorable conditions (2.5 $mg{\cdot}l^{-1}$ JA, 8h photoperiod), only 15% of the explants developed tubers, and no differential appearance was revealed in the tuber-forming explants, although they were so few that no conclusions can be drawn. Tubers were very small and they contributed in less than 1% to the total fresh or dry weight of the explants.

Explant development suffered important changes under the influence of the JA, with overall increases in all the parameters analyzed: explant length, branching, root number and thickness, total fresh and dry weight, and root fresh and dry weight. It seems that, at least under the experimental conditions tested, JA is a potent general growth promotor.

Table 2. Increase of elongation, number of branches, and fresh and dry weight, in jasmonic acid treated explants, after 45 days in culture under 8 h or 12 h photoperiod

	Elongation (mm)		Branch number		Total fresh weight (mg)		Total dry weight (mg)	
JA conc.	Photoperiod		Photoperiod		Photoperiod		Photoperiod	
($mg{\cdot}l^{-1}$)	8 h	12 h	8 h	12 h	8 h	12 h	8 h	12 h
0.0	132 a	103 a	1.9 a	1.2 a	104.9 a	119.7 a	8.0 a	10.3 a
0.1	389 d	422 c	4.1 c	4.6 b	648.0 bc	1413.1 b	54.4 b	130.6 b
1.0	307 c	401 c	2.8 b	3.8 b	735.1 c	1974.2 c	68.0 b	185.4 c
2.5	230 b	301 b	1.7 a	3.8 b	516.5 b	1221.6 b	51.3 b	116.0 b

Table 3. Increase of root fresh and dry weight, and root dry matter contribution to total dry weight, in jasmonic acid treated explants, after 45 days in culture under 8 h or 12 h photoperiod

JA conc.	Root fresh weight (mg)		Root dry weight (mg)		Dry matter in roots (%)	
	Photoperiod		Photoperiod		Photoperiod	
($mg \cdot l^{-1}$)	8 h	12 h	8 h	12 h	8 h	12 h
0.0	17.4 a	23.5 a	1.9 a	2.2 a	23.7	21.3
0.1	143.2 b	489.0 b	18.8 b	53.1 b	34.6	40.7
1.0	304.1 c	1080.7 c	36.9 c	110.8 c	54.3	59.8
2.5	225.2 bc	630.3 b	29.3 bc	68.3 b	57.2	58.9

Numbers within a column followed by the same letter are not significantly different (p = 0.05).

However, the JA effects were not equivalent in the aerial and in the "underground" part of the explant.

The increase of branching of the aerial parts is very interesting, because JA content has been reported to be higher in actively growing areas (Knöfel et al. 1984, Meyer et al. 1984), thus perhaps being related with a higher branching rate; and branching is associated with the potato juvenile stages. So, any tuber-inducing factor associated with senescence would be initially expected to be in connection with reduction in branching. Since tuber formation was very low for any treatment, no relationship between branching and tuberization rates could be established. The failure of JA to induce tuberization under any photoperiod poses some doubts on JA being itself the so called "tuberization hormone". However, the strong effectiveness of JA in the dark cannot be neglected. The coexistence of both facts indicates the presence of a JA-photoperiod interaction that deserves further investigation.

The JA effect in promoting development on the explant "underground" part was stronger than on the aerial part. JA caused a drastic change both in fresh and in dry matter distribution between shoot and root systems (Table 3). Roots of controls were 21–24% of the dry weight of the explant. The root contribution to the total dry weight of the explant increasing to 35–41% under the lower JA concentration, and went up to 54–60% with the remaining JA concentrations. Even though the system does not allow to completely separate the JA effects on shoots from those on roots, our results make the role of JA in enhancing root development very likely. Although there are some reports that contribute to the idea that JA has also direct effects on shoot development, it has not been possible to completely determine if the aerial effects reported here are due to JA itself or to its effect on root development. This remains to be dilucidated.

REFERENCES

Knöfel H.D., Brückner C., Kramell R., Sembdner G. and Schreiber K. 1984. A radioimmunoassay for jasmonic acid. Biochem. Physiol. Pflanzen 179:317–325.

Koda Y. 1992. The role of jasmonic acid and related compounds in the regulation of plant development. Int. Rev. Cytol. 135:155–199

Koda Y. and Okazawa Y. 1988. Detection of potato tuber-inducing activity in potato leaves and old tubers. Plant Cell Physiol. 29: 969–974.

Koda Y., Kikuta Y., Tazaki H., Tsujino Y., Sakamura S., and Yoshihara T. 1991. Potato tuber-inducing activities of jasmonic acid and related compounds. Phytochemistry 30: 1435–1438.

Meyer A., Miersch O., Buttner C., Dathe W. and Sembdner G. 1984. Occurrence of the plant growth regulator jasmonic acid in plants. J. Plant Growth Regul. 3:1–8

Murashige T. and Skoog F. 1962. A revised medium for rapid growth and bioassays with tobbaco tissue cultures. Physiol. Plant. 15: 473–497.

Parthier B., Brückner C., Dathe W., Hause B., Herrman G., Knöfel H.-D., Kramell R., Lehmann J., Miersch O., Reinbothe S.T., Sembdner G., Wasternack C. and Zur Nieden U. 1992. Jasmonates: metabolism, biological activities, and modes of action in senescence and stress responses. *in*: "Progress in Plant Growth Regulation," C.M. Karssen, L.C. van Loon and D. Vreugdenhil eds., Kluwer, The Netherlands, pp. 276–285.

Pelacho A.M. and Mingo-Castel A.M. 1991. Jasmonic acid induces tuberization of potato stolons cultured in vitro. Plant Physiol. 97, 1253–1255.

Ravnikar M., Vilhar B. and Gogala N. 1992. Stimulatory effects of jasmonic acid on potato stem node and protoplast culture. J. Plant Growth Regul. 11: 29–33.

Sembdner G. and Parthier B. 1993. The biochemistry and the physiological and molecular actions of jasmonates. Annu. Rev. Plant Physiol. Plant Mol. Biol. 44:569–89.

Steward F.C., Moreno U. and Roca W.M. 1981. Growth, form and composition of potato plants as affected by environment. Ann. Bot. 48, supplement No. 2, 45 pp.

Van den Berg J.H. and Ewing E.E. 1991. Jasmonates and their role in plant growth and development, with special reference to the control of potato tuberization: a review. Am. Pot. J. 68:781–794.

18

ROLE OF ROOT DERIVED ABA IN REGULATING EARLY LEAF GROWTH RESPONSES TO WATER DEFICITS

Peter Neumann,[1] Ofer Chazen,[1] Lev Bogoslavsky,[1] and Wolfram Hartung[2]

[1]Plant Physiology Lab.
Faculty of Agricultural Engineering
Technion I.I.T., Haifa
Israel
[2]Julius von Sachs Institute fur Biowissenchaften
Mittlerer Dallenburgweg 6
Univ. Wurtzburg, D97082
Germany

INTRODUCTION

Inhibition of leaf growth is usually the first whole plant response to soil water deficits (Hsaio, 1973; Chazen and Neumann, 1994). In the longer term, such inhibition will result in reduction in the leaf area available for photosynthesis and possible yield reductions. However, the chain of events by which soil water deficits in the root zones of plants are initially signaled to the leaves and then converted into a sustained inhibition of leaf growth, are still far from clear. In principle, the process could involve transmission of hydraulic, chemical or electrical signals from the stressed roots to the leaves.

There has been much recent interest in the possibility that increased transport from the roots to the leaves, of the hormone ABA (abscisic acid), may act as a long distance signal which regulates leaf adaptations to soil water deficits. For example, increased concentrations of endogenous ABA in the root to shoot xylem pathway of water stressed plants may correlate to some extent with reductions in leaf stomatal apertures and transpiration rates. Moreover, ABA may also inhibit leaf growth (see extensive reviews in Davies and Jones, 1991; Davies and Zhang, 1991).

However, the roots are not the only source of the additional ABA accumulated in plants responding to soil water deficits. Thus, leaf tissues responding to water deficits appear to be able to synthesize and transport ABA independently of the roots (e.g. Wright and Hiron, 1969; Wolf et al., 1990; Slovak et al., 1995). Furthermore, the possible role of root derived ABA in the inhibition of wheat leaf growth by water deficits has been critically questioned (Munns, 1992; Munns and Sharp, 1993). Similarly, Chazen and Neumann

Biology of Root Formation and Development, edited by Altman and Waisel.
Plenum Press, New York, 1997.

(1994) found that water deficits applied to maize seedlings with killed roots, caused rapid (within minutes) hardening of leaf cell walls and associated inhibition of leaf growth, even though the killed roots were presumably unable to synthesize and export ABA signals to the leaves. The authors suggested that a hydraulic signal (i.e. more negative water potential in the xylem), rather than a root derived hormonal signal, was involved in initiating the rapid leaf responses. Nonami and Boyer (1990) and Malone (1993) also provide evidence for the role of hydraulic signaling in early root to shoot communication.

This report describes the use of seedlings with live or killed roots, to further investigate the comparative roles of root to shoot ABA signals and /or hydraulic signals, in regulating early leaf growth responses to water deficits, in both wheat and maize. In this context, water deficits are defined as external water potentials which limit leaf growth rates to sub-optimal levels.

EXPERIMENTAL

Plant Growth

Germination and growth of wheat seeds (Triticum aestivum L. cv. Beit Hashita) and maize seeds (Zea mays L. cv. Halamish) were as previously described in detail by Chazen and Neumann (1994). Wheat seeds were germinated for 2d in the dark and then transferred to hydroponic culture in aerated nutrient solution in an illuminated growth chamber (16h photoperiod, light intensity $35 Wm^{-2}$ PAR, 27±2°C). The seedlings were assayed 2d later, when the mean length of the first true leaf of wheat was 47±3mm and that of the surrounding coleoptile 22±2mm.

Seedlings at this early developmental stage were chosen because (1) the elongation of the first leaf is inhibited by root water deficits and (2) effects on growth of possible treatment induced differences in rates of photosynthesis (or transpiration) are limited by the ability to utilize seed storage reserves to maintain growth, and by the relatively small area of mature leaf tissue directly exposed to the environment.

Leaf Elongation

Long term growth of leaves was followed by measuring leaf length at daily intervals. Water-deficits were induced by transferring plants from basal nutrient solutions to nutrient solutions containing polyethylene glycol PEG 6000. The solutions were vigorously aerated and care was taken to prevent any solution droplets from reaching the leaves. The PEG, supplied by Riedel-de Haën AG, Germany, was of analytical quality with a molecular weight range from 5000 to 7000, and was used without further purification. PEG concentration was 192 g/dm^3 of solution equivalent to an osmotic potential of −0.5MPa (Chazen et al., 1995).

Short term rates of leaf elongation were assayed using a computerised extensiometer consisting of linear variable displacement transducers (LVDTs) attached to the leaf tips of whole plants (Neumann, 1993; Chazen & Neumann, 1994). Each seedling was firmly fitted into a perspex holder so that the roots alone were bathed in appropriate aerated nutrient solution. Replacing the root medium with treatment solutions could be effected within 60s and without disturbing the shoot. The caryopsis was fixed above the solution by the holder, and the shoot (coleoptile and protruding primary leaf) extended vertically above it.

The leaf tip was stuck to an aluminium foil tab connected to a small alligator clip which was in turn connected to a thread looped over a low resistance pulley wheel and joined to the core of a LVDT supplied by Instruments and Control Inc., Haifa, Israel. Small range (±2mm) high sensitivity transducers (ST-3, 950 mv/mm) were used. An 0.4g weight was continuously applied in the direction of leaf growth in order to overcome frictional resistances. Up to sixteen plants could be monitored simultaneously. The electrical output of the LVDTs varied linearly with changes in the position of the tip of the growing leaves and was sampled at 1 min intervals. Output was amplified and fed into an A/D card (Das 1401, Kiethly Metrabyte, Taunton, MA, USA) in a PC 486 clone. Stored data was displayed graphically using Viewdac data acquisition and graphical display software (Viewdac 2.1, Kiethly Metrabyte, Taunton, MA, USA). The software was programmed to directly produce graphic plots of changes in individual leaf tip position and leaf elongation rates, against time. A moving average based on 7 data points was used to reduce background noise. Accuracy of leaf position measurements was within ±2μm.

Leaf ABA Analysis

Leaf segments were excised from the cell expansion zone of the first true leaf under humid conditions. Batches of expanding leaf tissue, containing 10 wheat leaf segments, or 3 to 6 maize leaf segments, were immediately frozen in liquid nitrogen and freeze dried. Wheat data are from 2 separate experiments, and maize data from 3 separate experiments. The dried leaf tissues were extracted overnight in 80% aqueous methanol at 4°C. Abscisic acid was analysed immunologically using an ELISA with monoclonal antibodies as described by Mertens et al. (1985). The aqueous fraction of the extract was placed on Waters C_{18} Sep Pak Cartridges and eluted with 70% methanol. The aqueous residue of that fraction was partitioned three times against ethyl acetate at pH 3.0. The organic fraction was reduced to dryness in vacuo, taken up in TBS buffer (Tris buffered saline: 50 mol · m^{-3} Tris, 1 mol · m^{-3} $MgCl_2$, 150 mol · m^{-3} NaCl; pH 7.8) and subjected to ELISA. The purification procedure was shown (by spike dilution assay) to remove cross reacting compounds from both maize and wheat tissues. Recovery of ABA during the purification procedure was checked routinely using radioactive ABA and was never less than 95%.

All the experiments reported here were repeated one or more times with similar results. Some of the comparative data for maize leaves is taken from earlier reports by Chazen and Neumann (1994), and Chazen et al. (1995).

RESULTS AND DISCUSSION

Root derived hormonal (or electrical) signals are presumably dependent on root metabolic responses to local water deficits. In contrast, root derived hydraulic signals (changes in pressure potential in the root to shoot xylem pathway) represent a simple physical response to water deficits i.e. the tendency for water potential in the xylem to equilibrate with that of the external water source. We used solutions of polyethylene glycol (PEG 6000), an inert water soluble osmolyte which does not penetrate through plant cell walls, to induce water deficits in the root media: Exposure of the live roots of control plants to PEG could induce transmission of physical, chemical or electrical signals to the leaves. However, PEG treatment of roots killed by a prior freeze thaw treatment, was expected to induce transmission of only hydraulic signals to the shoot (Chazen et al., 1995 and references therein).

We therefore exposed live or killed roots of wheat and maize seedlings to PEG solution (−0.5MPa water potential) in order to determine the relative importance of hydraulic and hormonal signals to early leaf growth responses.

ABA Accumulation in Leaf Expansion Zone

Previously un-stressed wheat and maize seedlings, with live or killed roots, were exposed to control or PEG solutions for 8min and 4h, and effects of PEG treatment on rates of ABA accumulation in expanding first leaf tissues, were then investigated. Although reductions in growth and wall extensibility were established after 8min of PEG treatment in both species, there were no significant increases in leaf ABA levels (not shown). However, 4h of PEG treatment caused significant increases in ABA accumulation (Table 1). Interestingly, the rate of ABA accumulation induced by PEG treatment in expanding maize leaf tissues, was considerably reduced (by 71%) when the roots were killed (i.e. unable to synthesize and export ABA to the shoot). In contrast, killing the roots of wheat plants resulted in a significant increase (+390%) in rates of ABA accumulation by the expanding leaf tissues, following exposure to equivalent water deficits. Killing the roots of otherwise un-stressed plants increased leaf ABA from 600 to 980 p mol ABA/gDW in wheat and from 320 to 580 p mol ABA/gDW in maize, over a 4h period. Thus, unlike the PEG treatment, killing the roots of otherwise un-stressed plants only had comparatively minor effects on levels of leaf ABA.

The comparison of patterns of ABA accumulation in maize seedlings with live or killed roots, suggest that much of the stress induced increase in rates of leaf ABA accumulation, was derived from the live roots. The remainder could reflect leaf synthesis. In contrast, the results for wheat seedlings suggest that the live roots of the water stressed plants acted as a sink to which leaf produced ABA was transported (Wolf et al., 1990). Thus, when the roots were killed, leaf produced ABA could no longer be transferred to the roots and leaf accumulation rates were therefore increased (+390%). A similar occurrence of leaf to root transport of ABA was predicted by a mathematical model simulating distribution of ABA in stressed plants (Slovik et al., 1995).

In other experiments (Figure 1 and unpublished results), we found that un-stressed maize leaves were rapidly (within minutes) susceptible to growth inhibition by injections of exogenous ABA solution into the leaf expansion zone. However, the concentrations of

Table 1. Effect of killing roots on PEG induced accumulation of ABA in the expanding leaf tissues of wheat and maize seedlings

	ABA accumulation rate: p mol ABA / g.DW / 4h	
	Maize leaf	Wheat leaf
4h PEG, live roots	25,700 a	740 b
4h PEG, killed roots	7,330 b (−71%)	3,660 a (+390%)

Three separate batches of 3–6 maize plants or of 10 wheat plants, were used to determine ABA levels for each treatment. Comparative rates of ABA accumulation were calculated from differences between control and water stressed plants after 4h. Water deficit was induced by exposure of roots to PEG solutions (−0.5MPa water potential). Where indicated, the roots were killed by freeze thaw treatment 1h prior to start of experiments. PEG does not penetrate the cell walls of killed roots and should therefore induce similar water deficits when applied to live or killed roots (Chazen et al 1995). Different letters in vertical columns indicate that within species differences are significant (p=0.05). Figures in brackets show percent changes in rates of leaf ABA accumulation caused by killing the roots.

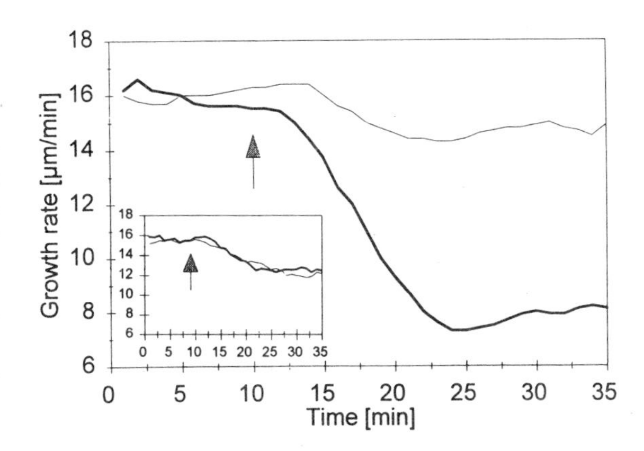

Figure 1. Typical early responses of growing wheat and maize leaves to ABA injection. 15μl of 100 μM ABA solution were injected directly into the expansion zone at the base of the first true leaf of an intact maize seedling (arrow). Control plant was injected with 15μl of water. Growth rate based on assay of position of leaf tip at 1 min intervals, using a position transducer. Upper thin line is leaf growth rate for control leaf; descending bold line is for ABA treated leaf. Inset shows lack of response to similar ABA injection in wheat leaves (thin line is water control).

injected ABA which effectively inhibited leaf growth (≥ 50μM) were far higher than levels of endogenous ABA found in maize leaf tissues after 4h of PEG treatment (2.9μM).

The wheat leaves were not inhibited by equivalent ABA injections (Figure 1, inset). In addition, ABA injection induced cell wall hardening (loss of plastic extensibility) in the expanding maize leaf tissues, but did not affect wheat leaves (not shown). Thus, it appeared that cell expansion processes in maize leaves were susceptible to inhibition by injections of high levels of exogenous ABA, whereas expanding wheat leaf cells, were relatively insensitive.

Effects of Root Water Deficits on Leaf Growth

The comparative effects of PEG induced water deficits on early leaf growth responses were also assayed. Figure 2 shows the kinetics of early inhibition of leaf growth rate induced by root application of PEG to wheat and maize seedlings with live roots. Table 2 shows that killing the roots had relatively little effect on the subsequent growth of leaves of un-stresssed wheat and maize seedlings, as measured after 4h. However, exposure of killed roots of either wheat or maize seedlings to PEG, consistently induced a substantial inhibition of leaf growth, despite the presumed absence of any root capacity to generate and export hormonal, or electrical signals.

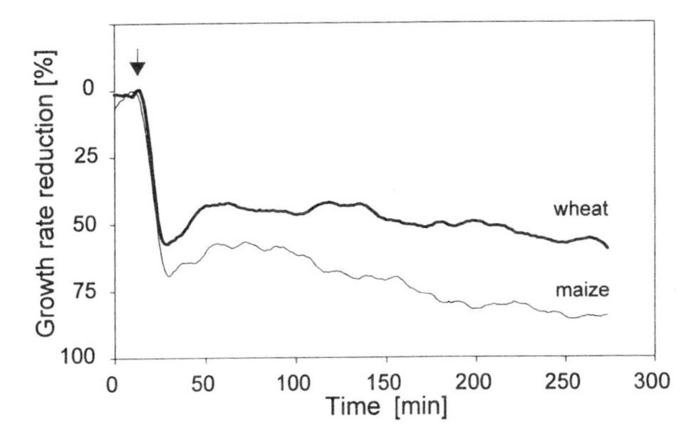

Figure 2. Kinetics of early inhibition by water deficits of leaf elongation rates in wheat and maize seedlings. Percent reductions in rate of elongation growth of newly emerged first true leaves on intact seedlings calculated as $(GR-GR_p/GR).100$, where GR and GR_p represent leaf growth rate before and after addition of PEG6000 (osmotic potential −0.5MPa) to the root medium (arrow). Growth rate based on assay of position of leaf tip at 1 min intervals using a position transducer. Maize leaf data adapted from Chazen et al., (1995). Rates of leaf growth (GR) for wheat and maize leaves before PEG addition, were 15 ± 1 μm/min and 19 ± 0 μm/min, respectively. Leaf elongation rates 4h after PEG addition (GR_p) were 6 ± 1 μm/min for wheat and 3 ± 1 μm/min for maize. Mean elongation rates of 13 wheat plants or 5 maize plants ± SE.

Table 2. Early reductions in leaf growth rates induced by PEG application to live or killed roots of maize and wheat seedlings

	Leaf growth rate, µm min^{-1}			
	Wheat		Maize	
	Live root	Killed root	Live root	Killed root
Control	14a	11a	16a	14a
4h PEG	6b	5b	4b	7b
% reduction in leaf GR	57	55	75	50

Maize data adapted from Chazen and Neumann 1994; wheat data from unpublished results. Each live root datum based on assay of 6 seedlings. Each killed root datum based on 7 wheat seedlings and 6 maize seedlings. Different letters indicate that differences within the vertical columns are significant (p=0.05).

It is nevertheless interesting that exposure of maize seedlings with live roots to PEG, caused a greater reduction in leaf growth (75% reduction) than exposure of seedlings with killed roots (only 50% reduction). Thus, metabolic responses of live maize roots to local water deficit may have generated an additional growth inhibitory factor (ABA ?). Such root supplied ABA might therefore have a secondary role in maintaining the inhibition of leaf growth, after its initiation by a hydraulic signal (Chazen et al., 1995).

In contrast to maize, exposure of either live or killed roots of wheat seedlings to PEG, induced nearly identical reductions in leaf growth (57% and 55% respectively). Moreover, the reductions in leaf growth were apparently unaffected by the large differences in leaf ABA accumulation by stressed plants with live or killed roots. It might be argued that the ABA levels in the wheat leaves after 4h of water stress, were above the threshold concentration required to inhibit leaf growth. However, the fact that injections of high concentrations of ABA into the growing tissues of wheat leaves had no early effects on their growth rates (Figure 2), supports another interpretation; i.e. that the wheat leaves of the species studied here, show relatively little short term sensitivity to growth inhibition by ABA (Blum and Sinmena, 1995). Unknown inhibitors of leaf growth, such as that found by Munns (1992) in the xylem sap of water stressed wheat plants, may be additionally involved in regulating leaf growth responses to water deficits in wheat.

Comparative Roles of Hormonal and Hydraulic Signals

Leaf growth in wheat seedlings with killed roots was rapidly inhibited (within minutes) by PEG induced water deficits.This result, together with previous findings for maize (Chazen and Neumann, 1994; Chazen et al., 1995) suggests that neither ABA nor electrical signals from the water stressed roots, were essential for the induction of early reductions in wheat or maize leaf growth by water deficits.

Although root derived hormonal signals did not appear to be essential for inducing early leaf growth responses to root water deficits, the involvement of a root to shoot hydraulic signal is clearly indicated (Nonami and Boyer, 1990: Malone, 1993). A hydraulic signal, i.e. induction of a more negative xylem water potential, could rapidly and directly affect water availability for ongoing cell expansion in the leaf.

A hydraulic signal might have additional effects. For example, it might affect the degree of opening of stretch activated ion channels in the plasma membranes of the expanding leaf cells. This might in turn affect the influx of secondary messengers (e.g. calcium ions) which could activate metabolic changes (e.g. associated decreases in the plastic extensibility of the expanding cell walls and onset of solute accumulation for osmotic adjustment). Regulated decreases in the plastic extensibility of cell walls in expanding leaf tissues would tend to inhibit growth, and together with increased rates of solute accumulation, act to minimize turgor loss (see review by Neumann, 1995). Further work is required to investigate these possibilities. In all events, the comparison of early leaf growth responses to PEG induced water deficits, in seedlings with live or killed roots, provides a useful tool for evaluating the relative roles of hormonal and hydraulic signaling.

CONCLUSIONS

1. The evidence did not indicate an essential requirement for root to shoot ABA signals in initiating the early inhibition of either wheat or maize leaf growth by root water deficits.
2. Rather, root to shoot transmission of a hydraulic signal appeared to be primarily involved in both cases.
3. The comparison of early leaf growth responses to PEG induced water deficits, in seedlings with live or killed roots, provides a useful tool for evaluating the relative roles of hormonal and hydraulic signals from the roots.

AKNOWLEDGMENTS

This research was supported by the fund for the promotion of research at the Technion.

REFERENCES

Blum A., and Sinmena B,. 1995, Isolation and characterization of variant wheat cultivars for ABA sensitivity. Plant Cell Environment 18: 77–83.

Chazen O., Hartung W., and Neumann P.M. 1995, The different effects of PEG 6000 and NaCl on leaf development are associated with differential inhibition of root water transport. Plant Cell & Environment 18:727–735

Chazen O. and Neumann P.M., 1994, Hydraulic signals from the roots and rapid cell wall hardening in growing maize (Zea mays L.) leaves, are primary responses to polyethylene glycol- induced water deficits. Plant Physiology 104: 1385–1392

Davies W.J. and Jones H.G., 1993, Abscisic acid physiology and biochemistry. Bios Scientific Publishers. Oxford UK pps 1–260

Davies W.J., and Zhang J., 1991, Root signals and the regulation of growth and development of plants in drying soil. Annu Rev Plant Physiol Plant Mol Biol 42: 55–76

Hsiao T.C., 1973, Plant responses to water stress. Ann. Rev. Plant Physiology 24:519–570

Malone M., 1993, Hydraulic signals. Phil. Trans. R. Soc. Lond. B. 341: 33–39

Mertens R.J., Deus-Neumann B., and Weiler E.W., 1985, Monoclonal antibodies for the detection and quantitation of the endogenous plant growth regulator, abscisic acid. FEBS Letters 160: 269–272

Munns R., 1992, A leaf elongation assay detects an unknown growth inhibitor in xylem sap from wheat and barley. Aust J Plant Physiol 19: 127–135

Munns R., Sharp R.E., 1993, Involvement of abscisic acid in controlling plant growth in soils of low water potential. Aust. J. Plant Physiol. 20: 425 -437.

Neumann P.M., 1993, Rapid and reversible modifications of extension capacity of cell walls in elongating maize leaf tissues responding to root addition and removal of NaCl. Plant Cell & Environment. 16: 1107–1114

Neumann P.M., 1995, The role of cell wall adjustment in plant resistance to water deficits. (Review and Interpretation) Crop Science 35: 1258–1266

Nonami H., and Boyer J.S., 1990,Primary events regulating stem growth at low water potentials. Plant Physiol 93: 1601–160

Slovik S., Daeter W., and Hartung W., 1995, Compartmental distribution and redistribution of abscisic acid (ABA) in roots as influenced by environmental changes. A biomathematical model. J. Experimental Botany 46: 881–894

Wolf O., Jeschke W.D.,and Hartung W., 1990, Long distance transport of abscisic acid in NaCl - treated intact plants of Lupinus albus. Journal of Experimental Botany 41: 593–600

Wright S.T.C., and Hiron R.W.P., 1969, (+)- Abscisic acid, the growth inhibitor induced in detached leaves by a period of wilting. Nature 224:719–720

MICROCUTTINGS ROOT SYSTEM MORPHOLOGY AND DEVELOPMENT UNDER *IN VITRO* AND *IN VIVO* CONDITIONS IN CITRUS ROOTSTOCKS (*Citrus macrophylla* West and Citrange Troyer—*Poncirus trifoliata* (L.) X *Citrus sinensis* (L.) OSB.)

M. Castro,[1] V. Miranda,[1] and B. Palma[2]

[1]Facultad de Agronomía. [2]Instituto de Biología, Universidad Católica de Valparaíso, Casilla 4059, Valparaíso, Chile.

INTRODUCTION

Macrophylla and Troyer are the most used rootstocks in Chile, for lemons and mandarins respectively. There are experimental evidence of differential root system morphology and anatomy, depending on their *in vitro* or *in vivo* origin, which could affect their survival potential during acclimation and the subsequent field performance. The objective of this research was to compare the adventitious root systems in microcuttings obtained from *in vitro* plantlets, under *in vitro* and *in vivo* conditions.

EXPERIMENTAL

Seedlings and microcuttings were obtained from seeds of *Citrus macrophylla* (Cm) and Citrange Troyer (CT), which were treated with hot water (52°C) for 10 min. Seed coats were removed, and then surface sterilized with 0.1% sodium hypochlorite plus 0.1 ml l^{-1} Tween 20. Germination medium was MS (Murashige and Skoog, 1962), plus 7.5 g l^{-1} agar. Tubes were kept in a growth chamber at 1600 lux, 16 hr light , at 25±2°C. Seedlings and microcuttings were obtained after 4 weeks of culture; microcuttings were prepared by selecting nodal sections with 1 node and 2 cm long. They were dipped in 200 mg l^{-1} IBA and then transplanted to minirhizotrons (Riedaker, 1974), to compare with *in vitro* rooted microcuttings. Peat moss + perlite was used for Cm and peat moss + sand for CT. Light microscopy was used to observe the stem-root union zone, and the medial and terminal parts of the root.

RESULTS AND DISCUSSION

In both species, the number of roots and root lengths were higher in the *in vitro* rooted plantlets. Cm and CT showed 100% of *in vitro* rooting, with an average of 2 roots per microcutting. *In vivo* rooted microcuttings presented 60 and 100% rooting in CT and Cm, respectively. A better quality of root system was obtained *in vitro*. Stem development was higher in plantlets rooted *in vitro*, whose length was 1–2 cm, while no stem growth (or very incipient) was observed in microcuttings rooted *in vivo*. This is opposite to the results obtained by Rogers and Smith (1993) and Apter *et al.* (1993). Histological analyses showed that *in vitro* formed roots by microcuttings of Cm and CT had the functional elements all the required. Xylem and phloem in microcuttings are hexarc, while seedlings presented heptarc elements. The medial zone of microcuttings roots of Cm and CT had lower diameters that seedlings. A continuous xylematic connection between the *in vitro* neo-formed root and the stem was observed in both species, as reported by Apter *et al.* (1993) and Harbage *et al.* (1993). Root growth rate was initially higher in seedlings than in *in vitro* rooted microcuttings of both species. After the fourth week this situation was reverted. This is another reason to prefer the use of *in vitro* rooted microcuttings instead of seedlings, in addition to the higher number of plants that can be obtained from one seed by using microcuttings.

REFERENCES

Apter, R., Mc Williams, E. and Davies, F. 1993. *In vitro* and *ex vitro* adventitious root formation in Asian Jasmine *(Trachelospermum asiaticum)* I. Comparative morphology. *J Amer Soc Hort Sci* 118: 902–905.

Harbage, J., Stimart, D. and Evert, R. 1993. Anatomy of adventitious root formation in microcuttings of *Malus domestica* Borkh "Gala" *J Amer Soc Hort Sci* 118: 680–688.

Murashige, T. and Skoog, F. 1962. A revised medium for rapid growth and bioassays with tobacco tissue cultures. *Physiol Plant* 15: 473- 497.

Riedaker, A. 1974. Le minirhizotron. *Ann Sci Forest* 31: 129–134.

Rogers, R. and Smith, M. 1993. Consequences of *in vitro* and *ex vitro* root initiation for miniature rose production. *J Hort Sci* 67: 535–540.

ROOT AND SHOOT GROWTH OF OAT PLANTS TREATED WITH CCC AND ETHEPHON

Ari Rajala and Pirjo Peltonen-Sainio

University of Helsinki, Department of Plant Production, P.O. Box 27, FIN-00014, Helsinki, Finland.

ABSTRACT

The plant growth regulators (PGRs) are used on cereals primarily to reduce stem height and hence, prevent lodging. Recently interest has grown to modify canopy structure, root performance, and root/shoot ratios by use of PGRs. However, this may result in a need for reconsideration of the application time of the PGRs. Two preliminary trials were carried out in a greenhouse, in which three oat cultivars were used to evaluate the effect of early application (at 2–3 leaf stage) of chlormequat chloride (CCC) (1%) and ethephon (0.5%) on root and shoot growth. The cultivars that were included in this study represent differences in growth habit: *Jalostettu maatiainen* is an old, long-strawed and lodging-sensitive landrace cultivar, *Salo* is a modern high yielding cultivar with relative short and stiff straw, and *Pal* is a Minnesota-adapted dwarf cultivar. Oat seeds were sown in 5 L pots containing clay illitic top soil from the experimental fields. Fifteen seeds were sown per pot, but when the seedlings had emerged, plants were thinned to ten per pot. Cultivars were set as a main plot and treatments were split across them with five replicates. Pots were placed in a container (120x180cm), and hence they received continuos ground watering. At two to three leaf stage, plants were sprayed with a battery operated small scale atomizer at the rate of 10 ml of PGR per pot. Control plants were excluded. Two weeks after treatments, the above ground plant parts were cut and roots were washed carefully to remove the soil. Root and shoot samples were dried overnight at 100EC to determine their dry weight (mg/plant). CCC as well as ethephon reduced root and shoot growth of all three cultivars. This may have resulted from a too high application rate (I ml per plant). Neither CCC nor ethephon have affected the root/shoot ratio. CCC had more deleterious effects when compared to ethephon. Root growth of the landrace and modern cultivar, was more sensitive to CCC application, at early growth stages, as compared to the response of the dwarf cultivar *Pal*.

ODC-MEDIATED BIOSYNTHESIS AND DAO-MEDIATED CATABOLISM OF PUTRESCINE INVOLVED IN THE INDUCTION OF ROOTING BY LEAF EXPLANTS OF *Chrysanthemum morifolium* Ramat *IN VITRO*

Josette Martin-Tanguy,[1] Marcel Aribaud,[2] Monique Carré,[2] and Thomas Gaspar[3]

[1]Lab. de Biologie végétale, Univ. de Rennes I, F. 35042 Rennes, France. [2]Lab. de Physiopathologie, INRA, F-21034 Dijon, France. [3]Institut de Botanique B 22, Sart Tilman, B-4000 Liège, Belgium.

Foliar discs (8 mm diameter) from expanding leaves of the middle part of vegetative shoots of *Chrysanthemum morifolium* Ramat raised *in vitro*, rooted on their basal side from the 6th day on when cultured *in vitro* on a Murashige and Skoog medium containing IAA (1 mg l^{-1}). Every two days measurement of free putrescine indicated a continuous level increase from the beginning. Its hydroxycinnamoyl form (75% of the conjugated polyamine pool) raised from zero at the beginning of culture to peak at day 4. No ODC (ornithine decarboxylase) or ADC (arginine decarboxylase) activity was detected at day 0 but a high activity of both enzymes was measured at day 2 (ODC > ADC) before a decline. DFMA (α-difluoromethylarginine), an inhibitor of ADC, in the rooting medium from time 0 (10^{-3}M), although completely inhibiting ADC activity, reduced the level of free putrescine by only 25% in the leaf explants without affecting the conjugated pool, and it had no effect on rooting. DFMO (α-difluoromethylornithine), an inhibitor of ODC at the same concentration, completely inhibited ODC activity with a reduction of putrescine accumulation and a hindrance of the formation of polyamine conjugates. Root formation was totally inhibited by DFMO alone. A simultaneous application of putrescine (10^{-3}M) to DFMO restored complete rooting.

There was no DAO (diamine oxidase) activity at day 0 in the foliar explants but it increased rapidly to a maximum at day 2 on the rooting medium. PAO (polyamine oxidase) activity remained at a low level in the same conditions. DAO did not develop any activity on the root-non forming medium containing DFMO β-OH hydrazine, an inhibitor of DAO, hindered the development of activity of this enzyme in the leaf explants on the rooting medium. Putrescine and its conjugate accumulated in such treated explants. Rooting was completely inhibited in the presence of β-OH hydrazine but an important callus grew around the disc. Such a callus did not develop in the presence of DFMO.

All together considered, the results thus pointed to the involvement of ODC-mediated biosynthesis and DAO-mediated catabolism of putrescine in the induction of rooting of this material. Implication of DAO-mediated putrescine catabolism in the inductive phase of rooting already has been argumented for another material in chapter of the present book (Gaspar et al.).

POLYAMINE METABOLISM, AND ROOT FORMATION IN DEVELOPING TOBACCO PLANTS

J. Martin-Tanguy,[1] B. Pasquis,[2] C. Dreumont,[2] D. Tepfer,[3] and F. Leach[3]

[1]Lab. Biologie Végétale, Université de Rennes I, F-35042 Rennes, France. [2]Station de Génétique et d'Amélioration des Plantes, INRA, F-21034 Dijon, France. [3]Laboratoire de la Rhizosphère, INRA, F-78026 Versailles, France.

The content of water-insoluble amine conjugates (di-feruloylputrescine, di-feruloylspermidine, feruloytyramine) decreased drastically during tobacco germination, concomitantly with a rapid increase in free polyamines (putrescine, spermidine, tyramine). Neither DFMA nor DFMO inhibited the emergence of the radicle, which is considered to be the end of the germination process, nor where the putrescine and spermidine titers modified during the first two days. We present evidence that ADC and ODC regulate putrescine biosynthesis at different stages of tobacco root development, with ADC being active early in development and ODC later. Arginine decarboxylase reached a peak on day 3 of germination (Fig. 1) and was required for root elongation. Ornithine decarboxylase activity peaked 3 days after emergence of the radicle (Fig. 2), and was required for subsequent root devel-

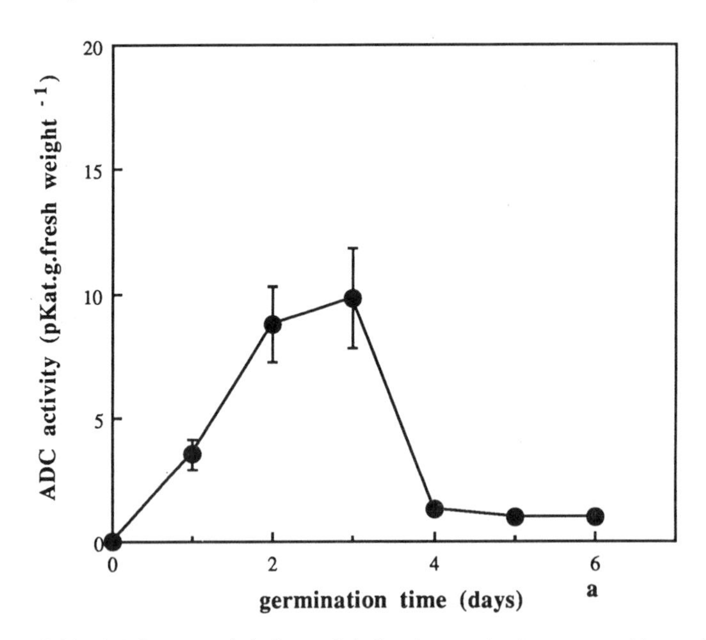

Figure 1. ADC activities in tobacco seeds before and during the germination process. Means of 2 experiments, each point representing 200 to 250 seeds. a. emergence of the radicle (end of germination).

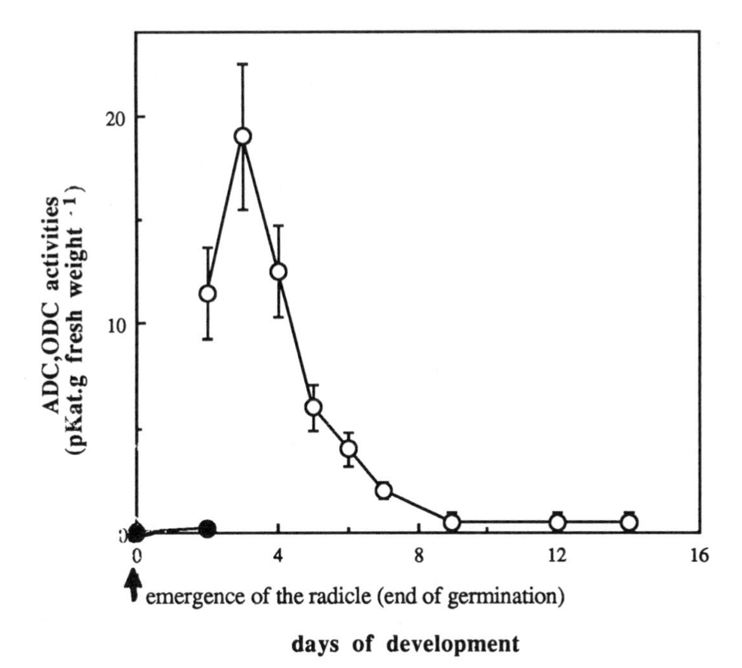

Figure 2. ADC (●) and ODC (○) activities in tobacco roots during seedling development. Means ± SD of 2 experiments, each point representing 200 to 250 explants.

opment and branching. When ADC action was blocked by DFMA, polyamine diminished only during the first few days of development; when agmatine or putrescine were added, normal polyamine titers and growth were restored. The effects of DFMA were concentration dependent. DFMO promoted root elongation during the first few days of seedling development. This effect of DFMO is related to the observed enhancement of plant ADC and accumulation of polyamines. Fundamental changes in growth and tobacco root development were associated with inhibition of ODC by DFMO (root elongation was stopped, and no, or only limited ramification, was observed during the whole period of development). They correlated with the inhibition of the production of free polyamines, and with inhibition of putrescine conjugates. Our results lead to the following conclusions: ADC leads to the production of free polyamines which are required for growth by cell expansion; ODC is required for cell division and differentiation; putrescine conjugates are important in regulating the branching of the root system.

2,4-D REVERSAL OF DFMO-INDUCED INHIBITION OF ADVENTITIOUS ROOT FORMATION IN LEAFY SPURGE (*Euphorbia esula* L.) *IN VITRO*

David G. Davis

U. S. Department of Agriculture, Agricultural Research Service, Biosciences Research Laboratory, State University Station, Fargo, North Dakota 58105-5674.

Indoleacetic acid (IAA) stimulates adventitious root formation and partially reverses the effects of difluoromethylornithine (DFMO), a specific suicide inhibitor of putrescine biosynthesis, in etiolated aseptically-grown hypocotyl segments of the perennial weed, leafy spurge. DFMO also is a strong inhibitor of adventitious root formation in the hypocotyl segments. Dichlorophenoxyacetic acid (2,4-D) has auxin-like activity and was also tested to determine if it could counteract DFMO-induced inhibition of root formation. When 2,4-D was applied at 45 to 450 nM in the nutrient medium, root formation was inhibited. Unexpectedly, 450 nM 2,4-D applied simultaneously with 500 μM DFMO reversed the DFMO-induced inhibition of adventitious root formation, forming root numbers similar to controls. This reversal occurred under growth conditions with complete B5 nutrient medium or with B5 medium containing salts and vitamins reduced to 10% of their normal concentrations. This action of 2,4-D resembles that of indoleacetic acid (IAA), although 2,4-D did not stimulate root formation in the absence of DFMO, as did IAA (at 230 to 1100 nM).

ROLE OF ENDOGENOUS CYTOKININS IN ROOTING OF ETIOLATED AND LIGHT-GROWN OAK CUTTINGS

Y. Eshed and J. Riov

Department of Horticulture, Faculty of Agriculture, The Hebrew University of Jerusalem, P.O. Box 12, Rehovot 76100, Israel.

Endogenous cytokinins have long been associated with inhibition of adventitious root formation in cuttings. This inhibitory role of cytokinins is supported by the observations that application of cytokinins usually inhibits adventitious root formation (Van Staden and Harty, 1988). However, the regulatory role of endogenous cytokinins in this process is less clear. There are several reports which observed a negative correlation between endogenous cytokinin level and adventitious root formation (Bollmark and Eliasson, 1990) while in other reports no such correlation was found (Smalley et al., 1991).

Etiolation of stock plants has long been known to effectively enhance rooting of difficult-to-root woody plant cuttings (Maynard and Bassuk, 1988). In view of some reports on the regulation of endogenous cytokinin level by light, the aim of the present work was to examine the possibility that etiolation enhances rooting by affecting cytokinin levels and/or cytokinin metabolism in *Quercus ithaburensis* cuttings.

Cuttings from oak mother plants grown under natural light rooted poorly, whereas rooting of cuttings from etiolated mother plants was significantly higher. Rooting of the two types of cuttings was negatively correlated with endogenous cytokinins levels per cutting as determined by antibodies against isopentenyl adenosine and trans-zeatin riboside. However, there was no difference in cytokinin levels between the two types of cuttings on a fresh weight basis.

The elution pattern of various cytokinins from an HPLC column was significantly different in the two types of cuttings. More bound forms, presumably glycosides and monophosphates of the zeatin group were found in etiolated cuttings, whereas light-grown cuttings contained higher levels of zeatin and zeatin riboside-like compounds.

The inhibitory effect of zeatin glycoside on adventitious root formation which was tested in the mung bean rooting assay was significantly lower than that of zeatin or zeatin riboside.

It is suggested that the enhanced rooting ability of etiolated cuttings results from increased conversion of active cytokinins to bound forms with reduced biological activity. As to the question of the role of endogenous cytokinins in adventitious root formation, the present data indicate that the nature of the cytokinins present in the tissues must be considered in addition to the determination of cytokinins levels.

REFERENCES

Bollmark M. and, Eliasson, L. 1990. A rooting inhibitor present in Norway spruce seedlings grown at high irradiance - a putative cytokinin. *Physiol. Plant.* 80:527.

Maynard, B.K. and Bussuk N.L., 1988. Etiolation and Banding Effects on Adventitious Root Formation. In: Adventitious Root Formation in Cuttings. T.D. Davis, B.E. Haissig, and N. Sankhla, eds., Adv. in plant Sci. Ser. vol 2, Dioscorides Press, Portland, pp. 29–43.

Smalley, T.J., Dirr, M.A., Armitage, A.M., Wood, B.W., Teskey, R.O., and Severson, R.F. 1991. Photosynthesis and leaf water, carbohydrate, and hormone status during rooting of stem cuttings of *Acer rubrum*. *J. Amer. Soc. Hort. Sci.* 116:1052.

Van Staden J., and Harty A.R. (1988) Cytokinins and adventitous root formation. in:Adventitious Root Formation in Cuttings. T.D. Davis, B.E. Haissig, and N. Sankhla, eds., Adv. in plant Sci. Ser. vol 2, Dioscorides Press, Portland, pp. 185–209.

INVOLVEMENT OF PUTRESCINE AND OF ITS CATABOLIC PATHWAY IN THE INDUCTION OF ROOTING OF WALNUT SHOOTS *IN VITRO*

Claire Kevers,[1] Jean-Francois Hausman,[2] and Thomas Gaspar[1]

[1]Univ. of Liege, Institute of Botany, B 22, Sart Tilman, B–4000 Liège, Belgium.
[2]CRP–CU, 162A, av. de la Faiencerie, L–1511 Luxembourg, Luxembourg.

Walnut *(Juglans regia)* shoots raised *in vitro* did not root on a single agar rooting medium, with or without auxin, in the light or under darkness. About 100% rooting was achieved through the use of two successive rooting media: a gelrite gelified MS "inductive" medium with auxin (3 mg l^{-1} IBA for 5 days in the dark), followed by a gelrite-vermiculite "expressive" medium (1/4 DKW nutrients) without auxin (in the light). Using peroxidase as marker, it was determined that the time spent on the "inductive" medium corresponded to the physiological inductive and initiative phases of rooting. The level of endogenous IAA underwent a typical transient increase at inductive phase before decreasing at the initiative one.

Three types of arguments pleaded in favour of putrescine playing an essential role in the inductive phase of rooting:

- putrescine (not the other polyamines) was able to bring about up to 31% rooting when applied in the induction medium, in the absence of auxin

- the level of endogenous putrescine (not the other polyamines) underwent a temporary increase at inductive phase
- rooting could be perturbed (enhanced or decreased) using inhibitors of polyamine metabolism in the induction medium.

Aminoguanidine, an inhibitor of putrescine catabolism into the Δ^{-1}-pyrroline-GABA (γ-aminobutyric acid) pathway, inhibited the rooting of walnut shoots. This is an additional indication that putrescine catabolism is an essential event in the inductive phase of rooting, as discussed elsewhere in this book (papers by Gaspar *et al.* and by Hausman *et al.*).

19

AUXIN-INDUCED GENE EXPRESSION DURING ROOTING OF LOBLOLLY PINE STEM CUTTINGS

Barry Goldfarb, Zhigang Lian, Carmen Lanz-Garcia, and Ross Whetten

Department of Forestry
North Carolina State University
Raleigh, North Carolina 27695

INTRODUCTION

Vegetative propagation is becoming an important tool for mass producing elite genotypes to establish forest plantations. A number of vegetative propagation techniques have been researched, but at the present time, rooting stem cuttings is the method that is most commonly employed in operational programs (Ritchie, 1994). In many tree species, especially conifers, operational use of rooted stem cuttings is limited by maturation. The time required to field test individual trees precludes propagation of those trees, because rooting ability declines with age of the donor plant (Hackett, 1988).

The North Carolina State University Loblolly and Slash Pine Rooted Cutting Project is conducting research to develop operational methods for producing rooted cuttings of *Pinus taeda* and *P. elliottii* (Weir and Goldfarb, 1993). One focus of this research is to understand the biochemical and molecular events during root initiation and how they are affected by maturation.

Although loblolly pine is often classified as a difficult-to-root species, it could alternatively be viewed as a species in which the onset of maturation occurs very early in development. For example, cuttings, prepared from 2-week-old seedlings by severing the primary root in the hypocotyl region, root at high frequencies when exposed to an exogenous auxin treatment (Diaz-Sala et al., 1996). Exposing pine hypocotyl cuttings to 1.6 mM (300 mg/L) 1-naphthaleneacetic acid (NAA) for 5 to 10 minutes causes nearly 100% of cuttings to form roots and the number of roots per cutting is typically greater than twenty (Goldfarb et al., 1992). Adventitious root primordia form and develop rapidly and synchronously throughout the hypocotyl portion of the cutting in a manner that is morphologically similar to lateral root formation in roots (Goldfarb et al., 1993). Cuttings prepared from the epicotyl portions of 7-week-old seedlings already exhibit a decreased rooting response as compared with hypocotyl cuttings (Diaz-Sala et al., 1996). Although most of these cuttings will form roots if placed in suitable propagation conditions, root formation is considerably delayed as compared with hypocotyl cuttings and many fewer roots are formed. Cuttings obtained from

Biology of Root Formation and Development, edited by Altman and Waisel.
Plenum Press, New York, 1997.

2-year-old, continuously hedged stock plants root at a similar rate and frequency as epicotyl cuttings, but cuttings from unhedged, 2-year-old trees root at low frequencies.

Because root formation in hypocotyl cuttings is induced by auxin treatment, we have chosen to study auxin-induced gene expression. There are several classes of genes in plants that are induced by auxin (reviewed in Napier and Venis, 1995; Abel and Theologis, 1996). One class has been referred to as "early auxin-induced genes" and members of this class have been cloned from four angiosperm species (soybean, Ainley et al., 1988; pea, Oeller et al., 1993; mung bean, Yamamoto et al., 1992; and *Arabidopsis*, Conner et al., 1990, Abel et al., 1995). Some members of this gene family have attributes of primary response genes: (1) they are induced rapidly following auxin exposure, (2) the gene products have short half-lives, (3) the gene products have been localized to nuclei, and (4) the deduced amino acid sequences bear structural similarities to bacterial transcriptional regulators (Abel et al, 1994).

To determine if some members of this class of early auxin-induced genes are involved in adventitious root initiation, we have cloned auxin-induced cDNAs from pine that are homologous to the early auxin-induced genes in pea, mung bean, soybean and *Arabidopsis* (Goldfarb et al., 1995). This paper reports progress on the cloning, sequence analysis, and expression of the *L*oblolly *P*ine *E*arly *A*uxin-induced (LPEA) genes.

EXPERIMENTAL

Total RNA was purified from loblolly pine hypocotyls 10 minutes after the cuttings had received a 10-minute NAA treatment (as described above). Single-stranded cDNA was synthesized from the RNA using reverse transcriptase and oligo dT primers. Degenerate oligonucleotide primers were designed to be complementary to the conserved sequence domains of the eight available angiosperm early auxin-induced gene sequences. These primers were used to amplify DNA fragments from the cDNAs. Amplification products were ligated to the pGem-T vector (Promega, Madison, Wisconsin, USA) and transformed into *E. coli*. Cloned fragments were used as radiolabelled probes to screen a cDNA library. The library was synthesized from mRNA purified from hypocotyls treated as above using the UniZap-XR vector and cDNA synthesis kit (Stratagene, La Jolla, California, USA).

cDNAs were sequenced at the sequencing facilities at Iowa State University and the University of North Carolina, Chapel Hill. Sequence analyses were conducted using the computer software package provided by the Genetics Computer Group (Madison, Wisconsin, USA), version 8.0. Phylogenetic analysis was conducted using corrected distances calculated with the Kimura correction by the Distances program and assembled into a phylogenetic tree with the neighbor joining method by the GrowTree program. Analysis of protein structure was performed with the PeptideStructure and Motifs program. Database searching for sequences similar to known genes was conducted using the Blast e-mail server of the National Center for Biotechnology Information.

Expression of the genes represented by the cloned cDNAs was tested by northern analysis in four experiments. The first experiment tested the timing of auxin induction in NAA- and control- treated hypocotyl cuttings. The second experiment compared the timing of expression in hypocotyl cuttings, epicotyl cuttings and cuttings from 2-year-old, hedged stock plants. The third experiment examined expression in different tissues and organs: hypocotyls, cotyledons, and roots of two-week-old seedlings; and leaves and stems from cuttings taken from scions of mature trees that had been grafted onto seedling rootstock and then hedged. These tissues and organs were sampled 24 hours after treatment with NAA or a control solution. The fourth experiment tested for induction of the cloned genes in hypocotyl cuttings sampled 24 hours after treatment for 10 minutes with NAA (1.6 mM), gibberellic acid (GA_3, 1.6 mM), N^6-benzyladenine (BA, 1.6 mM), abscisic acid (ABA, 3.2 mM), the translation inhibitor cycloheximide (0.08 mM), and a combination of NAA (1.6 mM) and cycloheximide (0.8 mM).

RESULTS AND DISCUSSION

PCR amplification with the degenerate primers resulted in the cloning of four unique fragments. When these fragments were used as probes to screen the cDNA library, five unique clones (LPEA1–5) were isolated, including four that had identical sequences with the four original fragments. All five LPEAs contain a complete open-reading frame, except LPEA3 which is missing a short segment of the 5′ end. The open-reading frames range in size from 211 amino acids in LPEA1 to 303 amino acids in LPEA2. There are 211 deduced amino acids in the cDNA of LPEA3 which is missing approximately 20 amino acids.

Phylogenetic analysis grouping the 5 pine and 8 angiosperm clones revealed that LPEA2 is most closely related to GmAux28 from soybean, whereas the other four LPEA clones are more closely related to each other than to any of the angiosperm genes. This pattern holds true for the angiosperm genes as well; the loci within a species are more closely related to genes in other species than they are to other loci within the same species. Our analysis, however, should be regarded as preliminary, because Abel et al. (1995) have shown there are at least 15 members of this gene family in *Arabidopsis*. We are currently comparing our genes to these newly available sequences and screening a pine genomic library to determine if there are additional gene family members in loblolly pine.

All thirteen genes (five pine and eight angiosperm) contained four domains with a high degree of sequence conservation. In domains I and II, 78% of the deduced amino acids were identical in all thirteen sequences and the identity percentages for domains III and IV were 63% and 57%, respectively. Several of the angiosperm sequences contain two conserved potential tyrosine kinase activation sites in domain III and both of these sites are also present in LPEA1–4, with one site present in LPEA5. All five LPEAs contain the beta-alpha-alpha helix motif reported for PS-IAA4/5 and PS-IAA6 that is similar to motifs in bacterial transcriptional regulators (Abel et al., 1994).

While the functions of the LPEAs are not yet known, sequence similarities of portions of the genes suggest possible roles. LPEA1, 2 and 4 contain potential serine and/or threonine kinase sites, in addition to the tyrosine kinase sites mentioned above. All five LPEAs contain predicted glycosylation sites, as well as a hydrophobic beta sheet conformation in domain IV. This hydrophobic region may be important for the interaction of the proteins with membranes or other proteins. LPEA2 contains sequence similarity with several myristoylated protein kinases. This similarity could indicate a functional role for this product involving membrane association, as well. LPEA5 contains similarities to G-proteins along much of its length. Taken as a whole, these similarities to other genes of known function suggests that the LPEAs may be part of a signal transduction pathway with each other and/or other proteins. We will be testing this hypothesis in future research.

We are studying the auxin-inducibility and expression of these genes during root initiation. All five LPEAs are induced above background levels in auxin-treated hypocotyls. The timing of expression, however, differs among the LPEA cDNAs. LPEA2 and 3 have been detected above background levels as rapidly as 10 minutes following the 10-minute NAA treatment. LPEA1 and 4 have not been detected before one hour and LPEA5 not before five hours following NAA treatment. Four of the five LPEAs reach their maximal level of expression 24 hours after treatment, but LPEA4 is maximally expressed at 5 hours after treatment. All five genes remain at levels above background for at least 5 days after auxin treatment. This is longer than the duration of expression after auxin induction reported for any of the angiosperm genes that have been studied in this way. It is not known whether this difference is the result of differences in the genes themselves; differences in uptake, distribution, or persistence of auxin; or differences in the tissues studied in the various species.

Expression of LPEA2 and 3 has been tested in cuttings from the epicotyl portions of seedlings and cuttings from hedged stock plants. Both genes are induced by NAA treatment in both cutting types. The timing of expression is similar for both genes, but is somewhat slower in epicotyl and hedge cuttings than in hypocotyl cuttings. Maximal expression in the hypocotyls occurs 1 day after treatment, but 3 days after treatment in the epicotyl and hedge cuttings.

Expression of LPEA2 and 5 has been examined in different tissues and organs, with and without NAA treatment. Both genes were induced by auxin in hypocotyls and cotyledons. LPEA2 appeared to have higher levels of transcripts in these tissues without auxin treatment, although both genes were strongly expressed when these tissues were treated with NAA. LPEA2 was detected in both needles and stems from mature cuttings, although auxin did not induce higher levels of transcript abundance than controls. LPEA2 was not detected in roots, but LPEA5 was detected in auxin-treated roots. Thus, it appears that the LPEAs exhibit a complex pattern of tissue- and organ-specific expression as has been reported for the early auxin-induced gene family in *Arabidopsis* (Abel et al., 1995).

We also tested the specificity of auxin induction by treating hypocotyl cuttings with other plant growth regulators. LPEA3 was not induced above basal levels by ABA, BA or GA. It was induced by treatment with cycloheximide and the effect of treatment with both NAA and cycloheximide simultaneously was induction at higher levels than with either treatment alone. Cycloheximide induction has been previously reported for the early auxin-induced genes in pea (Koshiba et al., 1995). These authors hypothesized that this indicates control of expression of these genes by a labile repressor protein. When translation is inhibited by cycloheximide treatment, the repressor is depleted and the gene is available for transcription.

CONCLUSIONS

Loblolly pine contains a multi-gene family of auxin-induced genes with homology to the "early auxin-induced genes" in angiosperms. Five members of the pine family have been cloned as cDNAs and all are induced by auxin treatment in hypocotyl cuttings. The timing of induction and pattern of expression in different plant parts is complex. We are conducting additional research to more fully characterize the expression of these genes in cuttings undergoing adventitious root initiation, to identify other possible auxin-induced genes in loblolly pine, and to determine functions of the genes, their role in root initiation, and whether expression is affected by maturation.

ACKNOWLEDGMENTS

The research reported in this paper was supported by the industrial supporters of the North Carolina State University Loblolly and Slash Pine Rooted Cutting Project, the

North Carolina Agricultural Research Service and the Southern Experiment Station of the US Forest Service.

This article was submitted in July 1996.

REFERENCES

Abel, S., Nguyen, M.D. and Theologis, A., 1995 The PS-IAA4/5-like family of early auxin-inducible mRNAs in *Arabidopsis thaliana*, J. Mol. Biol. 251:533.

Abel, S., Oeller, P.W. and Theologis, A., 1994, Early auxin-induced genes encode short-lived nuclear proteins, Proc. Natl. Acad. Sci., USA 91:326.

Abel, S., and Theologis, A., 1996, Early genes and auxin action, Plant Physiol. 111:9.

Ainley, W., Walker, J. Nagao, R., and Key, J., 1988, Sequence and characterization of two auxin-regulated genes from soybean, J. Biol. Chem. 263:10658.

Conner, T.W., Goekjian, V.H., LaFayette, P.R. and Key, J.L., 1990, Structure and expression of two auxin-inducible genes from *Arabidopsis*, Plant Molecular Biology 15:623.

Diaz-Sala, C., Hutchison, K.W., Goldfarb, B., and Greenwood, M.S., 1996, Maturation-related loss in rooting competence by loblolly pine stem cuttings: the role of polar auxin transport and tissue sensitivity to auxin, Physiol. Plant. 97:481.

Goldfarb, B.,. Hackett, W.P., Furnier, G.R., Howe, G.T., Plietsch, A., and Mohn, C.A., 1992, Molecular approaches to understanding adventitious root initiation in eastern white pine (*Pinus strobus*), in: "I.U.F.R.O. Workshop on the Molecular Biology of Forest Trees," June 15–18, 1992 Carcans - Maubuisson, France.

Goldfarb, B., Hackett, W.P., Howe, G.T., Furnier, G.R., and Mohn, C.A., 1993, Analysis of adventitious root initiation in eastern white pine cuttings, in: "First International Symposium on the Biology of Adventitious Root Formation," April 18–22, 1993, Dallas, Texas.

Goldfarb, B., Whetten, R., and Hackett, W.P., 1995, Cloning of auxin-induced genes from loblolly pine, in: "Keystone Symposium: Frontiers of Plant Morphogenesis," March 29-April 4, 1995, Hilton Head Island, South Carolina.

Hackett, W.P., 1988, Donor plant maturation and adventitious root formation. in: "Adventitious Root Formation in Cuttings," T.D. Davis, B.E. Haissig, and N. Sankhla, eds., p. 11, Dioscorides Press, Portland.

Koshiba, T., Ballas, N., Wong, L.-M., and Theologis, A., 1995, Transcriptional regulation of PS-IAA4/5 and PS-IAA6 early gene expression by indoleacetic acid and protein synthesis inhibitors in pea (*Pisum sativum*), J. Mol. Biol. 253:396.

Napier, R.M., and Venis, M.A., 1995, Auxin action and auxin-binding proteins, New Phytol. 129:167.

Oeller, P.W., Keller, J.A., Parks, J.E., Silbert, J.E., and Theologis, A., 1993, Structural characterization of the early indoleacetic acid-inducible genes, PS-IAA4/5 and PS-IAA6, of pea (*Pisum sativum* L.), J. Mol. Biol. 233:789.

Ritchie, G., 1994, Commercial application of adventitious rooting to forestry, in: "Biology of Adventitious Root Formation," T.D. Davis, and B.E. Haissig eds., Plenum Press, New York. p. 37.

Weir, R.J., and Goldfarb, B., 1993, Loblolly and slash pine rooted cutting research at North Carolina State University, in: "22nd Southern Forest Tree Improvement Conference, June, 1993, Atlanta, Georgia, p. 434.

Yamamoto, K.T., Mori, H., and Imaseki, H., 1992, cDNA cloning of indole-3-acetic acid regulated genes: Aux22 and Saur from mung bean (*Vigna radiata*) hypocotyl tissue, Plant Cell Physiol. 33:93.

20

THE USE OF MUTANTS TO UNDERSTAND COMPETENCE FOR SHOOT-BORNE ROOT INITIATION

W. P. Hackett, S. T. Lund, and A. G. Smith

Department of Horticultural Science
University of Minnesota
St. Paul, Minnesota 55108

INTRODUCTION

Many research approaches have been used with an attempt to explain the basis for differences in competence for root initiation among species, clones and developmental phases within clones. Because of the long reproductive cycle in most woody perennial species, in which competence for root initiation is a limiting and practically important variable, a genetic research approach has not been feasible for them. With the rather recent development of molecular biological techniques, methods of generating and screening for monogenic mutants, and methods of tagging and cloning affected genes, a genetic approach to understanding the basis of competence, may now be the method of choice. Because shoot-borne rooting is likely to be a complex, multi-step process, a series of single gene mutants with, developmental blocks in various steps in the process, would provide a powerful tool for understanding competence for root initiation.

MUTANTS FOR ROOT INITIATION

A few monogenic mutants for altered root initiation have been identified (Table 1). They can be categorized into three general groups: 1) those which have an excess of lateral and shoot-borne roots (Boerjan, *et al.*, 1995; Celenza, *et al.*, 1995; King *et al.*, 1995; Neil Olszewski, personal communication) 2) those in which auxin does not induce or has a small affect on lateral and shoot-borne root initiation (Cheng *et al.*, 1995; Lund *et al.*, 1996; Celenza *et al.*, 1995) and 3) those in which there is aberrant development of lateral root initials (Cheng, *et al.*, 1995; Celenza *et al*, 1995). For those in which there is an excess number of roots it has been shown that these mutants have elevated levels of indoleacetic acid and its conjugates (Boerjan *et al.*, 1995; King *et al.*, 1995). The affected gene has been tagged and has been cloned in one of these mutants (Neil Olszewski, personal

Table 1. Monogenic mutants for root initiation and development that may be useful for analyzing the basis of competence for shoot-borne root initiation

Species	Mutant name	Mutation type	Root phenotype	Root response to auxin	Endogenous auxin levels	Gene tagged or mapped	Gene cloned	Citation
Arbidopsis thaliana	superrot, sur 1-1 to 1-7	recessive	excess lateral roots; roots on hypocotyl	excised roots are auxin autropic	elevated	mapped to chromosome 2	no	Boerjan et al., 1995
Arbidopsis thaliana	rooty, rty-1 to 4	recessive	excess lateral roots; roots on hypocotyl	excised roots are auxin autotropic	elevated	mapped to chromosome 2; rty 3 T-DNA tagged	yes	King et al., 1995; Joe Ecker, Janson Reed, & Neil Oszewski, personal communication
Arbidopsis thaliana	alf 1-1	recessive	excess lateral roots; roots on hypolotyl	unknown	unknown	no	no	Celenza et al., 1995
Arbidopsis thaliana	rml 1	recessive	primary & lateral roots cease growth	shoot-borne roots induced by auxin, root growth not rescued by auxin	unknown	mapped to chromosome 4	no	Cheng et al., 1995
Arbidopsis thaliana	alf 4-1	recessive	no lateral roots	shoot-borne & lateral roots *not* induced by auxin	unknown	no	no	Celenza et al., 1995
Arbidopsis thaliana	rml 2	recessive	no lateral roots, primary roots cease growth	shoot-borne & lateral roots *not* induced by auxin		mapped to chromosome 3	no	Cheng et al., 1995
Nicotiana tabacum 'Xanthii'	rac	dominant	primary roots cease growth	shoot-borne roots *not* induced by auxin	auxin conjugates normal	no	no	Muller et al., 1985; Lund et al., 1996

communication). One of the mutants which has no response to auxin for root initiation has been proposed to be an auxin sensitivity mutant (Barbier-Brygoo *et al.*, 1990). The nature of the gene involved and the basis of the mutant phenotype has not been unequivocally determined for any of the mutants. The mutants in which auxin treatment does not induce or has a small effect on root initiation may be most useful for analyzing the basis for competence of root initiation. The rml 2 and alf 4-1 *Arabidopsis* mutants might be particularly useful. Ernst (1994) listed and discussed characteristics of a model system or species for studying causal mechanisms in shoot-borne rooting. He concluded that *Arabidopsis thaliana* was superior in many ways to any other species. Most of the mutants listed in Table 1 are of *A. thaliana.*

Although it doesn't have some of the desirable characteristics listed by Ernest, 1994 we've begun and analysis of the developmental and molecular basis of rooting competence using the monogenic mutant *rac* of *Nicotiana tabacum* v. Xanthii.

GENETIC AND DEVELOPMENTAL BASIS OF THE LACK OF ROOTING IN THE *RAC* MUTANT

Muller *et al.* (1985) identified an auxin-resistant mutant, *rac*, of tobacco (*N. tabacum* v. Xanthii) which is impaired in primary root development. They generated the mutant genotype via ultraviolet irradiation of wild-type mesophyll protoplasts; mutant protoplasts were selected that had the ability to proliferate in media containing levels of naphthaleacetic acid (NAA) that were toxic to wild-type protoplasts. A mutant line was recovered that confers a ten-fold increase in the resistance of cell suspensions to auxin and a cosegregating impaired primary root development phenotype in seedlings. *Rac* plantlets from seed or stem cuttings from mutant plantlets do not form shoot-borne roots. Mutant shoots grafted onto wild-type rootstock are fertile and produce viable seed upon selfing. A karyotype analysis of mutant petal cells revealed that mutant plants are diploid, even though they were derived from mutagenized haploid protoplasts. Progeny tests showed that the auxin resistance and the impaired primary root development phenotypes are dominant and caused by a mutation in a single gene. *Rac* plantlets are not impaired in auxin transport and do not contain altered levels of auxin conjugates (Caboche *et al.*, 1987).

Our experiments (Lund *et al., 1996) demonstrate that shoot-borne root formation does not occur when heterozygous and homozygous Rac* micro shoot-derived stem cuttings are treated *in vitro* with indole-3-butyric acid (IBA) concentrations ranging from 0.5 μM to 500.0 μM. Optimal root initiation was obtained in wildtype cuttings at 5.0–10.0 μM IBA. Histological analysis showed that some phloem parenchyma or inner cortical parenchyma cells in wild-type stem cuttings undergo shoot-borne root morphogenesis when treated with 5.0 μM IBA. The same cell types in heterozygous and homozygous *rac* stem cuttings undergo mitoses in response to auxin, but never form shoot-borne root meristems. We conclude that *rac* mutants are incompetent for shoot-borne root initiation in response to auxin. The lack of shoot-borne root initiation in *rac* stem cuttings is phenotypically distinct from the aberrant primary root development in *rac* stem seedlings in which case a root meristem is formed but aborts during germination (Lund *et al.*, 1996). The *rac* mutation appears to block an essential process for auxin induction of shoot-borne root initiation but not cell division in phloem parenchyma or inner corical parenchyma cells. Comparisons of *rac* heterozygous and homozygous seedling primary root length and callus formation in response to auxin in stem cuttings indicate that *rac* copy number is correlated to the degree of expression of these two phenotypes.

Based on responses of *rac* and wildtype protoplast and cell suspension to auxins Barbier-Brygoo *et al.* (1990) concluded that expression of the dominant *rac* mutation causes lowered sensitivity to auxin by reducing the number of functional auxin binding sites at the plasmalemma. However, our results show that formation of shoot-borne root meristems is completely blocked in *rac* stem cuttings even at high auxin concentrations, while cell divisions occur readily in response to auxin. Assuming that Barbier-Brygoo's conclusion is correct, our results suggest that the cellular incompetence of *rac* stem parenchyma cells, specifically for shoot-borne root initiation but not for cell division, may reflect a higher threshold of auxin binding required for proper signaling for root initiation than for cell division. Alternatively, cellular incompetence of *rac* stem cells for shoot-borne root initiation but not for cell division may be due to the existence of different auxin receptors for the two responses.

MOLECULAR ANALYSIS OF THE BASIS OF ROOT INITIATION INCOMPETENCE IN THE *RAC* MUTANT

Our developmental analysis indicated that the *rac* mutation blocks shoot-borne root initiation but not unorganized cell divisions in response to auxin (Lund *et al.*, 1996). This work also suggests that the *rac* mutation blocks shoot-borne root initiation prior to the first organized cell divisions that normally lead to the formation of an shoot-borne root meristem. This difference in competence for root initiation but not cell division in wildtype and *rac* cuttings provides an excellent experimental system for analysis of the molecular basis of competence for root initiation.

We have studied the effect of the *rac* mutation on the temporal and spatial expression patterns of three genes, *HRGPnt3*, *iaa4/5*, and *gh3* previously shown to be expressed in shoot-borne root meristems (Vera *et al., 1994; Ballas et al., 1993; Hagen et al.*, 1991). This was done by comparing the temporal and/or spatial expression patterns of the three genes during early stages of cell division and root initiation in cuttings with wild type versus *rac* background. Three genes were selected from previously isolated, auxin-responsive or root-associated genes because they all have been shown to be expressed in shoot-borne root initials and transgenic tobacco lines with promoter-*GUS* fusions for each of the genes were available.

iaa4/5 and *gh3* are thought to be early auxin-responsive genes because accumulation of each of their transcripts is rapid, specific for biologically active auxins, and independent of protein synthesis. Transcriptional activation of both the *iaa4/5* and *gh3* promoters is detectable in a variety of auxin-responsive cell types in addition to shoot-borne and lateral root meristematic cells. Activation of HRGPnt3 promoter sequences is only detectable in primary root meristematic cells or cells deterimined for secondary (shoot-borne or lateral) root formation. Since the *rac* mutation had been defined on a phenotypic but not on a molecular basis, the effects of *rac* on the transcriptional induction of *iaa4/5*, *gh3*, and *HRGPnt3* should provide information regarding *rac*'s role in auxin signal transduction for cell divisions and/or specifically for shoot-borne root initiation.

Using histochemical staining and quantitative analyses of *iaa4/5-GUS* and *gh3-GUS* transformant, wild-type and *rac* cuttings, we determined that *rac* does not repress the auxin activation of the promoters of these two genes. This indicates that activation of the *iaa4/5* and *gh3* promoters is not limiting for shoot-borne root initiation in *rac* cuttings. Based on a fluorometric analyses of *gh3*-GUS activity in *gh3*-GUS hemizygous cuttings versus *gh3*-GUS hemizygous, *rac* heterozygous cuttings treated with auxin for one day, we concluded that the *rac* mutation does not cause any reduction in IBA sensitivity at the level of *gh3*-GUS activity. Thus, these results do not support the conclusion of Barbier-Brygoo *et al;* (1990) that the *rac* mutation causes a general reduction in auxin sensitivity.

Using histochemical staining analyses of auxin-treated, *HRGPnt3- GUS* transformant wild-type and *rac* cuttings, we demonstrated that the *rac* mutation blocks activation of the 1.3 kb, 5′ upstream region of the *HRGPnt3* promoter. Thus, activation of a region within the *HRGPnt3* promoter occurs specifically during shoot-borne root initiation in tobacco cuttings. RNA blot analyses, however, suggest that *HRGPnt3* expression is regulated both developmentally and environmentally. Based on the *HRGPnt3-GUS* analyses, we concluded that differential expression in response to auxin treatment occurs during shoot-borne root initiation in the wild type versus callus formation in *rac* cuttings.

If shoot-borne root initiation is regarded as an organized form of cell divisions and auxin-induced shoot-borne root initiation occurs via auxin reception followed by transduction of the auxin signal, our data place *iaa4/5* and *gh3* expression upstream from *rac*, and *HRGPnt3* expression downstream from *rac* in a simple linear model. Although early, auxin-responsive gene expression is likely a component of shoot-borne root initiation, it is doubtful that expression of genes such as *iaa4/5* and *gh3* have specific roles in the organization of the cell divisions that are required for shoot-borne root initiation. Furthermore, it is unlikely that expression of upstream, auxin signal transduction genes has a role in the determination of competence for shoot-borne root initiation in cuttings. Conversely, it is plausible that genes such as *HRGPnt3* that code for structural proteins might be up-regulated prior to, or coincident with, the organization of cell divisions that occurs during shoot-borne root initiation, and thus, be a factor in the determination of the competence for shoot-borne root initiation in tobacco cuttings. Since the *rac* mutation has a major phenotypic effect of blocking shoot-borne root initiation prior to the first organized cell divisions, it is likely that *rac* has an important regulatory role during the early stages of shoot-borne root initiation in the phloem parenchyma or inner cortical parenchyma cells in tobacco cuttings.

CONCLUSIONS

Monogenic mutants of *Arabidopsis thaliana* that do not form shoot-borne roots in response to auxin mutants, may be most useful in analyzing the basis of shoot-borne rooting competence. Of the mutants currently identified rml 2 and alf 4–1may be particularly useful.

In the *rac* mutant of tobacco which is incompetent for auxin induced root initiation but not cell division, activation of the promoters of the early auxin-responsive genes is not limiting for shoot-borne root initiation based on histochemical GUS staining and quantitative analyses of appropriate genetic transformants.

Based on histochemical staining of appropriate genetic transformants, we also conclude that the rac mutation blocks activation of the *HRGP nt3* promoter and that there is differential expression of the promoter in response to auxin treatment during shoot-borne root initiation in wild type versus callus formation in *rac* cuttings.

REFERENCES

Ballas, N., Wong, L.M. and Theologis A., 1993. Identification of the auxin-responsive element Aux RE in the primary indoleacetic acid-inducible gene, PS-Iaa 4/5, of pea (*Pisum sativum*). *J. Mol. Biol* 223:580.

Barbier-Brygoo H., Maurel, C., Shen, W-H., Ephritikhine, Delbarre, A. , Guern, A. & C., 1990. Use of mutants and transformed plants to study the action of auxins. in: "Hormone Perception & Signal Transduction in Animals and Plants," Soc. Expt. Biol. Symp., R. Hooley & J. Roberts, eds., Company of Biologists, Cambridge. p.67.

Boerjan, W., Cervera, M.T., Delarue, M., Beeckman, T., Dewitter, W., Bellini, C., Caboche, M., Van Onckelen, H., van Montagu, M., Inze, M. &. D., 1995. *Superroot*, a recessive mutation in *Arabidopsis*, confers auxin overproduction. *Plant Cell* 7:1405.

Caboche, M., Muller, J.F., Chanut, F., Aranda, G. and Cirakoglu, S., 1987. Comparison of the growth promoting activities and toxicities of various auxin analogs on cells derived from wild-type and a non-rooting mutant of tobacco. *Plant Physiol.* 83: 795.

Celenza, J.L., Grisafi, P.L. and Fink, G.R., 1995. A pathway for lateral root formation in *Arabidopsis thaliana. Genes & Devel.* 9:2131.

Cheng, J.C., Seeley, K.A. and Sung, Z.R., 1995. rml. 1 and rml. 2, *Arabidopsis* genes required for cell proliferation at the root tip. *Plant Physiol.* 107:365.

Ernst, S.G. 1994. Model systems for studying adventitious root initiation. in: "Biology of Shoot-borne Root Formation," T.D. Davis and B.E. Haissig, eds., Plenum Press, New York & London. p. 77.

Hagen, G., Martin, G., Li, Y. and Guilfoyle, T.J., 1991. Auxin-induced expression of the soybean *gh3* promoter in transgenic tobacco plants. *Plant Mol. Biol.* 17:567.

King, J.J, Stimart, D.P., Fischer, R.H. and A.B. Bleecker, 1995. A mutant altering auxin homeostasis and plant morphology in *Arabidopsis. Plant Cell* 7:2023.

Lund, S.T., Smith, A.G. and Hackett, W.P., 1996. Cuttings of a tobacco mutant, *rac*, undergo cell division but do not initiate adventitious roots in response to exogenous auxin. *Physiol. Plant* 97:372.

Muller, J.F., Goujaud, J. and Caboche, M., 1985. Isolation *in vitro* of naphthalene acetic acid-tolerant mutants of *nicotiana tobacum* which impair root morphogenses. *Mol. Gen. Genet.* 199:194.

Vera, P., Lamb, P.J., and Doerner, P.W., 1994. Cell-cycle regulation of hydroxyproline-rich glycoprotein *HRGPnt3* gene expression during initiation of lateral root meristems. *The Plant J.* 6:717.

21

TRANSGENIC ROOTING IN CONIFERS

Roland Grönroos, Anders Lindroth, Haile Yibrah, and David Clapham

Department of Forest Genetics
Uppsala Genetic Centre
Swedish University of Agricultural Sciences
Box 7027, S-750 07 Uppsala
Sweden

INTRODUCTION

Genetic transformation of conifer tissues has been difficult due to low transformation frequencies, problematic detection and gene inactivation. Genetically transformed roots were obtained in seedlings of *Picea abies*, *Pinus sylvestris* and *Pinus contorta* by infection with *Agrobacterium rhizogenes* (Magnussen *et al.*, 1994). However, the transformation technique was far from optimal. Here we present a recently developed technique to obtain a high frequency of transgenic roots from a conifer, *P. contorta* (Yibrah *et al.*, 1996).

EXPERIMENTAL

Plant Material

Pinus contorta Dougl. ex. Loud seeds were surface sterilized. After imbibition, the seeds were sown in Magenta GA7–3 vessels containing a sterile mixture of vermiculite and 25% Hoagland nutrient solution. Another jar was placed upside-down over the first as a lid. The seeds were germinated in a growth room at 22°C under continuous light and allowed to grow for 2–4 weeks before inoculation.

Bacterial Strains and Binary Vectors

The bipartite strains of *Agrobacterium rhizogenes* used to inoculate the hypocotyls were constructed from LBA 9402 (Spano *et al.*, 1982) or A4RSII (from Dr. J. Tempé). The use of bipartite strains entail that transformed cells may obtain either T-DNA from the Ri plasmid or from the binary vector or from both. LBA 9402 carried either the binary vector pMOG6GUS (Yibrah *et al.*, 1993) or p35S GUS INT (Vancanneyt *et al.*, 1990). A4RSII

Biology of Root Formation and Development, edited by Altman and Waisel.
Plenum Press, New York, 1997.

carried pMOG6GUS. Both binary vectors contained the *uidA* gene fused between the 35S promoter and 35S terminator. In p35S GUS INT the coding region contains a plant-derived intron which prevents its expression in bacteria (Vancanneyt et al. 1990). The bacteria were grown at 28°C for 2–3 days on solid YMB containing kanamycin. If not otherwise stated the experiments were carried out with LBA 9402 carrying the binary plasmid p35S GUS INT.

Inoculation and Root Development

A vertical piercing infection cut of about 2–3 mm was made straigth through the middle of each hypocotyl with a sterile eye-surgery knife (Beaver® MVR Unitome® knife cat 375560) or a 0.4 mm diam injection needle. A piece of agar containing the bacterial lawn was placed hanging on the hypocotyl at the cut.

After 6–8 weeks the inoculated seedlings were cut on the hypocotyl about 5 mm below the site of inoculation and then planted in vermiculite containing 25% Hoagland's medium. After 13 months of *in vitro* culture, the inoculated plants were transferred to *ex vitro* culture, where a continuously circulating nutrient solution was supplied. The plants were cultivated by this method for one year.

5-Azacytidine Treatment

In one experimental series, 96 putatively transformed roots were cut, two years after inoculation with LBA 9402 pMOG6GUS, into 30 segments of about 5–10 mm. These were either: (1) immediately stained for GUS activity, (2) cultured as control on medium containing 300 μg Claforan/ml, 0.1% Ronilan*FL (anti-fungal agent) and solidified with 0.3% Gelrite™ gelan gum, or (3) cultured as in (2) but with 30 μM 5-azacytidine in the medium. Root segments from treatments (2) and (3) were stained for GUS activity after 10 days.

GUS Assay

The tissues were histochemically stained with 5-bromo-4-chloro-3-indolyl glucuronide (X-GLU) (Jefferson, 1987) at pH 7.0 for 24–48 hours at 37°C and thereafter examined under a dissecting microscope.

Anatomical Studies

To study the pattern of GUS activity in inoculated hypocotyls, entire infection sites were sampled at 8, 15, and 22 days after infection. After staining histochemically for GUS activity, the samples were frozen in dry-ice-chilled ethanol, dehydrated at −80°C and embedded in paraffin as described by Stomp (1992). Sections were cut serially. Every section was examined for GUS activity using dark field microscopy.

Verification of *uidA* Integration by Southern Blotting

P. contorta DNA extracted from roots putatively transformed with the *uidA* gene and the pMOG6GUS plasmid were restricted with *Not*I and electrophoresed on an agarose gel. The fractionated DNA was blotted onto a membrane which was hybridized to the T-DNA region of pMOG6GUS.

RESULTS AND DISCUSSION

Development of Transgenic Roots in *Pinus contorta*

Roots started to develop from the inoculation site during the first month after inoculation. After about 2 months the primary root system was cut from the hypocotyl and new roots developed on many plants from the wound tissue formed at the cut surface. After one year of culture *ex vitro*, tumors had developed at the inoculation sites. New roots continued to emerge from the tumors at the inoculation site during the culture *ex vitro*.

No obvious morphological differences were found between most of the GUS-positive and negative roots. Most of the roots looked like normal *P. contorta* roots grown in liquid culture. A few of the GUS-positive roots showed extensive lateral branching and may have been transformed with one or both T-DNA segments from the Ri plasmid as well as the T-DNA from the binary vector.

The use of dark field microscopy allowed easy detection of single GUS-positive cells in thin tissue sections. In 43% of the hypocotyls, GUS activity was detected somewhere in the hypocotyl. No rooting or GUS activity was found unless the endodermis had been penetrated during inoculation. The proportion of GUS-positive cells seen in sections close to the inoculation site was high in the original tissue inside the endodermis of the stem at day 7, transient GUS activity was obtained in almost all cell types found inside the endodermis. By day 15 the GUS activity had diminished and was localised to small areas at the upper and lower ends of the inoculation scar.

As the roots developed and callus was formed the intensity of the GUS staining increased. Callus and roots developed from cells inside the endodermis. No GUS activity was found in roots before day 22. Anatomical examination of four hypocotyls with a total of 33 roots, 6 weeks after inoculation, revealed that all 7 GUS-stained roots had developed via callus tissue.

The rooting followed three main patterns: direct rooting from the inoculation site, indirect rooting from callus formed at the inoculation site and indirect rooting from basal wound tissue. The first two types of rooting were unique for *A. rhizogenes*-inoculated seedlings, while the third type occurred both in infected and in mock-inoculated seedlings. Direct rooting from the inoculation site resembles auxin-induced rooting (Grönroos and von Arnold 1987). From these studies it seemed that the transformed roots developed from callus that was derived from transformed cells inside the endodermis. Consequently both transformed and untransformed roots developed from the inoculation site. Furthermore, transformed roots developed from the basal wound tissue localized about 5 mm below the inoculation site. This could be explained by movement of *A. rhizogenes* within the hypocotyl.

A significantly higher proportion of the seedlings produced GUS-positive roots after inoculation with LBA 9402 (20%) than with A4RSII (4%). The positive seedlings had a similar mean number of GUS-positive roots per seedling, irrespective of which bacterial strain had been used.

The Southern blot with *Not*I-digested *P. contorta* showed that the probe hybridized to fragments in the *P. contorta* DNA which were substantially larger (>23 kb) than the linearized plasmid (13 kb). This indicates that the *uidA* gene has been incorporated into the *P. contorta* genome and that the signal is not due to residual bacteria. The number of copies of the *uidA* gene in DNA extracted from calli varied among transformants, from 1 up to 20 copies. Transgene copy number did not correlate with gene expression.

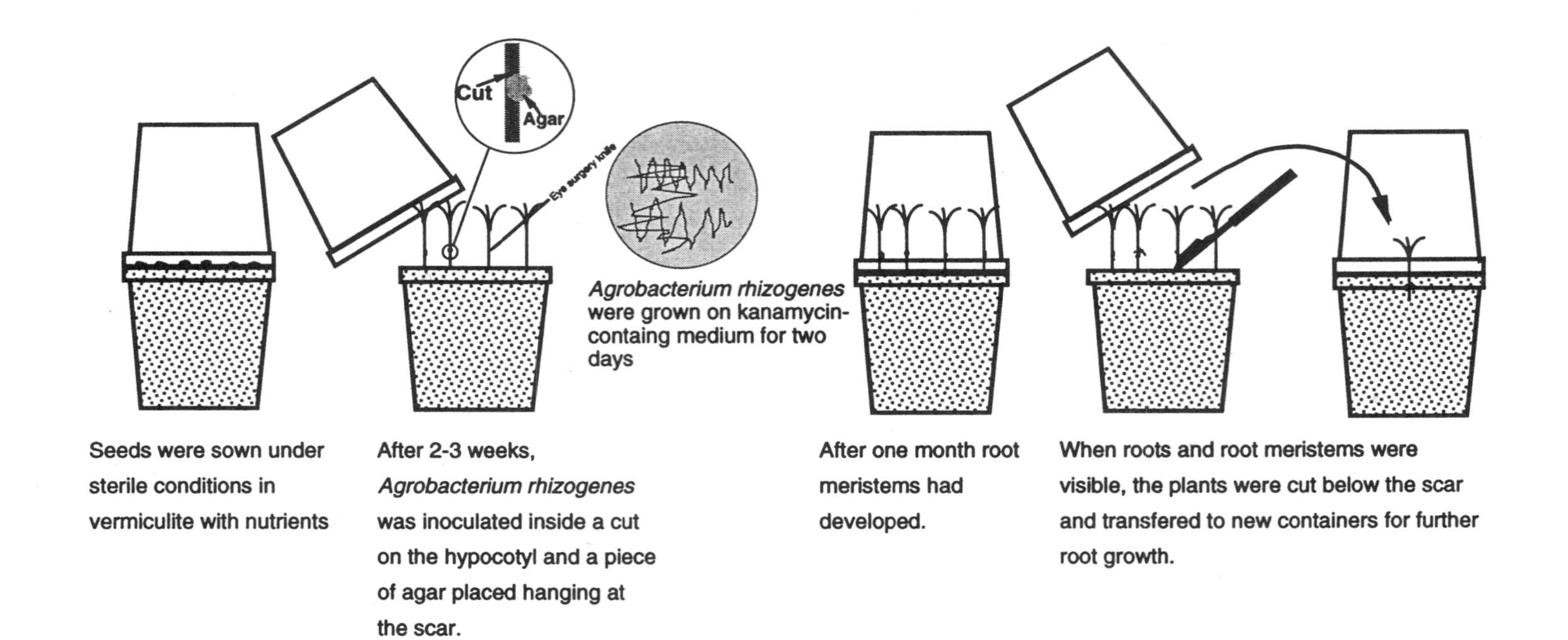

Figure 1. Transformation of *Pinus contorta* with *Agrobacterium rhizogenes*.

Expression Pattern in Roots of *P. contorta*

The uidA gene under 35S promoter control was expressed at the root apex, inside the endodermis or at the base of lateral roots. There was almost always a zone with no visible expression behind the apex of leader roots. Dichotomously branched lateral roots showed GUS activity at both apices. The most intensive GUS staining was found in the pith and around the resin ducts in the xylem at the base of primary adventitious roots after two years. At present we are studying how different promotors are expressed in roots (see the paper by Lindroth *et al.*, in this book).

Transgene Inactivation in *P. contorta*

To test whether more roots had been transformed than showed GUS activity, root segments initiated from putatively transformed roots were treated with 5-azacytidine in an attempt to reactivate silenced *uidA* genes. Initially 25% of the roots were GUS positive. The number of GUS positive roots rose to 55% after 10 days, as a result of 5-azacytidine treatment. The increase was highly significant ($p=0.0001$). No significant increase in expression in the untreated controls was obtained during 10 days of culture. When recalculated to the inoculated seedlings, GUS-expression could be detected in roots in 28% of the rooted seedlings after 14 weeks. However, after 5-azacytidine treatment of root segments, it was found that 48% of the rooted seedlings had produced transformed roots. Furthermore, an increase in percentage of GUS-positive roots per seedling was obtained.

Development of Transgenic Roots on *Picea abies* and *Pinus sylvestris*

In a previous study we showed that transgenic roots can be obtained from *P. abies* and *P. sylvestris*. However the transformation frequencies were lower than now obtained in *P. contorta*. The reason for this is not clear. We have used the new method on *P. abies* but in spite of 5 independent experiments with inoculation with LBA9402 p35SGUSINT, only a few GUS positive roots were obtanied (unpublished results). Some explanations for this may be that (1) the number of roots that develop from the inoculation cut is small, (2) there might be a more efficient gene silencing than in *P. contorta*—an indication of this is that the transient expression disappears within 24 h in *P. abies* compared to after more than 7 days in *P. contorta* (unpublished results).

CONCLUSIONS

There is now a method avaliable for producing transgenic roots in *Pinus contorta* (Fig. 1).

REFERENCES

Magnussen, D., Clapham D., Grönroos, R. and von Arnold, S. (1994) Induction of hairy and normal roots on *Picea abies*, *Pinus sylvestris* and *Pinus contorta* by *Agrobacterium rhizogenes*. *Scand. J. For. Res.* 9:46–51.

Spano, L., Pomponi, M., Costantino, P., van Slogteren, G.M.S. and Tempé, J. (1982) Identification of T-DNA in the root-inducing plasmid of the agropine type *Agrobacterium rhizogenes* 1855. *Plant Mol. Biol.* 1:291–300.

Stomp, A-M. (1992) Histochemical Localization of ß-Glucuronidase. In Gallagher, S.R. ed., *Gus Protocols: Using the GUS gene as a Reporter of Gene Expression*, pp.103–113. San Diego: Academic Press.

Vancanneyt, G., Schmidt, R., O'Connor-Sanchez, A., Willmitzer, L. and Rocha-Sosa, M. (1990) Construction of an intron-containing marker gene: Splicing of the intron in transgenic plants and its use in monitoring early events in *Agrobacterium*-mediated plant transformation. *Mol. Gen. Genet.* 220:245–250.

Yibrah, H. S., Clapham, D. H. and von Arnold, S. (1993) Antisense RNA inhibition of *uidA* gene expression in transgenic plants: Evidence for interaction between first and second transformation events. *Hereditas* 118:273–280.

Yibrah, H.S., Grönroos, R., Lindroth, A., Franzén, H., Clapham, D., and von Arnold, S. (1996) *Agrobacterium rhizogenes*-mediated induction of adventitious rooting from *Pinus contorta* hypocotyls and the effect of 5-azacytidine on transgene activity. Transgenic Research 5:75–85.

22

IMPROVED ROOTING ABILITY AND ROOT-SYSTEM PERFORMANCE IN TRANSGENIC ASPEN PLANTS

Tzvi Tzfira,[1] Christian Sig Jensen,[2] Alexander Vainstein,[1] and Arie Altman[1]

[1]The Otto Warburg Center for Biotechnology in Agriculture
The Hebrew University of Jerusalem
Rehovot 76-100, Israel
[2]The Royal Veterinary and Agricultural University
2000 Frederiksberg, Denmark

INTRODUCTION

Adventitious root formation and root-system performance are important aspects in the clonal propagation of different plant species, both from cuttings and in vitro. Root formation from cuttings is affected by anatomical, physiological and environmental factors and by their interaction. In addition, it is largely dependent on selecting a suitable organ (e.g. stem, leaf or root cuttings). This decision is influenced mainly by the inherent ability of the cuttings to regenerate root primordia. Several techniques have been proposed to improve rooting ability in difficult-to-root species, including heavy pruning (Scott, 1987) and etiolating stock plants to obtain easy-to-root juvenile material (Maynard and Bassuk, 1987), using specific rooting media (Sabalka, 1986), fog (Harrison-Murray et al., 1988) or mist atmospheres (Rosenberg et al., 1992), and applying growth regulators directly to the target tissue.

In recent years another "tool" for improving root formation and root-system performance has become available, i.e. transformation or inoculation with root-inducing bacteria. The phytopathological soil bacterium *Agrobacterium rhizogenes* triggers root formation in infected plants by transforming the T-DNA fragment of the Ri (root-inducing) plasmid into plant cells (Tepfer, 1984). Using molecular insertion and deletion analysis methods, four different loci have been identified in the T-DNA of *A. rhizogenes* (White et al., 1985), termed *rol* genes. Two of these genes are known to directly affect auxin (*rolB*) and cytokinin (*rolC*) content and/or tissue sensitivity to these hormones, thus stably modifying the inherent root regeneration ability (Cardarelli et al., 1987; Spena et al., 1987; Schmulling et al., 1988), and leading to enhanced root formation in difficult-to-root plants. The action of *rolA* and *rolD* genes is not as clear, but their presence in transgenic plants does not seem to interfere with the main effects of the *rolB* and *rolC* genes (Spena et al., 1987).

Biology of Root Formation and Development, edited by Altman and Waisel.
Plenum Press, New York, 1997.

Integration and expression of the *rol* genes does not prevent the regeneration of normal plants from hairy roots, as evidenced in many species (Tepfer, 1984; Cardarelli et al., 1987; Spena et al., 1987; Otani et al., 1996; Tzfira et al., 1996a). However, regeneration is often accompanied by several unique morphological traits resulting from the *rol* gene expression, e.g. wrinkled leaves (Tepfer, 1984), bushy plants, and reduced apical dominance (Tepfer, 1984; Spena et al., 1987; Schmulling et al., 1988, 1993; Tzfira et al., 1996b). Extensive root formation and modified root architecture are other *rol*-expression-related traits in transformed plants. Thus, the *rol* genes can be harnessed, at least theoretically, to improve root performance in transgenic plants. The use of *A. rhizogenes* to enhance root formation in cuttings was suggested before (Hartmann et al., 1990), and has been effected in several species (Strobel and Nachmias, 1985; Patena et al., 1988; McAffe et al., 1993; Caboni et al., 1996).

In the following, we present evidence of the effect of *A. rhizogenes* on root formation in genetically engineered aspen (*Populus tremula*) plants. Transgenic plants, as confirmed by Southern blot analysis, were tested for their rooting ability and root-system performance. The use of wild-type *A. rhizogenes* to transform and alter root performance of this woody plant is expected to have practical implications, because plant-rooting ability was substantially affected without evidence of shoot abnormalities in many of the transformants.

EXPERIMENTAL

Bacteria

Wild-type, agropine-type, *A. rhizogenes* strain LBA9402 was used for transformation. This strain carries the pRi1855 *vir*-oncogenic plasmid and the p35SGUSINT binary plasmid. pRi1855 carries the *rol* genes, while the binary plasmid carries genes coding for neomycin phosphotransferase II (NPTII) driven by a nopaline synthase (NOS) promoter, and the *uidA* gene coding for β-glucuronidase (GUS) driven by a cauliflower mosaic virus (CaMV) 35S promoter. Bacteria were cultured at 28°C in liquid yeast mannitol broth (YMB) (Tzfira et al., 1996a).

Transformation

Aspen (*P. tremula*) in vitro-cultured plants were transformed as described by Tzfira et al., (1996a). Briefly, approximatly 10-mm long stem explants were soaked in an overnight bacterial suspension, and co-cultivated for 2 days on half-strength MS (Murashige and Skoog, 1962) hormone-free regeneration medium. The explants were then washed under sterile conditions and transferred to the same medium, supplemented with ampicillin. About 2–3 weeks after co-cultivation, roots had regenerated from the stem's basal cut surface. These roots were excised, transferred to half-strength liquid MS medium (without hormones) and placed on a rotary shaker for shoot induction. Two to three weeks later, shoots had emerged from these roots: they were tested for GUS expression, excised and transferred to half-strength MS medium (without hormones) for rooting. About 100 different transgenic lines (phenotypes) were established, and five lines were selected for further analyses. The transformed lines were designated T-4, T-5, T-26, T-27 and T-31. Transformed and non-transformed (control) plants were propagated in vitro using small stem segments containing two to three nodes, on half-strength MS medium without hormones.

Analysis of Transformants

Transient and stable transformation in aspen explants was monitored by histochemical GUS detection (Stomp, 1992). Kanamycin resistance was confirmed by the ability of stem segments to produce adventitious roots in its presence. DNA was extracted from cultured aspen plants and Southern blot analysis was performed to detect *uidA*, *nptII* and *rol* genes in the plant's digested genomic DNA. Rooting percentages, number of roots per stem explant, root fresh weight and root surface-area measurements were performed at the specified time points (20 replicates per line). Single-node segments were cultured in half-strength MS medium without hormones, and root number was counted for each segment. After 31 days in culture, the roots were excised and weighed, and their surface area was evaluated using image-analysis computer technique.

RESULTS

Five putatively transformed plants were selected based on phenotypic abnormalities relative to non-transformed (control) plants. GUS expression and Southern and northern blot analyses confirmed the *uidA* transgenic nature of the transformants, as described previously (Tzfira et al., 1996a, b). The *uidA* gene was present in four of the transgenic lines, but not in line T-27 (data not shown). Kanamycin resistance was observed, to various degreas, in the fully developed transformed plants of four lines, but again, not in line T-27 (data not shown). Southern blot analysis of the oncogenic *rol* genes demonstrated their presence in all transgenic lines except T-4, and their absence from the control line (data not shown). The same integration pattern was found in lines T-5, T-27 and T-31, the different pattern observed in line T-26 suggested that it harbors more than one oncogenic T-DNA copy.

Stem segments of transformed plants formed more roots, in a shorter period of time and at a higher rate, than the non-transformed (control) line. Whereas control stem segments maintained an average three roots per explant during the culture period (Fig. 1a), transgenic lines exhibited extensive root formation resulting in an average 10 roots per explant for line T-27 (Fig. 1b), and more than 12 roots per explant for line T-31 (Fig. 2a). Transgenic lines also exhibited higher rooting percentages during the culture period relative to the control line. This was reflected not only by the fact that stem segments of transformed plants had higher rooting percentages, but also by their reaching almost 100% rooting after 14 to 18 days in culture, as compared with the 23 days required for the control line. It is worth noting that transgenic plants maintained the capacity for 100% rooting throughout the year, whereas control plants had a year-round average of about 90%, with only 70–80% rooting during the winter season.

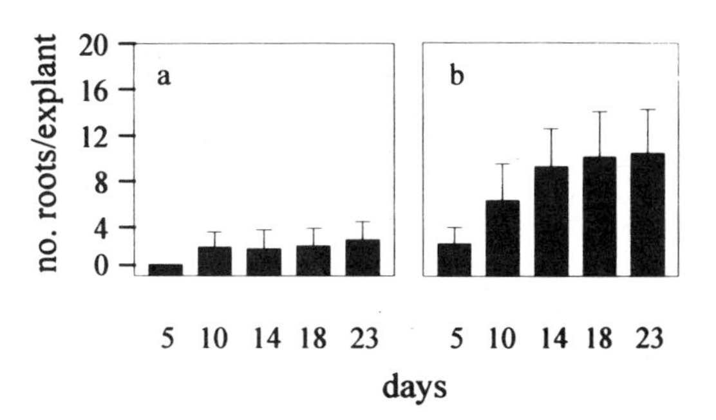

Figure 1. Kinetics of root formation in different aspen lines. Bars represent the average number of roots per explant (20 replicates) during the culture period. (a) Control line; (b) line T-27. SE of the mean is indicated.

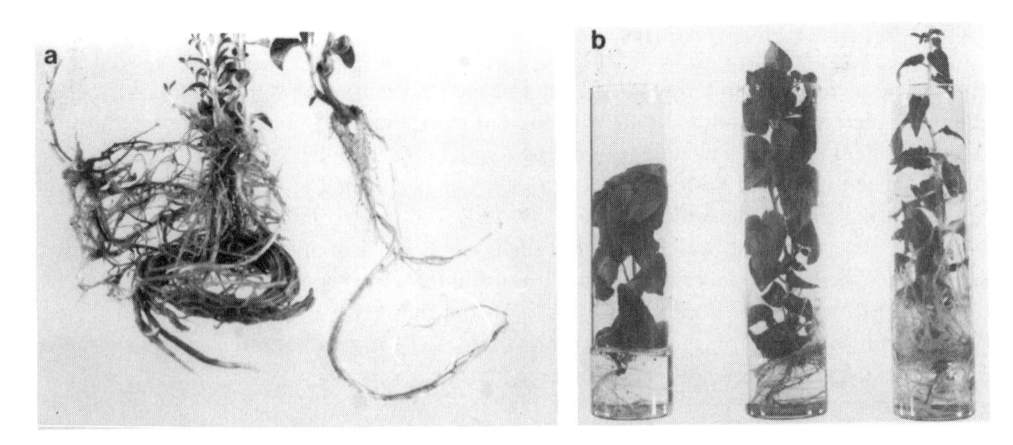

Figure 2. Phenotypic effect of *rol* genes in transgenic aspen plants. (a) More than 12 roots were observed in line T-31 transgenic plants (left), as compared to three roots in the control line (right). (b) Differences between three different lines of aspen plants (30 days of culture): control (left), T-4 (middle) and T-5 (right).

Significant differences were observed in the root surface area of specific transgenic lines and control plants, as estimated by image-analysis computer technique. All transformed plants presented a high average root surface area as compared to the control. Both quantitative and qualitative differences were observed: transgenic lines tended to branch more and to form a cluster of roots when grown in vials (Fig. 2b), whereas control plants formed only two to four roots, which did not develop many lateral roots, and usually continued to elongate. The increase in root number and the altered root-system architecture, as well as the rapid root formation in transgenic plants, led to accelerated growth in the transgenic plants.

Total root fresh weight was higher in transformed lines (e.g. up to 5.9-fold in line T-26 relative to control roots). The increment of root fresh weight was accompanied by an increase in total plant fresh weight, a substantial portion of which was root tissue (Fig. 3). Some transformed lines also tended to regenerate adventitious roots from the upper parts of the stem segment, from leaves and from stems of developed intact plants. However,

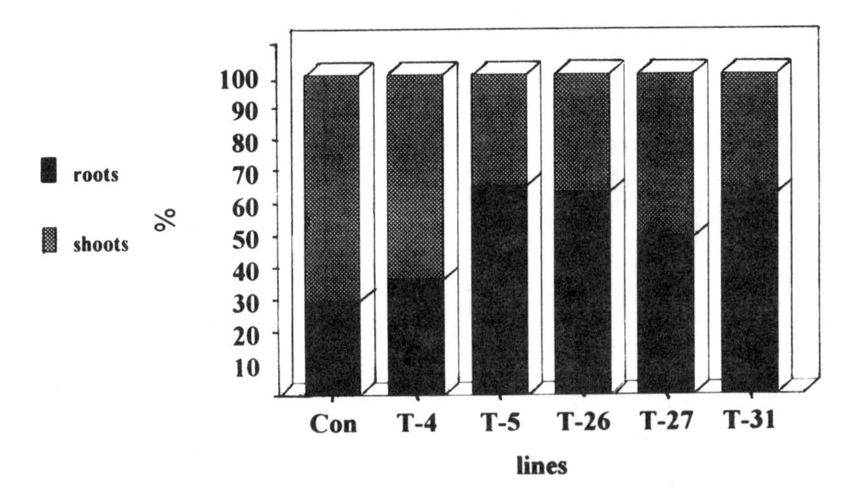

Figure 3. Distribution of plant part fresh weights in transgenic and control aspen lines. Blocks represent the proportion of root and shoot fresh weights out of the total plant fresh weight.

these abnormalities were observed only in vitro, or under high humidity (during plantlet hardening in a fog chamber), but never in greenhouse-grown plants.

CONCLUSIONS

In-vitro propagation can help overcome some rooting difficulties because controlled treatments can be easily applied in tissue culture. The use of different nutrient media, well-defined growth regulators, accurate hormone concentrations, well-controlled environmental conditions, and a juvenile plant source, can all improve adventitious root formation and root performance. However, the high cost of tissue-culture propagation (Mudge et al., 1986) has led to an ongoing search for better rooting performance combined with reduced culturing time and costs (Strode and Abner, 1986). Abolishing the need for additions to the medium, reducing the time required for adventitious root formation, and increasing rooting percentages could reduce propagation costs and increase the propagation coefficient.

High rooting percentages in transgenic plants, relative to non-transformed lines can reduce the culturing time required for root-system establishment in vitro. The use of genetically modified plants abolishes the need for additional hormones in culture. The increase in root number, as well as in root surface area, may eventually lead to better plantlet establishment and growth because the roots, particulary the root surface area, comprise the main site for the plant's interaction with its supporting medium (Tachibana and Ohta, 1983; Turner et al., 1985; Anderson et al., 1988; Bashan and LevAnony, 1989).

Indeed, the improved aspen root system found in this study led to improved plant performance, as reflected in the growth rate and development of in-vitro- and greenhouse-grown plants (data not shown). The transgenic plants were generally taller, with higher fresh weights than non-transformed plants. The fact that the proportion of root fresh weight was more than 50% of the total plant's fresh weight in most of the transgenic lines (Fig. 3), did not seem to interfere with their normal development. Nevertheless, the oncogenic root production that accompanied some of the transgenes under in-vitro or fog conditions was never observed in the greenhouse, after hardening. We believe that the high humidity induced the expression/activity of those genes, which later became silenced (Meins and Kunz, 1995) in the greenhouse. Thus, no abnormal root production is expected under normal growth conditions in nurseries, plantations and forests.

We should note that rooting improvment was uslly related to *rol*-gene expression, as it was accompanied with other *rol*-related phenotype (e.g. lose of apical dominance), in all transgenic lines, except T-4. We persume that in the case of the later line, a T-DNA mutagenesis or tagging event had randomly occured, in which, with the proper selection (Tzfira et al., 1996a) led to this phenotype.

Naturally, desirable plant traits also include shoot form, shoot growth rate, total plant weight and alterations in apical dominance. Evaluation of the *rol* genes' effects on shoots and shoot-related traits are currently being studied in our laboratory.

ACKNOWLEDGMENTS

This research was funded in part by the German-Israeli Agricultural Research Agreement for the benefit of the Third World (GIARA), by the Land Development Authority (Keren Kayemet Le'Israel) and by the Authority for Research and Development at The Hebrew University. The authors wish to thank Dr. D. Clapham (Upssala, Sweden) and Dr. R. Gardner (Auckland, New Zealand) for the bacterial strains.

REFERENCES

Anderson, A.J., Habibzadegah, T.P., and Tepper, C.S., 1988, Molecular studies on the role of a root surface agglutinin in adherence and colonization by *Pseudomonas putida*, *Appl. Environ. Microbiol.* 54:375.

Bashan, Y., and LevAnony, H., 1989, Factors affecting adsorption of *Azospirillium brasilence* Cd. to root hairs as compared with root surface of wheat, *Can. J. Microb.* 35:936.

Caboni, E., Lauri, P., Tonelli, M., Falasca, G., and Damiano, C., 1996, Root induction by *Agrobacterium rhizogenes* in walnut, *Plant Sci.* 118:203.

Cardarelli, M., Mariotti, D., Pomponi, M., Spano, L., Capone, I., and Constantino, P., 1987, *Agrobacterium rhizogenes* T-DNA genes capable of inducing hairy root phenotype, *Mol. Gen. Genet.* 209:475.

Harrison-Murray, R.S., Howard, B.H., and Thompson, R., 1988, Potential for improved propagation of cuttings through the use of fog, *Acta Hort.* 227:205.

Hartmann, H.T., Kester, D.E., and Davies, F.T., 1990, Vegetative propagation: Source selection and management in vegetative propagation, *in*:

Maynard, B.K., and Bassuk, N.L., 1987, Stock-plant etiolation and blanching of woody plants prior to cutting propagation, *J. Amer. Soc. Hort. Sci.* 112:273.

McAffe, B.J., While, E.E., Pelcher, L.E., and Lapp, M.S., 1993, Root induction in pine (*Pinus*) and larch (*Larix*) spp. using *Agrobacterium rhizogenes*, *Plant Cell, Tissue and Organ Culture* 34:53.

Meins, F.J., and Kunz, C., 1995, Gene silencing in transgenic plants: A heuristic autoregulation model, *in*:

Mudge, K.W., Borgman, C.A., Neal, J.C., and Weller, H.A., 1986, Present limitations and future prospects for commercial micropropagation of small fruits, *Proc. Inter. Plant Prop. Soc.* 36:538.

Murashige, T., and Skoog, F., 1962, A revised medium for rapid growth and bioassays with tobacco tissue culture, *Physiol. Plant.* 15:473.

Otani, M., Shimoda, T., Kamada, H., Teruya, H., and Mii, M., 1996, Fertile transgenic plants of *Ipomoea trichocarpa* Ell. induced by different strains of *Agrobacterium rhizogenes*, *Plant Sci.* 116:169.

Patena, L., Sutter, E.G., and Dandeker, A.M., 1988, Root induction by *Agrobacterium rhizogenes* in a difficult-to-root woody species, *Acta Hort.* 227:324.

Rosenberg, D., Altman, A., and Bravdo, B., 1992, A physiological analysis of the differential rooting performance of Calamondin and Pelargonium cuttings under mist and high humidity, *Acta Hort.* 314:257.

Sabalka, D., 1986, Propagation media for flats and direct sticking: what works? *Proc. Inter. Plant Prop. Soc.* 36:409.

Schmulling, T., Fladung, M., Grossmann, K., and Schell, J., 1993, Hormonal content and sensitivity of transgenic tobacco and potato plants expressing single *rol* genes of *Agrobacterium rhizogenes* T-DNA, *The Plant J.* 3:371.

Schmulling, T., Schell, J., and Spena, A., 1988, Single genes from *Agrobacterium rhizogenes* influence plant development, *EMBO J.* 7:2621.

Scott, M.A., 1987, Management of hardy nursery stock plants to achieve high yield of quality cuttings, *HortSci.* 22:738.

Spena, A., Schmulling, T., Koncz, C., and Schell, J.S., 1987, Independent and synergistic activity of *rol A*, *B* and *C* loci in stimulating abnormal growth in plants, *EMBO J.* 6:3891.

Stomp, A.M., 1992, Histochemical detection of GUS, *in:*

Strobel, G.A., and Nachmias, A., 1985, *Agrobacterium rhizogenes* promotes the initial growth of bare rootstock almond, *J. Gen. Microbiol.* 131:1245.

Strode, R.A., and Abner, G., 1986, Large-scale tissue culture production for horticultural crops, *in*:

Tachibana, Y., and Ohta, Y., 1983, Root surface area, as a parameter in relation to water and nutrient uptake by cucumber plant, *Soil Science and Plant Nutrition* 29:387.

Tepfer, D., 1984, Transformation of several species of higher plants by *Agrobacterium rhizogenes*: sexual transmission of the transformed genotype and phenotype, *Cell* 37:959.

Turner, S.M., Newman, E.I., and Campbell, R., 1985, Microbial population of ryegrass root surfaces: influence of nitrogen and phosphorus supply, *Soil Biol. Biochem.* 17:711.

Tzfira, T., Ben-Meir, H., Vainstein, A., and Altman, A., 1996a, Highly efficient transformation and regeneration of aspen plants through shoot-bud formation in root culture, *Plant Cell Rep.* 15:566.

Tzfira, T., Ben-Meir, H., Yarnitzky, O.,Vainstein, A., and Altman, A., 1996b, Highly efficient transformation and regeneration of transgenic aspen plants through shoot-bud formation in root culture, and transformation of *Pinus halepensis*, *in:* Proceedings of the IFURO Conference, Somatic Cell Genetics and Molecular Genetics of Trees, M.R. Ahuja, ed., Kluwer Academic Press, Gent, Belgium (in press).

White, F., Taylor, B., Huffman, G., Gordon, M., and Nester, E., 1985, Molecular and genetic analysis of the transferred DNA regions of the root-inducing plasmid of *Agrobacterium rhizogenes*, *J. Bacteriol.* 164:33.

23

ROOT-DIRECTED EXPRESSION OF ALIEN GENES IN TRANSGENIC POTATO

Sarcotoxin and *Gus*

Yael Mahler-Slasky,[1,2] Shmuel Galili,[3] Avihai Perl,[3] Radi Aly,[3] Shmuel Wolf,[2] Dvora Aviv,[1] and Esra Galun[1]

[1]Department of Plant Genetics
The Weizmann Institute of Science
Rehovot, Israel
[2]The Faculty of Agriculture
The Hebrew University
Rehovot, Israel
[3]Agricultural Research Organization
Bet Dagan and Neve Yaar, Israel

ABSTRACT

Bacterial pathogens of potato, e.g. *Pseudomonas solanacearum*, are known to infect potato roots causing severe losses in relatively warm climates. Our intention was to express, in potato roots, a bactericidal peptide that was identified in the larvae of the flesh fly (*Sarcophaga peregrina*) by Natori and associates in 1977. The cDNA coding for this peptide was subsequently isolated and it was termed sarcotoxin IA (*sarco*) by these investigators. The respective protein was also characterized and the mature proteins mass is about 5kDa. We used the *Tob* promoter that is root specific, to direct *sarco* expression in roots. In parallel we used the *Gus* gene as a reporter gene for this (*Tob*) promoter activity. Thus, two constructions of fusion genes were made. In one we inserted *Tob* up-stream of *sarco* and in the second *Tob* was inserted up-stream of *Gus*. In both cases the coding region was followed by a terminator. Both transformation cassettes contained also a kanamycin (*kana*) resistance gene (*nptII*) that was inserted as a selective marker under the 35S (CaMV) promoter. *Agrobacterium* mediated genetic-transformation was performed with potato tuber discs. Five potato cultivars and breeding lines were used: Desiree, Achirana INTA, LT-9, TS-10, TS-15. Potato plants that regenerated from *Agrobacterium* infected tuber-discs and rooted on selective medium were regarded putative transformants and were further analyzed. We found that putative transformants that resulted from transformation of a vector that contained *Gus* driven by *Tob*, indeed expressed the reporter

Biology of Root Formation and Development, edited by Altman and Waisel.
Plenum Press, New York, 1997.

gene in their roots. This verified the potency and specificity of the chimeric genes in the transformation vector. Polyclonal anti-*sarco* antibodies were produced and used to evaluate the expression of *sarco* in the putative transgenic potato plants. Preliminary western-blot assays indicated that indeed the roots of some of the plants that were transformed with the chimeric-gene that contained the *Tob* promoter and the *sarco* cDNA, showed bands that reacted with the anti-*sarco* antibodies.

INTRODUCTION

Potato plants sensitivity to bacterial pathogens became prevalent and has an increasing economic impact in recent years. This is due to successful breeding of warm-weather tolerant cultivars and the expansion of potato cultivation into areas with warm and humid climatic conditions. While breeding activity, including the use of molecular-genetic tools, is going on, to produce potato cultivars that are tolerant to other pathogens (e.g. viral, fungal), very little, if at all, breeding was conducted to reduce potatoes sensitivity to bacterial diseases. One important pathogenic bacterium is *Pseudomonas solanacearum* that mainly infects potato roots (brown rot) (Rich, 1983). We wished to use the potency of an antibacterial protein to confer tolerance agaisnt pathogenic bacteria infecting potato roots. For this purpose we focused on one antibacterial toxin, sarcotoxin IA, that is a member of a group of sarcotoxins studied extensively by Natori and associates at the University of Tokyo, Japan (e.g. Natori, 1977; Kanai and Natori, 1989; Matsumoto et al., 1986; Nakajima et al., 1987). These proteins are produced by the larvae of the flesh fly *Sarcophaga peregrina*, especially when the body of the third-instar is injured or when either living or dead bacteria are injected. The toxic sarcotoxins are accumulating in the haemolymph. Natori and associates isolated three such sarcotoxin I proteins (IA, IB & IC) that are very similar and belong to the cercopin type toxins. Sarcotoxin IA was purified and sequenced. The respective cDNA and the genomic DNA that codes for this toxin were also fully sequenced. The latter was found to contain one intron. The cDNA of sarcotoxin IA codes for a total of 63 amino acids. Of these, 40 are in the mature protein, while 23 N-terminal amino acids constitute the putative signal peptide that is cleaved off upon transfer to the haemolymph. The primary target of sarcotoxin IA is the cytoplasmic membrane of bacteria. Studies with *Escherichia coli* showed that upon exposure of the bacteria to sarcotoxin IA the membrane potential almost instantaneously disappeared and they lost the ability to actively transport amino acids and to generate ATP. Sarcotoxin IA interacts with liposomes constituted from acidic phospholipids. The toxin produces holes that enable the release of glucose from these liposomes. Interestingly, adding cholesterol to these liposomes interfered with this interaction, explaining why sarcotoxin IA, has a bacterial specificity. Contrary to antibiotic drugs there should be no selection to resistance in bacteria against insect antibacterial proteins.

Our goal was to direct sarcotoxin IA expression to potato roots. Since the cDNA of sarcotoxin was available, our intention was to construct chimeric genes that include a root specific promoter, the cDNA for the toxin, a plant terminator, and a selective marker. There are only few promoters which were demonstrated to direct root specific expression. One of them is the *Tob* promoter (Yamamoto et al., 1991) that was not yet demonstrated in potato roots. In order to be able to check the activity of the *Tob* promoter in potato we constructed another chimeric gene having the *Gus* reporter gene replacing the sarcotoxin cDNA. The constructs were cloned into binary vectors and the latter were transformed into an *Agrobacterium* strain that also had a helper plasmid. Transgenic potato plants should be derived after due genetic transformation and selection on selective medium. Root cultures from putative transgenic potato plants, containing the sarcotoxin IA cDNA were initiated. Since the toxin's expression is expected to be revealed in the protein-fraction of roots, we initiated root-cultures to increase the yield of the analyzed material. These cultures from putatively *Gus* containing plants were thus instrumental in checking the activity of the *Tob* promoter in potato roots. Transgenic potato plants with a high level of sarcotoxin IA in the roots should then be tested for their tolerance against the pathogenic bacteria.

MATERIALS AND METHODS

Construction of the Transformation Vector

We constructed two transformation vectors. In one we included the coding sequence of the *Gus* reporter gene (for β-glucuronidase) and in the other the cDNA of *sarco*. Both vectors contained also the *npt*II coding sequence for kanamycin resistance driven by the 35S CaMV promoter. The vector with *Gus* as well as that with *sarco* contained the *Tob* promoter for expression in roots. For using this promoter, we had to create appropriate cleavage sites. This was achieved by PCR (polimerase chain reaction) with synthesized primer-oligonucleotids containing the desireable restriction sites. Details of the engineering shall be provided elsewhere. The genes to be expressed as well as the selective gene with their promoters and terminators were cloned into the pGA492 binary vector (An, 1986). The two constructs (without the nptII cassette) are schematically shown in Fig. 1.

Genetic Transformation

We used the *Agrobacterium*-mediated transformation procedure according to Sheerman and Bevan (1988) with some modifications, as described by us previously (Perl et al., 1993). Briefly we used potato tubers that were stored at least one month after harvest and had a size of a hen-egg or smaller. After surface sterilization and washings with sterilized water, slice-pieces ("discs") were cut out (about 2–3 mm thick, about 1 × 1 cm in surface area). The slices were maintained (ca 20 min) in an overnight culture of the *Agrobacterium* strain that contained the transformation vector and then blotted and cultured in solidified MS medium supplemental with 2 mg/L zeatin riboside and 1mg/L indole-3-aspartic acid. After 48h the slices were blotted again and transferred to the same solidified

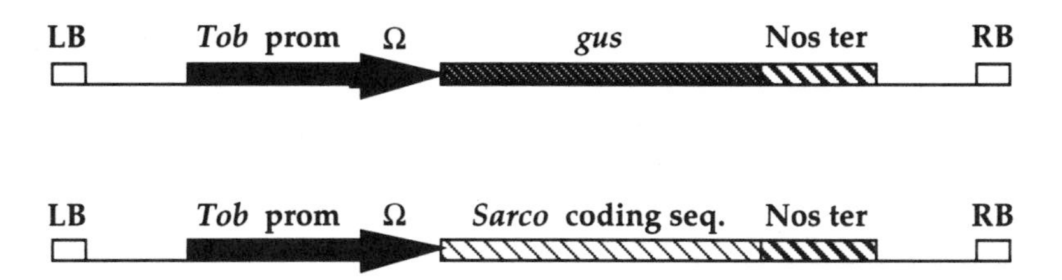

Figure 1. Chimeric genes that were constructed for genetic transformation of potato discs via the binary procedure of *Agrobacterium*-mediated transformation. These genes were cloned into a modified pGA492 binary vector (An, 1986). *Tob* prom is root specific promoter. Omega (W) is a translational enhancing sequence (Gallie et al., 1989). LB and RB are left border and right border of the *Agrobacterium* binary plasmid, respectively.

medium supplemented with 500 mg/L carbenicillin and 100 mg/L kanamycin and maintained in dim light. Shoots emerged during 3 or more weeks. The latter were transferred to the same medium and maintained under stronger light. Well developed shoots were rooted in MS medium containing 250 mg/L carbenicillin, 100 mg/L kanamycin and 1 mg/L indole-3-butyric acid. Rooted plantlets were transferred to plastic boxes with the same medium. Well developed plantlets that rooted normally were assigned designation numbers and further propagated *in vitro* in solidified MS also containing 100 mg/L kanamycin.

Five cultivars/breeding lines of potato were used: (1) cv Desiree, (2) Achirana INTA; (3) TS-10; (4) TS-15; (5) LT-9. The latter four were obtained from Dr. Ali Golmirzaie (CIP, Lima, Peru).

Root Cultures

Root sections (1–2 cm) from the putative transgenic potato plants were cultured for 3 days in 9 cm Petri dishes containing 6 ml of MS medium supplement with 3 mg/L indole-3-butyric acid (IBA). The root cultures were harvested and transferred to new Petri dishes (cd 50 mg per dish) containing 6 ml MS medium (pH 7.0) supplemented with 1 mg/L IBA. Cultures were kept at 25°C on a gyratory shaker (100 RPM) for 7 days and then 5 ml of the same medium were added and cultures were kept on the shaker for another 3–5 days. The roots were then blotted on a filter paper and their fresh weight was determined (Perl et al., 1993).

Staining for *Gus* Expression

Roots and leaves from intact, *in vitro* grown plantlets and roots from root cultures were collected. The plant tissues were stained with X-gluc as previously described (Perl et al., 1992). They were then fixed in an ethanol: acetic acid solution (3:1) and transferred to 70% ethanol. The latter was changed until the discoloration of the plant tissue and mounted on glass slides.

Production of Polyclonal Antibodies and Western Blots

Pure sarcotoxin IA served for the production of polyclonal anti-*sarco* antibodies. The production of antibodies and the procedure for western-blot hybridization were according to standard methods (Ausubel et al. 1995).

RESULTS

While the construction of the expression vectors and the genetic transformation of potato lines were completed, the assays for *sarco*-levels in the transgenic plants are in an initial phase. Therefore the following results should be considered as preliminary.

Genetic Transformation

Shoot and root regenerations from non-transformed (control) tuber slices were performed to verify the regeneration capability of these slices and the reliability of the kanamycin selection. When 24 and 13 such slices from Achirana INTA and TS-10 respectively were cultured as described but without kanamycin, all these slices regenerated rooted plantlets.

Table 1. Summary of genetic transformation[a]

	Tob: sarco			Tob: *Gus*		
Cultivar	Slices	Shoots	Rooted	Slices	Shoots	Rooted
Achirara	30	33	7	35	15	1
TS-10	40	26	5	40	18	5
LT-9	20	2	1	30	4	0
TS-15	30	0	0	40	0	0
Desiree	30	30	16	30	5	1

[a]Numbers indicate number of *slices* transformed, number of *shoots* regenerated from the transformed slices, and number of *rooted* shoots obtained, respectively. Two expression vectors served in the transformation: *Tob* promoter with the *sarco* gene, and *Tob* promoter with the *Gus* gene.

But when 100 mg/L kanamycin was added to the culture medium none of 12 slices from each of these lines produced shoots; not even callus formation was observed.

A total of 325 potato slices were transformed. These slices were derived from 5 cultivars and lines and they were transformed with 2 different expression vectors—as summarized in Table 1. Only rooted shoots were considered putative transgenic plants and were further analyzed. The cultivars Achirana INTA and Desiree as well as line TS-10 regenerated shoots and rooted plants, while LT-9 and TS-15 lacked root-regeneration capability under our conditions.

To verify the specificities and potency of the *Tob* promoter, we analyzed putative transgenic plantlets that were transformed with the expression vector that contained the *Gus* gene. Roots of intact plants showed X-gluc pigmentation. This pigmentation was restricted to the roots, (no pigmentation was observed in leaves) indicating the potency and specificity of the *tob* promoter in potato. We wished to find whether or not root cultures of transgenic plants harboring *Gus* behind the *Tob* promoter express the *Gus* reporter gene. Such root cultures were established from putative transgenic plants and a strong X-gluc pigmentation was observed.

Sarco Levels: Preliminary Results

To produce sufficient amounts of root tissue we initiated root cultures from putative transgenic plants that resulted from transformation with an expression vector that contained the *sarco* gene behind the *Tob* promoter. The roots were lyophilized, extracted in buffer and aliquotes were run on gels. The latter were processed as western-blots and reacted with anti-*sarco* antibodies. Root extracts of at least two such plants showed the presence of sarcotoxin IA.

CONCLUSIONS AND FUTURE

Our results up to present indicated that our expression vectors containing the *Tob* promoter were effective in directing gene-expression to the roots. We also found that at least some of the plants that resulted from transformation of a vector that contained the *sarco* gene behind the *Tob* promoter, expressed sarcotoxin IA in their roots. We are about to analyze additional putative root cultures in a procedure that should indicate the level of sarcotoxin IA that is produced. We shall then perform Southern-blot and northern-blot analyzes on selected trasngenic plants to correlate sarcotoxin IA levels with gene integration and transcript levels. Finally, the plants with highest sarcotoxin IA levels will be sent to the International Potato Center (CIP) in Lima, Peru, where these plants tolerance to bacterial pathogens shall be evaluated. Additional information on this system, especially on the expression of *Gus* in potato tubers and transformation vectors that should express *sarco* in there tubers, was presented elsewhere (Galun et al., 1996).

ACKNOWLEDGMENTS

This work is being supported by an USAID/CDR grant (no. 94-C12–245) and is part of a collaboration with Dr. Peter Gregory and Dr. Ali Golmirzaie of the International Potato Center (CIP), Lima, Peru. The authors are grateful to Prof. Shunji Natori, Faculty of Pharmaceutical Sciences, University of Tokyo for the cDNA of Sarcotoxin IA and to Dr. Satoshi Koikeda, Tsukuba Research Laboratories, Tsukuba, Japan, for a sample of Sarcotoxin IA as well as to Dr. David Granot, The Volcani Center, Bet Dagan, Israel, for his help in expressing Sarcotoxin IA in yeast.

REFERENCES

An, G. 1986. Development of plant-promoter expression vectors and their use for analysis of differential activity of nopaline synthase promoter in transformed tobacco cells. Plant Physiol. 81: 86–91.

Ausubel, F.M., Brent, R., Kingston, R.E., Moore, D.D., Seidman, J.G., Smity, J.A. & Struhl., K. 1995. Current Protocols in Molecular Biology. John Wiley & Sons, Inc., N.Y.

Gallie, D.R., Lucas, W.J. & Walbot, V. 1989. Visualizing mRNA expression in plant protoplasts: factors influencing efficient mRNA uptake and translation. Plant Cell 1: 301–311.

Galun, E., Galili, S., Perl, A., Aly, R., Aviv, D., Wolf, S. and Mahler-Slasky, Y. 1997. Defence against pathogenic bacteria in transgenic potato plants. Acta Horticulture (in press).

Kanei, A. & Natori, S.. 1989. Cloning of gene cluster for sarcotoxin I, antibacterial proteins of *Sarcophaga peregrina*. FEBS Lett. 258: 199–202.

Matsumoto, N., Okada, M., Takahashi, H., Ming, Q.X., Nakajima, Y., Nakanishi, Y., Komano, H., & Natori, S. 1986. Molecular cloning of a cDNA and assignment of the C-terminal of sarcotoxin IA a potent antibacterial protein of *Sarcophaga peregrina*. Biochem. J. 239: 717–722

Nakajima, Y., Qu X. M. & Natori, S. 1987. Interaction between liposomes and sarcotoxin IA, a potent antibacterial protein of *Sarcophaga peregrina* (flesh fly). J. Biol. Chem. 262: 1665–1669.

Natori, S. 1977. Bactericidal substance induced in the haemolymph of *Sarcophaga peregrina* larvae. J. Insect Physiol. 23: 1169–1173.

Perl, A., Kless, H., Blumenthal, A., Galili, G., & Galun, E. 1992. Improvement of plant regeneration and GUS expression in scutellar wheat calli by optimization of culture conditions and DNA microprojectile delivery procedures. Mol. Gen. Genet. 235: 279–284.

Perl, A., Perl-Treves, R., Galili, S., Aviv, D., Shalgi, E., Malkin, A. & Galun, E. 1993. Enhanced oxidative-stress defense in transgenic potato expressing tomato Cu,Zn auperoxide dismutases. Theor. Appl. Genet. 85: 568.

Rich, A.E. 1983. Potato Diseases. Acad. Press, New York, 238p.

Sheerman, S. & Bevan, M.W. 1988. A rapid trasnformation method for *Solanum tuberosum* using binary *Agrobacterium tumefaciens* vector. Plant Cell Rep. 7: 13–16.

Yamamoto, Y.T., Taylor, C.G., Acedo, G.N., Cheng, C.L. & Conkling, M.A. 1991. Characterization of *cis*-acting sequences regualting root-specific gene expression in tobacco. Plant Cell 3: 371–382.

24

QUANTITATIVE GENETIC ANALYSIS OF ADVENTITIOUS ROOT FORMING ABILITY IN *Populus trichocarpa* (TORR ET. GRAY)

D. E. Riemenschneider and E. O. Bauer

Forestry Sciences Laboratory
5985 Highway K
Rhinelander, Wisconsin 54501

INTRODUCTION

The ability of poplars (*Populus* sp.) and poplar hybrids to form adventitious roots is critical to the commercial deployment of intensively cultured poplar plantings, because the propagule of choice for such plantings is an unrooted dormant hardwood cutting. Given the importance of rooting ability, it is difficult to explain the shortage of quantitative genetic understanding of rooting compared to other commercially important traits. One reason for the aforementioned lack of knowledge is that genetic selection for rooting ability is mostly applied incidentally during serial propagation between stages of field testing for growth potential and disease resistance. Clones that root poorly are automatically eliminated from the testing process. However, such incidental selection provides little knowledge about selection efficiency, or possible correlated effects of selection for rooting ability, on other important traits. Likewise, incidental selection does not lead to the identification of more easily evaluated above-ground traits that could be used as indirect selection criteria for rooting ability.

Simultaneous selection for rooting ability and other important traits could be practiced in a more deterministic manner by use of various discriminant function selection indices. However, the use of such indices requires knowledge of phenotypic and genotypic covariances among traits. Estimates of phenotypic and genotypic covariances for rooting ability and other important traits, such as growth and disease resistance, could provide knowledge of the (1) indirect effects of selection for clone stem growth on the ability of clones to form roots and (2) strategies for improving rooting ability without the need to observe the trait directly. Indirect selection strategies would be particularly useful because the direct observation of root formation and development can be laborious. This research was conducted to: (1) determine broad-sense genetic and phenotypic covariances for rooting ability and shoot growth characteristics of hardwood cuttings from a population of 50 *Populus trichocarpa* clones and (2) determine broad-sense covariances between the devel-

Biology of Root Formation and Development, edited by Altman and Waisel.
Plenum Press, New York, 1997.

opment of roots and shoots, and previous measures of clone growth, phenology, and leaf and branch development made on the parent ortets.

EXPERIMENTAL

The experimental design in the original nursery test was two replications of a randomized complete block design with three-tree plots. Trees were grown for 5 years during which time various measurements of stem growth, branch and leaf morphology, and pest resistance were obtained. Trees were cut in January after the fifth growing season to a 10 cm stump and allowed to coppice. One-year-old coppice stems from a subsample of two trees per replication from each of 50 clones were harvested in March. The top and bottom-most 20 cm of each stem were discarded and the remaining section subdivided into 10 cm cuttings which were used for a greenhouse rooting experiment. Cuttings were soaked for 1 week in water at 21°C then planted in coarse sand in individual containers (folding book planters) in the greenhouse (22°C days, 16°C nights). The experimental design in the greenhouse was randomized, complete blocks with two replications of 12 cuttings from each nursery tree (ortet). Four developing cuttings from each replication and ortet were excavated at 1, 2 and 3 weeks after planting and dissected into the original cutting and the developing shoot, leaves and roots (8 cuttings per ortet per sample time). The dissected plants were photographed using a computerized imaging system and the images stored on high-resolution (High 8 format) video tape. Images were subsequently analyzed using an image analysis system (Optimus™) to determine cutting diameter; shoot length; leaf number and area; and root number, length, and area. Each component was also bagged separately, dried and dry weights were determined to the nearest mg. Data were analyzed using analyses of variance according to the all random effects model:

$$X_{ijkl} = m+R_i+C_j+RC_{ij}+T_{k(ij)}+e_{l(ijk)}$$

where X_{ijkl} is an experimental observation, m is the experiment mean, R_i is the effect if the ith replication, C_j is the effect if the jth clone, RC_{ij} is the clone x replication interaction effect, $T_{k(ij)}$ is the effect if the kth ortet within the jth clone and i^{th} replication, and $e_{l(ijk)}$ is experimental error (variation among cuttings within ortets). Variances and covariances attributable to each effect were estimated by equating expected mean squares to observed mean squares and solving the resultant set of simultaneous linear equations. Variance and covariance estimates were then used to compute heritabilities and genotypic correlations among rooting traits. Covariances and correlations among rooting traits and previously obtained measures of growth, morphology, and phenology of the parent ortets (Riemenschneider et al., 1993) were also estimated.

RESULTS AND DISCUSSION

Variance Analyses

Clones varied greatly in rooting ability. For example, mean clone root dry weight ranged from less than 1.0 to 26.9 mg and root number ranged from 2.1 to 11.4 by three weeks after planting. Such a large range in rooting ability suggested that further analysis of variability in rooting was warranted. Several results of the analyses of variance were

Table 1. Analyses of variance mean squares for shoot, leaf, and root dry weight (mg) in an experiment to test differences among 50 clones of *Populus trichocarpa* in their ability to develop new shoots and adventitious roots from dormant hardwood cuttings

		Week 1			Week 2			Week 3		
ANOVA	df	Shoot	Leaves	Roots	Shoot	Leaves	Roots	Shoot	Leaves	Roots
Replication	1	12680.3**	6560.8**	63.90*	1081.1 ns	264.5 ns	403.0**	0.91 ns	3086.2 ns	207.1 ns
Clone	49	3009.2**	3285.2**	54.76**	785.2**	6797.2**	129.5**	1971.4**	12407.8**	699.4**
Rep * clone	49	889.1 ns	765.8 ns	12.18 ns	270.4 ns	2455.7 ns	41.35 ns	780.9 ns	5137.2 ns	124.4 ns
Ortet	100	700.8**	751.5**	13.02**	240.6**	2407.7**	30.9**	765.7**	4317.7**	120.7**
Cutting	600	255.7	216.7	9.01	134.2	931.0	17.45	407.1	1554.3	57.2
Variance components										
Replication		29.	14.	0.13	2.0	0.	0.9	0.0	0.	0.2
Clone		132.	157.	2.66	32.2	271.	5.5	74.4	654.	35.9
Rep * clone		23.	2.	0.00	3.7	6.	1.3	1.9	102.	0.5
Ortet		111.	134.	1.00	26.6	369.	3.4	89.5	691.	15.9
Cutting		256.	217.	9.01	134.2	931.	17.5	407.1	1554.	57.2
Statistics										
Heritability		0.25	0.31	0.21	0.16	0.17	0.20	0.13	0.22	0.33
Mean		37.4	16.6	3.18	27.2	86.1	5.49	45.5	123.5	10.4
C.V. (%)		43.2	86.6	100.1	42.9	35.5	69.7	43.9	31.8	68.8

ns = not significant, * = significant at $0.01 < p < 0.05$, ** = significant at $p < 0.01$.

notable (Table 1). First, greenhouse replication effects on shoot and leaf weight were significant only at week one while replication effects on root weight were significant only at weeks one and two. Apparently, environmental gradients in the greenhouse were large enough to affect flushing and early growth of new shoots and roots, but not large enough to affect subsequent development. Likewise, replication x clone effects were mostly non-significant. Overall, results indicated that complete block (or more complex) designs may not be needed to achieve experimental efficiency in our greenhouse rooting experiments.

Second, clone effects were always significant regardless of trait or sampling time. Heritability for shoot, leaf and root dry weight tended to be low, especially for dry weights measured 3 weeks after planting. Root dry weight measurements were characterized by very high coefficients of variation (100.1%, 69.7% and 68.8% at 1, 2 and 3 weeks after planting, respectively) (Table 1). Root length was more highly heritable than root number (Table 2) or component dry weights (Table 1) at all measurement times (0.25, 0.23, and 0.28). And, coefficients of variation for root length and number were lower than for root weight (Tables 2 and 1, respectively). It has been our general experience in the conduct of rooting experiments that root weights are usually associated with high experimental errors. High experimental errors are commonly attributed to inevitable loss of roots during excavation. But, the root systems observed in the current study were relatively small and unbranched. Thus, we are confident that root systems were excavated without significant loss of tissue. Therefore, we conclude that high errors were an inherent attribute of root systems in this study and not attributable to incautious handling. Previous quantitative studies of root number, root length and root weight produced by *P. deltoides* cuttings demonstrated significant broad sense heritability ($0.33 < H^2 < 0.58$) for all characteristics (Wilcox and Farmer, 1977). Subsequent research resulted in higher broad sense heritability estimates for root number ($0.85 < H^2 < 0.91$) in the same

Table 2. Analyses of variance of cutting, shoot, leaf, and root linear dimensions in an experiment to test differences among 50 clones of *Populus trichocarpa* in their ability to develop new shoots and adventitious roots from dormant hardwood cuttings

		Week 1		Week 2		Week 3	
ANOVA	df	Root length (cm)	No. roots	Root length (cm)	No. roots	Root length (cm)	No. roots
Replication	1	964.02**	11.76 ns	4234.05**	117.82**	6044.15*	190.13*
Clone	49	456.77**	19.03**	1157.2**	32.3**	6744.29**	89.65**
Rep * clone	49	121.48**	7.59*	389.1*	14.17*	1662.58 ns	35.15 ns
Ortet	100	68.41*	4.51*	228.12**	8.21**	1598.95**	26.00 ns
Cutting	600	52.52	3.41	106.92	4.49	543.52	10.96
Variance components							
Replication		2.10	0.01	9.61	0.25	10.95	0.39
Clone		20.96	0.72	48.00	1.15	317.61	3.41
Rep * clone		6.63	0.38	20.12	0.74	7.95	1.14
Ortet		3.97	0.37	30.30	0.92	263.86	3.76
Cutting		52.51	3.41	106.92	4.49	543.52	10.96
Statistics							
Heritability		0.25	0.15	0.23	0.16	0.28	0.18
Mean		10.61	3.38	18.09	3.99	42.28	5.68
C.V. (%)		68.3	54.6	57.1	53.1	55.1	58.2

ns = not significant, * = significant at $0.01 < p < 0.05$, ** = significant at $p < 0.01$.

species (Ying and Bagley, 1977). Discrepancies among past tests and the current study could be attributed to a number of factors including: 1) ancestry and diversity of experimental populations, 2) method of accounting for C effects, if any, and 3) computational formulations.

Variance attributable to ortet within clone (i.e. "C"-effects) was consistently significant (Tables 1, 2). The component of variation due to ortet within clone in the current experiment was high in relation to variation due to clones (Table 1) and sometimes exceeded that due to clones (i.e.. shoot and leaf dry weights at 3 weeks after planting). Separation of genetic and environmental variation from C effects can be especially important because variation due to C effects can equal or exceed variation due to genetic effects (Wilcox and Farmer, 1968; Farmer et al., 1989; current study). Variation due to C effects can be confounded with genetic or environmental variation, depending on the experimental design (Wilcox and Farmer, 1968). The consistently large ortet effects in the current experiment support two conclusions. First, our genetics trial experiments should incorporate a strategy to estimate or eliminate ortet effects because, if ignored, those effects can be confounded with variance due to clone or error (depending on experimental design) which can lead to faulty predictions of selection efficiency. Second, significant ortet effects suggest that some experimental attention should be given to treatments, apart from genetic selection, that might enhance the expression of desired characteristics.

Correlations among Traits in Rooting Study

Shoot and leaf weights were strongly correlated both within and between sample times (Table 3). Correlations among shoot and leaf weights and root weight were not significant at week one but were significant thereafter. Overall, correlation analyses suggested that shoots and leaves were more tightly linked allometrically than were shoots and roots during very early development, but that an allometric relation was established among all developing plant parts by the third week after planting. The third week after planting was also associated with a decline in extraordinarily high shoot-root ratios of 18.0, 18.8, and 15.4 for weeks 1, 2 and 3, respectively. Various measures of rooting were

Table 3. Clone mean correlations (above diagonal) and genotypic correlations (below diagonal) among cutting, shoot, leaf, and root dry weights (g) at 1, 2 and 3 weeks after planting

	Week 1			Week 2			Week 3		
	Shoot	Leaves	Roots	Shoot	Leaves	Roots	Shoot	Leaves	Roots
Week 1									
Shoot		−0.52**	−0.03	**0.64****	0.15	−0.09	**0.34****	0.14	−0.22
Leaves	−0.65		0.11	−0.03	**0.46****	**0.31***	0.26	**0.30***	**0.54****
Roots	0.02	0.05		0.06	0.23	**0.56****	0.08	**0.42****	**0.51****
Week 2									
Shoot	0.79	−0.16	0.33		**0.47****	0.17	**0.70****	**0.42****	0.07
Leaves	0.03	0.53	0.06	0.48		**0.56****	**0.65****	**0.77****	**0.64****
Roots	−0.12	0.37	0.80	0.10	0.67		**0.38****	**0.61****	**0.81****
Week 3									
Shoot	0.31	0.33	0.27	0.46	0.79	0.56		**0.66****	**0.35***
Leaves	−0.08	0.44	0.68	0.94	1.00	0.91	0.66		**0.64****
Roots	−0.29	0.61	0.64	−0.03	0.74	0.86	0.45	0.82	

* = significant at $0.01 < p < 0.05$, ** = significant at $p < 0.01$.

Table 4. Clone mean (above diagonal) and genotypic (below diagonal) correlations among rooting characteristics at one week after planting in a population of 50 *Populus trichocharpa* clones

	Root dry weight (mg)	Root length (cm)	Number of roots
Root dry weight (mg)		0.81**	0.72**
Root length (cm)	0.86		0.89**
Number of roots	0.88	0.95	

* = significant at 0.01 < p < 0.05, ** = significant at p < 0.01.

highly correlated within sampling times (Tables 4, 5) and genotypic correlations tended to be higher than clone mean correlations (Tables 4, 5).

Cutting dry weights (data not shown) were correlated among sample times, as would be expected when sampling at random from a genetically variable population. But, cutting dry weights were uncorrelated with the dry weight of the developing shoots, leaves, or roots. Likewise, incorporating cutting dry weight as a covariate during analyses of variance of shoot, leaf, or root dry weight had little outcome on the results of the analyses, including the proportion of variation due to ortets within clones. Thus, an understanding of the nature of C effects on shoot and root development will probably not be gained by simple measurement of cutting mass.

Correlations among Rooting and Above-Ground Traits

Estimates of genetic and environmental covariances between rooting and above-ground plant characteristics are mostly lacking. However, past research has suggested that a genetic correlation exists between measures of rooting and date of bud break (Wilcox and Farmer, 1968; Farmer et al., 1989) suggesting that a potential indirect selection criterion might exist. Thus, we tested whether shoot, leaf and root dry weights measured in the current experiment were correlated with various measures of growth and morphology made on the same clones during a previous clonal nursery test (Riemenschneider et al., 1993). Cutting dry weight was significantly correlated with several measures of leaf size made on 2 year old clones. Correlations were strongest with leaf area on LPI 10 and 15 leaves when leaf measurements were made near the end of the growing season on September 1. Cutting dry weights were also correlated with clone height at age 1 year. Thus, several indirect selection criteria probably exist that would be useful in selecting for increased cutting weight. But, our observations of no correlation between cutting weight and the growth of roots and shoots from developing cuttings suggests that increased cutting weight would not be an important breeding objective.

Table 5. Clone mean (above diagonal) and genotypic (below diagonal) correlations among rooting characteristics at three weeks after planting in a population of 50 *Populus trichocharpa* clones

	Root dry weight (mg)	Root length (cm)	Number of roots
Root dry weight (mg)		0.87**	0.69**
Root length (cm)	0.89		0.90**
Number of roots	0.75	0.93	

* = significant at 0.01 < p < 0.05, ** = significant at p < 0.01.

Table 6. Clone mean and genotypic correlations between shoot, leaf and root dry weights (mg) in the current rooting experiment and several measures of growth and phenology in a previous nursery test of the same clones

	Shoot	Leaves	Roots	Shoot	Leaves	Roots	Shoot	Leaves	Roots
Height at 1 year									
Clone mean	−0.33	−0.17	−0.01	−0.05	−0.02	0.20	0.07	0.08	0.16
Genotypic	−0.27	−0.06	0.06	−0.08	0.07	0.17	0.45	0.20	0.21
Height at 2 years									
Clone mean	−0.14	−0.09	−0.09	−0.02	0.07	0.14	−0.01	0.05	0.14
Genotypic	−0.38	0.26	0.14	0.26	0.23	0.35	0.24	0.35	0.42
No. leaves July 1									
Clone mean	−0.09	0.14	**−0.29***	0.11	0.21	0.13	0.20	0.15	0.15
Genotypic	−0.59	0.88	−0.82	0.48	1.01	0.35	0.69	0.76	0.35
Height at 5 years									
Clone mean	0.14	−0.02	−0.06	**0.36****	0.16	0.03	0.24	0.12	0.12
Genotypic	0.13	0.15	0.12	0.70	0.35	0.12	0.48	0.23	0.20
Diameter at 5 years									
Clone mean	0.18	−0.01	−0.08	**0.31****	0.13	0.02	0.22	0.14	0.11
Genotypic	0.12	0.26	0.34	0.60	0.72	0.50	0.52	0.44	0.47
Bud-set date									
Clone mean	0.25	**−0.31***	0.10	**−0.45****	**−0.40****	−0.05	**−0.48****	−0.22	−0.15
Genotypic	−0.29	−0.30	−0.07	−0.62	−0.46	−0.05	−0.68	−0.35	−0.15
Bud-break date									
Clone mean	0.17	**−0.47****	0.01	−0.10	**−0.35***	−0.10	**−0.34***	−0.26	**−0.49****
Genotypic	0.07	−0.57	−0.09	−0.20	−0.47	−0.41	−0.48	−0.41	−0.54
LPI 5 leaf area									
Clone mean	0.01	0.03	−0.07	0.13	0.00	−0.11	0.01	−0.10	−0.06
Genotypic	0.01	0.19	−0.31	0:28	0.11	−0.28	0.04	0.01	−0.07
LPI 10 leaf area									
Clone mean	−0.03	0.08	−0.21	0.10	0.06	−0.15	0.03	0.13	−0.11
Genotypic	−0.15	0.39	−0.32	0.45	0.35	−0.16	0.26	0.39	0.07
LPI 15 leaf area									
Clone mean	0.18	0.02	0.00	0.12	0.19	0.05	0.06	**0.29***	0.11
Genotypic	0.84	0.21	0.52	0.82	0.93	0.72	0.96	1.55	0.49

* = significant at $0.01 < p < 0.05$, ** = significant at $p < 0.01$.

Shoot and leaf weights in the current test were often correlated with age 2 year bud-set date, with date of budbreak at the start of the third growing season, and with the area of a recently matured leaf (Table 6). Root dry weight was negatively correlated with bud-break date (Table 6), a result that has been observed in other *Populus* populations (Wilcox and Farmer, 1968; Farmer et al., 1989). We also found that root length and number of roots three weeks after planting were negatively correlated with date of budbreak (Table 7). Some results suggest that some increase in rooting ability might be achieved by selecting for clones that flush rapidly at the beginning of the growing season.

Our estimates of genetic correlations based on components of variance and covariance due to clones suggests that genetic correlations may be stronger than clone mean correlations when comparisons are made between the rooting experiment and the original nursery test from which cuttings were obtained (Table 6). For example, genetic correlations between number of leaves formed early in the growing season (July 1) in a nursery clonal test and the dry weight of leaves and roots produced by the developing cuttings obtained from the same clones several years later were greater, although of the same sign, than corresponding clone mean correlations. The largest differences between clone mean

Table 7. Clone mean and genotypic correlations among root length (cm) and root number in the current experiment and several measures of growth and phenology in a previous nursery test of the same clones

	Root length	No. roots	Root length	No. roots	Root length	No. roots
Height at 1 year						
Clone mean	−0.02	0.14	0.12	0.25	0.08	0.06
Genotypic	−0.20	0.14	0.04	0.19	0.05	−0.02
Height at 2 years						
Clone mean	−0.05	0.06	0.09	0.25	0.13	0.15
Genotypic	−0.04	−0.20	0.25	0.55	0.35	0.36
No. leaves July 1						
Clone mean	−0.15	−0.02	0.03	0.15	0.13	0.09
Genotypic	−0.88	−0.69	0.06	0.19	0.40	0.08
Height at 5 years						
Clone mean	−0.08	−0.06	−0.06	−0.11	0.08	0.08
Genotypic	−0.13	0.13	−0.04	0.16	0.10	0.02
Diameter at 5 years						
Clone mean	0.02	0.11	−0.02	0.08	0.04	0.07
Genotypic	0.03	0.35	0.21	0.43	0.25	0.36
Bud-set date						
Clone mean	0.22	0.20	0.08	0.10	−0.11	0.01
Genotypic	0.29	0.26	0.11	0.18	−0.08	0.11
Bud-break date						
Clone mean	0.06	−0.06	−0.08	−0.20	**−0.41****	**−0.32***
Genotypic	0.08	−0.10	−0.14	−0.26	−0.47	−0.42
LPI 5 leaf area						
Clone mean	−0.10	−0.13	−0.08	−0.04	−0.11	−0.13
Genotypic	−0.40	−0.63	−0.36	−0.47	−0.21	−0.41
LPI 10 leaf area						
Clone mean	−0.21	−0.16	−0.17	−0.16	−0.06	−0.12
Genotypic	−0.56	−0.53	−0.35	−0.29	−0.05	−0.32
LPI 15 leaf area						
Clone mean	0.02	−0.12	0.00	−0.03	0.05	−0.04
Genotypic	0.25	−0.02	0.38	0.07	0.25	−0.33

* = significant at $0.01 < p < 0.05$, ** = significant at $p < 0.01$.

and genotypic correlations observed in our tests was between mature leaf area and shoot, leaf and root dry weights (Table 6). These results suggest that clone mean correlations may be inadequate for the purpose of predicting indirect response to selection.

CONCLUSIONS

We come to several conclusions based on the results of the current study of adventitious rooting ability in *Populus trichocarpa*. First, clones differed significantly in all rooting traits regardless of sampling time. Thus, rooting was always a heritable trait in the broad sense and subject to improvement through selection. Second, "C"-effects were large, suggesting that differences among ortets within clones were important and that stock plant treatment should be considered in any applied program to improve rooting ability. Third, rooting was genetically correlated among sampling times, suggesting that experiments to determine adventitious root forming ability can be conducted quickly. Fourth, measures of rooting were correlated which indicates that a reduction in the number of variables measured would not result in the loss of much biological information. Fifth, roots and shoots become allometrically balanced very early in development. Sixth, rooting is genetically correlated with an above-ground trait, i.e. date of budbreak, suggesting that indirect selection for rooting ability based on an above-ground trait would be feasible. Last, early leaf development from cuttings is correlated with some above-ground traits, i.e. date of budbreak, date of budset and mature leaf area, suggesting that selection for the above-ground development of newly planted cuttings could be conducted indirectly based on observations of ortet development.

Overall, current results suggest that genetic selection for rooting ability could increase rooting at reasonable selection intensities, and that additional increases could be achieved through a more complete knowledge of the effects of ortet preconditioning.

ACKNOWLEDGMENTS

This research was supported by the U.S. Department of Energy, Biofuels Feedstock Development Program, Oak Ridge National Laboratory, under Interagency Agreement #DE-A105-940R22197.

REFERENCES

Farmer, R.E., Jr., Freitag,, M. and Garlick, K., 1989, Genetic variance and "C" effects in balsam poplar rooting, *Silv. Genet.* 38:62.

Riemenschneider, D.E., McMahon, B.E., and Ostry, M.E., 1993, Population-dependent selection strategies needed for 2-year-old black cottonwood clones, *Can J. For. Res.* 24: 1704.

Wilcox, J.R., and Farmer, R.E., Jr., 1967, Variation and inheritance of juvenile characters of eastern cottonwood, *Silv. Genet.* 16: 162.

Wilcox, J.R., and Farmer, R.E., Jr., 1968, Heritability and C effects in early root growth of eastern cottonwood cuttings, *Heredity* 23: 239.

Ying, C.C., and Bagley, W.T., 1977, Variation in rooting capability of *Populus deltoides*, *Silv. Genet.* 26:204.

25

DIFFERENTIAL GENE EXPRESSION DURING MATURATION-CAUSED DECLINE IN ADVENTITIOUS ROOTING ABILITY IN LOBLOLLY PINE (*Pinus taeda* L.)

Michael S. Greenwood,[1] Carmen Diaz-Sala,[3] Patricia B. Singer,[2] Antoinette Decker,[2] and Keith W. Hutchison[2]

[1]Department of Forest Ecosystem Science
University of Maine
Orono, Maine 04469-5755
[2]Department of Biochemistry
Microbiology and Molecular Biology
University of Maine
Orono, Maine 04469-5755
[3]Departmento de Biologica Vegetal
Universidad de Alcala de Henares
28871 Alcala de Henares
Madrid, Spain

1. INTRODUCTION

Loss of adventitious rooting ability occurs very quickly in loblolly pine seedlings, where hypocotyl cuttings from 50-day-old seedlings root readily in response to auxin, while epicotyl cuttings from the same seedling do not, despite their anatomical similarity (Diaz-Sala et al., 1996a). In hypocotyl cuttings, auxin causes root meristems to organize in the cambial region centrifugal to the former primary xylem poles over a 12-day period. In epicotyl cuttings, the cambium dedifferentiates in response to auxin, but roots rarely form, and then only after 50 or more days. A brief exposure (a 5 min pulse) to auxin is sufficient to promote rooting. While both types of cutting transport auxin in a polar manner, a pulse of N-(1-napthl)phthalamic acid (NPA), which inhibits auxin efflux, also delays rooting in hypocotyls only if applied within the first 3 days after the auxin pulse; thereafter it has no effect. These observations show that key events that determine root meristem formation occur in the first 72 hours, and that auxin binding to the efflux carrier that mediates polar auxin transport, may be necessary for rooting (Diaz-Sala et al., 1996a).

Based on these observations, our approach to understanding the loss of rooting competence by epicotyls is to compare gene expression with and without auxin in both types

Biology of Root Formation and Development, edited by Altman and Waisel.
Plenum Press, New York, 1997.

of cutting during the first few days of the rooting process. We have already reported that the expression of genes for PAL (phenylalanine ammonia lyase) and actin are different in hypocotyls and epicotyls, and that differential display detected about 35 PCR-produced sequences whose response varied with cutting type and auxin treatment (Diaz-Sala et al., 1996b). In this report we further describe differences in the expression of actin, a cyclin dependant kinase (CDK), and expansin (the latter was identified among the products of the differential display). Expansin is a newly characterized gene which causes cell wall loosening non-hydrolytically (Shcherban et al., 1995).

2. EXPERIMENTAL

Loblolly pine seedlings were grown in vermiculite as described by Greenwood and Weir (1995). RNA was extracted from the bases of cuttings at times ranging from 0 to 120 hours as described by Hutchison et al. (1990). Homologous probes were prepared for actin, PAL and CDK genes, and expression was followed using Northern blots. Expression was normalized relative to ribosomal RNA production, assuming that production of the latter is more constant than other RNAs. These techniques will be described in detail in Diaz-Sala et al. (1996c) and Decker (1996). Differential display screening of gene expression was carried out according to Liang and Pardee (1992) and is briefly described in Diaz-Sala et al., 1996b). About 30 of the differential display products were subjected to a BLASTX search to determine similarity to sequences of known genes. One gene, expansin, was fully characterized by Southern blot analysis, and full sequences of several family members were prepared by making both 3′ and 5′ RACE clones (Hutchison et al., 1996). Probes were created for Northern blot analysis of gene expression.

3. RESULTS

Expression of 4 genes (after 24 hours in both hypocotyl and epicotyl cuttings, with and without auxin) is shown in Table 1.

PAL expression was not affected by the auxin indole-3-butyric acid (IBA) at any time during the first 48h of auxin treatment. PAL expression was increased about 8-fold over all time points in hypocotyls following vertical incisions to the cutting base (results not shown). CDK expression may be slightly increased by IBA but further verification is required. Actin expression increases after epicotyl cuttings are made and decreases thereafter, while it increases over time in hypocotyls (Figure 1).

Table 1. Expression of several genes in the bases of stem cuttings from hypocotyl and epicotyls with or without IBA treatment, 24 hours after the cutting was made[a]

Hypocotyls	PAL	Actin	CDK	Expansin
10μM IBA	0.4	9.0	0.3	3.1
0 IBA	0.5	2.3	0.2	0
Epicotyls				
10μM IBA	0.5	2.0	1.7	0.9
0 IBA	0.6	1.9	0.4	0

[a]Homologous probes were used for each gene, and expression was normalized to ribosomal RNA production.

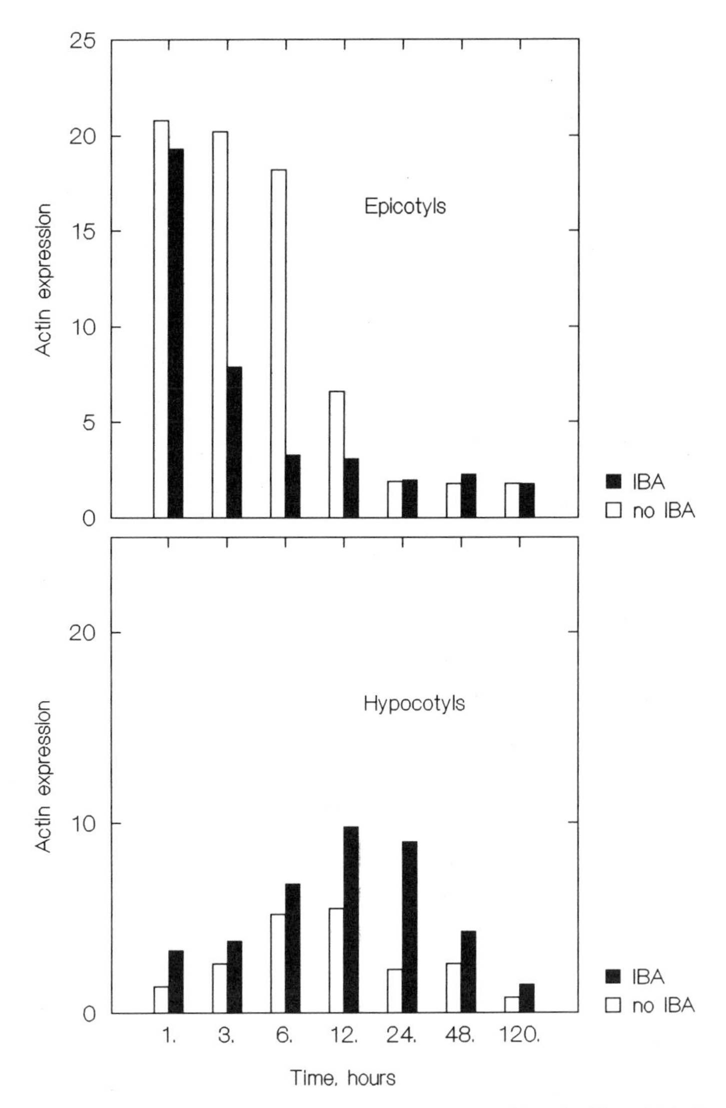

Figure 1. Time course of actin expression in epicotyls and hypocotyls, with and without IBA. Expression is normalized to ribosomal RNA production.

IBA appears to increase actin expression in hypocotyls, but is slightly inhibitory during the first 12h in epicotyls. NPA decreased actin expression but had no effect on PAL or CDK expression.

About 35 PCR products from the differential display were sequenced and subjected to a BLASTX search. One clone showed a high and significant level of similarity to the previously reported expansins of *Arabidopsis*, cucumber and rice. Expansin is of interest because its gene product clearly may play a key role in the early stages of root meristem formation (Hutchison et al., 1996). Southern blots reveal that in loblolly pine there are 8–10 members of the gene family. Clones obtained from 5′ RACE indicate at least 2 members of the family are expressed during the early stages of rooting. In addition, the gene family is only expressed in response to the auxin IBA, and appears to be more strongly expressed in hypocotyls (see Table 1). NPA slightly decreased expansin expression.

4. DISCUSSION

Because both actin and expansin expression is affected by auxin, the role of these genes in loss of rooting competence will be discussed further. Both epicotyls and hypocotyls show the same initial responses to auxin, so that signal transduction pathways are present in cuttings from both tissues. We hypothesize that lack of rooting competence in epicotyls is the result of a defect in the signal transduction pathway that is involved in the later stages of root meristem organization.

Although the importance of auxin binding prior to auxin response is well documented, there are few if any characterized auxin binding proteins that could participate in the early steps of such a pathway (Jones, 1994; Hertel, 1995). One such protein, called ABP1, is a 22 kDa protein from corn which may be a functional auxin receptor (Venis et al., 1996). Using antibodies to this protein (provided by Alan Jones, University of North Carolina, Chapel Hill, NC, USA), western blots of protein extracts from both epicotyls and hypocotyls revealed no trace of a band at 22 kDa in either tissue, while similar extracts of 4-day-old corn seedlings yielded a band at exactly 22 kDa (Greenwood, Morelli and Vayda, 1996). Hertel (1995) has recently questioned the evidence that ABP1 is a functional receptor, and proposes that the auxin efflux carrier which is involved in polar auxin transport may be the functional auxin receptor. The fact that NPA binds to the efflux carrier and delays rooting would support the possibility that auxin binding to the efflux carrier is required for rooting. Although the function of an efflux carrier in polar auxin transport has been thoroughly demonstrated, it has not been isolated and characterized, despite numerous attempts.

Cox and Muday (1994) have recently proposed that the efflux carrier may be a complex of proteins, consisting of auxin and NPA binding sites bound together by actin, which would explain why characterizing the efflux carrier may be so difficult. In epicotyls, auxin seems to slightly inhibit actin expression, while promoting it in hypocotyls. Thus it is tempting to speculate that the failure of auxin to upregulate actin expression in epicotyls may result in qualitative or quantitative differences in actin synthesis which will affect the efflux carrier or some other aspect of the rooting process. The observation that auxin transport is strongly polar in both epicotyls and hypocotyls (Diaz-Sala et al., 1996a) suggests there are no major differences in the concentration of efflux carrier in hypocotyls and epicotyls. Nonetheless, the remote possibility that the increased actin synthesis in hypocotyls may result in synthesis of efflux carrier which is qualitatively different or located at a unique cellular site that is specific to rooting can be considered, but would be very difficult to test experimentally. A more likely scenario is that failure to root results from alteration of the signal transduction pathway downstream from the efflux carrier (or an as yet uncharacterized auxin binding protein). This alteration will prevent actin synthesis that is needed for morphogenetic processes involved in root meristem organization, such as the orientation of cellulose microfibrils which will determine cell shape. Shibaoka (1994) has proposed that actin may stabilize the association between microtubules and the plasma membrane; microtubules in turn affect cellulose microfibril orientation in the primary wall.

Given the absence of well characterized auxin binding proteins, combined with the likelihood that failure to root does not involve an initial auxin receptor, the search for the cellular basis for loss in rooting competence cannot at present begin at the start of the auxin signal transduction pathway. Instead, emphasis on understanding the basis for the regulation of differential gene expression in hypocotyls and epicotyls during rooting will be a potentially more productive approach. Ideally, studies on genes that are induced by auxin only in tissues that are competent to root, and which produce a product with an understandable role in the rooting process would be the most informative. Expansin meets some of these criteria since as an agent which causes loosening of primary cell walls, it may have an obvious role in the cellular dedifferentiation which precedes rooting. Expansin is auxin inducible in both types of cuttings, with expression detectable after 6 hours and with a peak at 24 hours in hypocotyls. It is more strongly expressed over a longer period of time in hypocotyls than in epicotyls. NPA appears to slightly decrease the relative level of expression of expansin as was also observed with actin. Further studies are needed to see if this decrease is significant. Furthermore, since there seem to be 8–10 expansin genes in loblolly pine, the possibility that one of the members of the family is expressed in a strictly rooting specific manner can also be tested. We have shown that there are differences in expression of all the genes we have studied between epicotyls and hypocotyls, and that some of these differences are rooting specific to some degree. This leads to the unsurprising conclusion that the genetic regulation of rooting competence is complex, which renders the search for a single controlling factor, if one exists, very difficult. Goldfarb et al. (1996) have shown that some early auxin inducible genes which have been previously characterized in pea, bean and *Arabidopsis* are present in loblolly pine, and at least 2 are expressed in hypocotyl cuttings after auxin treatment. Expression can be detected within 1 hour of exposure to auxin for at least one of the genes, which raises questions about the relationship between auxin-induced expansin expression and the products of genes which may be auxin induced prior to expansin. Ballas et al. (1993) argue that since cycloheximide causes early auxin gene expression, that auxin inactivates a protein repressor which binds to an auxin-responsive element of early auxin genes. We can hypothesize that the early auxin genes are the first to respond to auxin, and produce a transcription factor needed for subsequent expression of expansin genes.

REFERENCES

Ballas, N., Wong, L.M. and Theologis, A., 1993, Identification of the auxin-responsive element, *AuxRE*, in the primary indoleacetic acid-inducible gene, *PS-IAA4/5*, of pea (*Pisum sativum*). *J. Mol. Biol. 233*: 580–596.

Cox, D.N., and Muday, G.K., 1994, NPA binding activity is peripheral to the plasma membrane and is associated with the cytoskeleton. *The Plant Cell* 6: 1941–1953.

Decker, A. 1996. A cyclin dependant kinase-like sequence from *Pinus taeda* L. and its expression during adventitious root formation. Ph.D. dissertation, Department of Microbiology, Biochemistry and Molecular Biology, University of Maine, Orono, ME.

Diaz-Sala, C., Greenwood, M.S. and Hutchison, K.W., 1996c, Actin and PAL gene expression associated with maturation-related loss in adventitious rooting ability by loblolly pine (*Pinus taeda* L.) stem cuttings: effect of auxin, wounding and NPA treatments. Unpublished manuscript.

Diaz-Sala, C., Hutchison, K.W., Goldfarb, B. and Greenwood, M.S., 1996a, Maturation-related loss in rooting competence by loblolly pine stem cuttings: The role of auxin transport, metabolism and tissue sensitivity. *Physiol. Plant.* 97: 481–490.

Diaz-Sala, C., Singer, P.B., Hutchison, K.W. and Greenwood, M.S., 1996b, Molecular approaches to maturation-caused decline in adventitious rooting ability in loblolly pin (*Pinus taeda* L.). In M.R. Ahuja et al. (eds.), Somatic Cell Genetics and Molecular Genetics of Trees, pp. 57–61. Kluwer Academic Publishers, Dordrecht.

Goldfarb, B., Lian, Z., Lanz-Garcia, C. and Whetten, R., 1997, Auxin-induced gene expression during rooting of loblolly pine stem cuttings. In A. Altman and Y. Waisel (eds): Biology of Root Formation and Development. New York: Plenum.

Greenwood, M.S., Morelli, J. and Vayda, M., 1996, Unpublished data. University of Maine, Orono, Me, USA.

Greenwood, M.S. and Weir, R.J., 1995, Genetic variation in rooting ability of loblolly pine cuttings: effects of auxin and family on rooting by hypocotyl cuttings. Tree Physiology *15*: 41–45.

Hertel, R., 1995, Auxin binding protein 1 is a red herring. *J. Exptl. Bot. 46*: 461–462.

Hutchison, K.W., Diaz-Sala, C., Singer, P.B. and Greenwood, M.S., 1996, Expansins are conserved in conifers and expressed in response to exogenous auxin. Unpublished manuscript.

Hutchison, K.W., Sherman, C.B., Weber, J., Smith, S.S, Singer, P.B. and Greenwood, M.S., 1990, Maturation in Larch II. Effects of age on photosynthesis and gene expression in developing foliage. *Plant Physiol. 94*: 1308–1315.

Jones, A.M. 1994. Auxin-binding proteins. *Ann.Rev.Plant Physiol. Plant Mol. Biol. 45*: 393–420.

Liang, P., and Pardee, A.B., 1992, Differential display of eukaryotic messenger RNA by means of the polymerase chain reaction. *Science 257*: 967–971.

Shcherban, T.Y., Shi, J., Durachko, D.M., Guiltinan, M.J., McQueen-Mason, S.J., Shieh, M. and Cosgrove, D.J., 1995, Molecular cloning and sequence analysis of expansins-a highly conserved, multigene family of proteins that mediate cell wall extension in plants. *Proc. Natl. Acad. Sci. 92*: 9245–9249.

Shibaoka, H., 1994, Plant hormone-induced changes in the orientation of cortical microtubules: alternation in the cross-linking between microtubules and the plasma membrane. *Ann. Rev. Plant Physiol. Plant Mol. Biol. 45*: 527–544.

Venis, M.A., Napier, R.M., and Oliver, S., 1996, Molecular analysis of auxin-specific signal transduction. *Plant Growth Regulation 18*: 1–6.

26

DIRECT REGENERATION AND SELECTION OF *Populus tremula* L. TRANSGENIC SHOOTS FROM *Agrobacterium tumefaciens*-TRANSFORMED STEM EXPLANTS

Christian Sig Jensen,[1] Tzvi Tzfira,[2] Alexander Vainstein,[2] and Arie Altman[2]

[1]Department of Agricultural Science
The Royal Veterinary and Agricultural University
2000 Frederiksberg, Denmark
[2]The Kennedy-Leigh Centre for Horicultural Research and The Otto Warburg Center for Biotechnology in Agriculture
The Hebrew University of Jerusalem
Rehovot 76100, Israel

INTRODUCTION

Transformation of *Populus* has been reported previously for different species. However, most of the transformation procedures using disarmed *Agrobacterium* strains relied on regeneration of transformed shoots from callus (Tsai et al., 1994; Klopfenstein et al., 1991, Confalonieri et al., 1995). This has some disadvantages, since callus formation may lead to increased rate of somaclonal variation and regeneration of chimeric plants (Lee & Phillips, 1988). Here, we describe an efficient procedure for regeneration of transformed adventitious shoots directly from stem explants, following inoculation with a disarmed *A. tumefaciens*, strain EHA105. Three different selection schemes were assesed for improving transformation efficiency.

EXPERIMENTAL

The leaves were removed from *in vitro* grown *P. tremula* plants, and 2 types of explants were prepared from the stem: (1) Whole 0.5–1.0 cm long segments or (2) segments were further excised longitudinally into two halves. The explants were soaked in a liquid Luria-Bertinin (LB) dilution (OD550nm = 0.1–0.25) of an overnight culture of *A. tumefaciens*, strain EHA105 for 30 min. After inoculation, explants were placed on a Woody Plant Medium (WPM) (Llyod and McCown, 1980) supplemented with TDZ (N-phenyl-

Biology of Root Formation and Development, edited by Altman and Waisel.
Plenum Press, New York, 1997.

N′-1,2,3-thiadiazol-5-yl-urea, thiadiazuron) for co-cultivation; the whole segments for 48 hours, and the half segments for 72 hours. Following co-cultivation, they were washed 3 times in sterilised water, blotted dry on filter paper, and transferred to a WPM supplemented with 0.04 mg/l TDZ and 300 mg/l carbenicillin. Three different selection schemes were evaluated: (1) Explants were subjected to kanamycin selection of bud development (50 mg/ml) immediately, or (2) after 10–14 days, or (3) after regeneration of plantlets from segments (100 mg/ml kanamycin). Adventitious shoots regenerated directly from the cut-surface after 2–3 weeks. Leaf samples from green regenerated shoots were tested for GUS-expression. Kanamycin resistance of roots from transgenic plants as judged by bud regeneration, was tested in a liquid medium. PCR and Southern blot analysis of kanamycin resistant plants were performed as described by Tzfira et al. (1996).

RESULTS

Application of kanamycin immediately after co-cultivation yielded only green resistant shoots while kanamycin selection after 10–14 days resulted in development of both white non-resistant shoots, and green resistant shoots. When kanamycin was applied to the plantlets, non-resistant plantlets formed pale-green leaves and failed to root, whereas resistant plantlets formed dark-green leaves and well-developed roots. The number of both white and green and GUS expressing shoots per explant was counted for the different selection types (Table 1). The differences between white and green shoots are discussed elsewhere.

A co-cultivation period of 72 h was found to be most effective. Resistant, GUS-expressing plants were assumed to be transgenic, as confirmed by PCR and Southern blot analysis (data not shown) for several randomly chosen plants. Transgenic roots regenerated shoots in liquid culture in the presence of up to 50 mg/l kanamycin, while shoot formation from control roots was inhibited by 10 mg/l kanamycin (data not shown), thus confirming the kanamycin resistance in transgenic plants.

DISCUSSION

A. tumefaciens-mediated transformation of *P. tremula*, as described above, proved to be very easy and efficient. Direct regeneration of shoots from the explants prevented formation of chimeras, which is very desirable. We found that explants selection immediately after infection was most efficient, both when considering the time required for selection, and the transformation efficiency (up to 25 transgenic plant per one cultured mother plant). Molecular analysis confirmed integration of bacterial T-DNA into host genome. Transformed plants were propagated in vitro, and no phenotypic changes were observed.

Table 1. Effect of three selection schemes on transformation efficiency of *P. tremula*. Both white and green shoots were recorded

Time of selection	Average number of shoots/explant	Average number of GUS expressing shoots/explant
Immediately	2.9	1.8
After 10–14 days	5.2	0.6
Plantlet stage	25.4	0.05

REFERENCES

Confalonieri, M.; Balestrazzi, A.; Bisoffi, S. & Cella, R., 1995. Factors affecting Agrobacterium tumefaciens-mediated transformation in several black polar clones. Plant Cell, Tissue and Organ Culture 43: 215–222

Klopfenstein, N.B.; Shi, N.Q.; Kernan, A.; McNabb, H.S.; Hall, R.B.; Hart, E.R.; Thornburg, R.W., 1991. Transgenic Populus hybrid expresses a wound-inducible potato proteinase inhibitor II-Cat gene fusion. Can. J. Forest. Res. 21: 1321–1328

Lee, M. & Phillips, R.L., 1988. The chromosomal basis for somaclonal variation. Annual Rev. Plant. Physiol. Plant. Mol. Biol. 39: 413–437

Llyod, G.B. & McCown, B.H., 1980. Commercially feasible micropropagation of mountain laurel (Kalmia latifolia) by use of shoot-tip culture. Proc. Int. Plant Propagators Soc. 30: 421–437

Tsai, C-J.; Podila, G.K.; Chiang, V.L., 1994. Agrobacterium-mediated transformation of quaking aspen (Populus tremuloides) and regeneration of transgenic plants. Plant Cell Reports 14: 94–97

Tzfira, T.; Ben-Meir, H.; Vainstein, A. & Altman, A., 1996. Highly efficient transformation and regeneration of aspen plants through shoot-bud formation in root culture. Plant Cell Reports 15: 566–571

27

STUDY OF ADVENTITIOUS ROOT FORMATION THROUGH LATERAL ROOT-SPECIFIC mRNA AND ROOTING-ASSOCIATED PROMOTERS IN *Pinus contorta*

Anders Lindroth, Roland Grönroos, and Sara von Arnold

Department of Forest Genetics
Swedish University of Agricultural Sciences
P.O. Box 7027, S-750 07 Uppsala
Sweden

INTRODUCTION

Adventitious root formation, by auxin treatment, on 20-day-old hypocotyls of *Pinus contorta* has been extensively studied at our department in recent years (Grönroos et al., 1993; Grönroos et al., 1987). Those studies have shown that there are great morphological similarities between lateral root formation and auxin-induced adventitious root formation. We have isolated cDNA clones presumably involved in early processes of lateral root formation, and their significance for adventitious root formation are under investigation. By a wound and inoculation procedure with *Agrobacterium rhizogenes* on hypocotyl cuttings (see Grönroos et al. elsewhere in this book), we have been able to study different promoter-GUS constructs in developing transformed roots. The armed strain we use to induce root formation gives almost no alteration in root morphology compared to normal adventitious roots.

EXPERIMENTAL

Plant Material

Pinus contorta Dougl. ex Loud., Lodgepole Pine seeds were surface-sterilised, imbibed, sown and cultured as described by Grönroos et al. (1993).

IBA Treatment, mRNA Preparation and Heterologous Screening of Root cDNA Library

When the seedlings were 20 days old, they were cut at the middle of the hypocotyl and treated with 1.23 mM IBA for 6h as described by Grönroos et al. (1993), and har-

Biology of Root Formation and Development, edited by Altman and Waisel.
Plenum Press, New York, 1997.

vested after 0, 3 and 6 days. RNA was prepared from the hypocotyls and subjected to RT-PCR with clone-specific primers. A cDNA library (named RI-library) was made via mRNA prepared from root tissue where lateral root initiation was in progress, namely between the root cap and the first lateral root. By heterologous screening with the coding region (1833bp) of HRGPnt3 (Keller et al.) as probe, we isolated 17 cDNA-clones which we sequenced. Another cDNA library (named C-library) was made as a control from root tissue where lateral root initiation was finished.

Transformation Technique, Bacterial Strain and Plasmid Constructs

The transformation technique, described in Yibrah et al. (1996), involves wounding and inoculation of *Agrobacterium rhizogenes* of the strain LBA9402. The plasmid constructs tested were promoter–GUS fusions of 35S–GUS INT, UBI–GUS INT, HRGPnt3–GUS and RSI–GUS in derivatives of pBIN19. The first two contain an intron in the GUS-gene.

RESULTS AND DISCUSSION

The only morphological event seen in the zone from where we made the RI-library is the formation of lateral root primordia. By heterologous screening of the RI-library with HRGPnt3, none of the 17 cDNA-clones showed high sequence similarity with the probe. The clones were instead tested for specificity to the RI-library by PCR with clone-specific primers. One of the clones was specific to the RI-library as compared to C-library, subsequently, possibly involved in the initiation of lateral roots. This mRNA was also induced by IBA, since it was detected with RT-PCR after 3 but not 0 and 6 days after treatment. Two other clones were shown to be specific to the RI-library, but not induced by IBA. Constitutively expressed clones, non IBA-induced, have also been isolated.

For the transformation studies we used both specific (HRGPnt3–GUS, RSI–GUS) and common (35S–GUS INT, UBI–GUS INT) constructs to study the expression pattern in lateral roots. The genes of the specific ones have been isolated in tobacco and tomato, respectively, and shown to be expressed in the site of incipient lateral root primordia (Vera et al., 1994, and Taylor, et al., 1994). In *P. contorta*, we have seen their expression exclusively in the vascular tissue and the endodermis, and they might be involved in similar processes as in tobacco and tomato. The common constructs show strong expression in almost all tissues of *P. contorta*, except for the cortex and the epidermis. The promoter of UBI seems to be stronger than even 35S, which is commonly considered to be stronger than most others.

REFERENCES

Grönroos, R., von Arnold, S., 1987. Initiation of roots on hypocotyl cuttings of *Pinus contorta* in vitro. *Physiol. Plantarum*, 69:227–236.

Grönroos, R., Flygh, G., von Arnold, S., 1993. Growth analyses on *in vitro, ex vitro* and auxin-rooted hypocotyl cuttings of *Pinus contorta* Dougl. ex Loud. *New Phytol.*, 125:829–836.

Keller, B., Lamb, C., 1989. Specific expression of a novel cell wall hydroxyproline- rich glycoprotein gene in lateral root initiation. *Genes Dev.*, 3:1639–1646.

Taylor, B., Scheuring, C., 1994. A molecular marker for lateral root initiation: The RSI-1 gene of tomato (*Lycopersicon esculentum* Mill) is activated in early lateral root primordia. *Mol. Gen. Genet.*, 243:148–157.

Vera, P., Lamb, C., Doerner, P., 1994. Cell-cycle regulation of hydroxyproline-rich glycoprotein HRGPnt3 gene expression during the initiation of lateral root meristems. *The Plant Journal*, 6(5):717–727.

Yibrah, H., Grönroos, R., Lindroth, A. Franzén, L., Clapham, D., von Arnold, S., 1996. *Agrobacterium rhizogenes*-mediated induction of adventitious rooting from *Pinus contorta* hypocotyls and the effect of 5-azacytidine on transgene activity. *Transgenic Research*, 5:75–85.

GENETIC VARIABILITY AND HERITABILITY OF HARDWOOD CUTTINGS PROPAGATION OF EUROPEAN PLUM

G. Bartolini and G. Roselli

Institute for the Propagation of Woody Plants - C.N.R., 50018 Scandicci (Firenze), Italy.

The estimation of quantitative genetic parameters (genetic variance, heritability, correlation between characters) enables the plant breeder to choose an appropriate scheme for selection and to predict the gain from it (1, 2).

Although propagation by hardwood cuttings is of significant agronomic and nursery importance, an estimation of quantitative genetic parameters for this trait has never been attempted with a vegetatively propagated species.

In the present investigation, the european plum (*Prunus domestica* L.) genetic variability and heritability of hardwood cutting propagation were estimated.

In order to quantitatively estimate rooting and establishment of various plum cultivars, root inducing hormone (IBA) concentration and the duration of time (30, 60 or 90 days), that the cuttings were kept in basal heated (20°C) perlite substrate, were determined first.

The optimal conditions determined in this experiment (IBA 3000 ppm, basal heating for 60 days), were applied to evaluate genetic and environmental variability and heritability. Self-rooting and establishment of 26, 20-year-old, european plum cultivars and of 33 progenies of cv "Precoce di Ebersucier" obtained through open pollination were tested with IBA 0 and 3000 ppm treatment and establishment once over the juvenile phase.

The heritability in broad sense (H) for the *rooting capacity* of the control treatment is 0.11 ± 0.08, and of the indole-3-butyric acid (IBA) 3000 ppm treatment is 0.21 ± 0.05.

The H for *establishment* of the control treatment is 0.29 ± 0.03 and for the IBA 3000 treatment of 0.05 ± 0.06.

For the traits where H value was relatively high an improved genetic performance could be based on mass selection. In the cases where H was low it is advisable to employ methods based on specific combining ability (S.C.A.).

The wide variability of the rooting and establishment traits, shown by the progenies, lead to the conclusion that selection of cultivars for good propagation performance, by wood cutting, must be based on establishment and not only on root formation.

The data show a wide variability among individuals (half-sib) for the 2 traits and the 2 treatments. It is interesting to note the following aspects:

1. The number of individuals that have increased rooting and establishment with IBA in respect to the control is similar for the 2 treatments (seedling 3/17).
2. The number of individuals that behave as above, but with an inferior establishment with IBA (seedling 5/13).
3. The number of individuals that have increased rooting with IBA and a full establishment (seedling 5/5).
4. The number of individuals that root well with IBA but then do not establishment (seedling 4/2–4/19).

5. The number of individuals that do not root either with, or without IBA, but establishment is noticeably superior to the control (seedling 3/2). For the selection of cultivar it is of utmost importance that rooting ability is followed by relative establishment.

REFERENCES

1. Hansche, P.E., V. Beres and R.M. Brooks, 1986. Proc. Amer. Soc. Sci., 88: 173–183.
2. Hansche, P.E., C.O. Hesse and V. Beres, 1972. J. Amer. Soc. Hort. Sci., 97(1): 76–79.

RECOMBINANT CELLULOSE BINDING DOMAIN (CBD) MODULATES *IN VITRO* ELONGATION OF DIFFERENT PLANT CELLS

Etai Shpigel, Levava Rois, and Oded Shoseyov

The Kennedy Leigh Centre for Horticulture Research and The Otto Warburg Center for Agricultural Biotechnology, The Faculty of Agriculture, The Hebrew University of Jerusalem, P.O. Box 12, Rehovot 76100, Israel.

INTRODUCTION

Shoseyov and Doi (1990) isolated a unique cellulose binding protein (CBPA) from the cellulolytic bacterium *Clostridium cellulovorans*. It was found that this major subunit of the cellulase complex has no hydrolytic activity and is essential for the degradation of crystalline cellulose. This protein has an extremely strong affinity to cellulose and chitin. More recently we cloned and sequenced cbpA (Shoseyov et al., 1992). Using PCR primers that flank the cellulose binding domain (cbd), we successfully PCR amplified and cloned cbd into an overexpression vector (pET8c) that enables us to overproduce the 17 kDa CBD in *Escherichia coli* (Goldstein et al., 1993). The recombinant CBD has a very strong affinity to cellulose. Here we report the plant cell and tissue modulation of elongation by CBD protein.

RESULTS AND DISCUSSION

Recombinant cellulose binding domain (CBD) modulates in vitro elongation of different plant cells. In peach pollen tubes, maximum elongation was observed at 50 μg/ml. Pollen tube staining with calcofluor indicated loss of crystallinity at the tip zone of CBD treated pollen tubes. At low concentrations (1×10^{-2} to 1 μg/ml) CBD enhance elongation of *Arabidopsis thaliana* roots and root hairs. At high concentrations (100–600 μg/ml) CBD inhibited dramatically roots and root hair elongation in a dose responsive manner. Gold-immunolabling of CBD revealed its presence predominantly at the tip zone of the pollen tubes and the root hairs. CBD compete with xyloglucan (XG) on binding to cellulose. When CBD was added first to the cellulose, increasing concentrations of CBD resulted in increasing amounts of unbound XG. However, when XG was added first increasing concentrations of CBD did not affect the level of unbound XG. The level of unbound CBD was higher when XG was added first to the cellulose.

CONCLUSIONS

It is accepted that a prerequisite for cell elongation is a loosening of the cross-linked cellulose network. CBD competes with XG for binding to cellulose. It is therefore possible that this competition results in a temporary loosening of the cell wall and consequently enhanced elongation. The inhibitory effect can be explained by steric hindrance of the cellulose fibrils by excess amounts of CBD, which block access for enzymes or other proteins that modulate cell elongation via loosening of the rigid cellulose-fibril network.

REFERENCES

Goldstein, M.A., Takagi, M., Hashida, S., Shoseyov, O., Doi, R.H., and Segel, I.H., 1993, Characterization of the cellulose-binding domain of the Clostridium cellulovorans cellulose-binding protein A. J. Bacteriol. 175, 5762–5768.

Shoseyov, O., and Doi, R.H.,1990, Essential 170-kDa subunit for degradation of crystalline cellulose by Clostridium cellulovorans cellulase. Proc. Natl. Acad. Sci. USA 87, 2192–2195.

Shoseyov, O., Takagi, M., Goldstein, M.A., and Doi, R.H., 1992, Primary sequence analysis of Clostridium cellulovorans cellulose binding protein A. Proc. Natl. Acad. Sci. USA 89, 3483–3487.

BUD REGENERATION AND GROWTH FROM TRANSGENIC AND NON-TRANSGENIC ASPEN (*Populus tremula*) ROOT EXPLANTS

B. Vinocur,[1] T. Tzfira,[1] M. Ziv,[2] A. Vainstein,[1] and A. Altman[1]

[1]The Kennedy-Leigh Center for Horticultural Research. [2]Department of Agriculture Botany, The Hebrew University of Jerusalem, Rehovot 76100, Israel.

INTRODUCTION

Poplars, including aspen, (*Populus tremula*), are fast growing trees employed in the wood industry and have a potential value for biomass and energy production. Vegetative propagation is carried out by cuttings, root suckers and grafting, but an increase of the propagation efficiency is highly desired. *In vitro* procedures are becoming available for efficient large-scale clonal propagation of poplar genotypes. These procedures can benefit from large-scale root culture in liquid media, based on the natural ability of aspen for shoot bud formation on roots (Carmi, 1994). The effect of different types of root explants upon bud regeneration of aspen was studied using root segments from adventitious roots that formed on stem sections of two transgenic lines that harbored the *rolB* and *rolC* genes, and one non-transgenic line.

MATERIALS AND METHODS

Adventitious root segments from *in vitro* stem sections of two transgenic lines (T-26 and T-27) and one non-transgenic line (control) were used. The roots were cut in 1.5 cm length segments, classified to proximal, middle and distal sections, cultured with respect to their position in the root, and were compared with whole non-sectioned roots. Ten root segments from each position (15 cm total root length) were cultured in half strength MS mineral basal liquid medium, with increasing benzyladenine (BA) concentrations (0 μM, 0.44 μM, 0.89 μM and 1.78 μM). Explants were cultured under a 16-h photoperiod of cool white fluorescent light (photon flux rate of 50–60 μmol/m^{2}s) at 25±1°C. Data were taken after 30 days culture.

RESULTS

Culture of root segments of T-26 and T-27 aspen lines in a medium with 0 μM of BA resulted in an increment in root formation expressed by the higher root growth value, as compared with roots from the control plants. When root segments were cultured under increasing concentrations of BA, root growth value of all lines decreased, as compared with roots cultured with 0 μM of BA. Yet, in the presence of increasing BA concentration T-26 and T-27 transgenic lines still showed a higher root formation than the control line. It was also found that the number of buds per explant increased in relation to root polarity in all treatments. Bud regeneration of all lines, in all treatments, was highest at the proximal zone, decreasing toward the root apex (Fig. 1). Root segments of T-26 and T-27 lines cultured in a medium with 0 μM of BA resulted in a slightly lower bud differentiation, as compared with roots from the control line. Only at the proximal zone of T-27, was bud regeneration higher than the other two lines. Under increasing concentrations of BA, T-26 and T-27 transgenic lines showed a higher bud regeneration than the control line. Southern blot analysis revealed the presence of *rolC* and *rolB* genes (from the *Agrobacterium rhizogenes* Ri plasmid) in lines T-26 and T-27. While T-27 resulted from a single insertion event, line T-26 appears to harbor more than one copy of the *rol* genes (Tzfira et al.1996). The number of regenerated buds, in all lines, was lower in whole non-sectioned roots than in sectioned ones. The control line showed high bud regeneration at 0.44 μM of BA and its number decreased with increasing concentrations of BA. While, with increasing BA concentrations, a general increase in bud regeneration in lines T-27 and T-26 was found; the highest bud differentiation was at 1.78 μM BA level. In contrast to bud regeneration, root biomass of sectioned explants decreased with increasing BA concentrations.

CONCLUSIONS

1. Bud regeneration from sectioned roots of aspen (combining all three zones) was considerably higher than in whole, non-sectioned roots in each treatment.
2. Benzyladenine enhanced bud regeneration of sectioned roots, in comparison with whole non- sectioned roots
3. Due to their higher root biomass, bud regeneration of T-26 and T-27 transgenic lines is potentially higher than in the control line
4. Bud differentiation in aspen root explants can be enhanced by optimizing both media and gene expression.

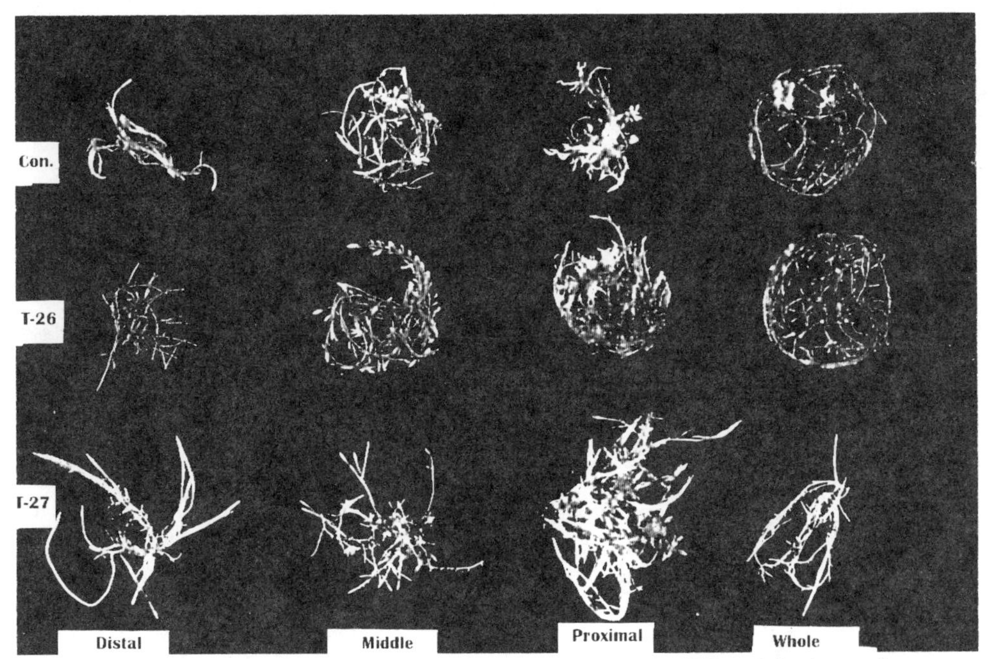

Figure 1. Regeneration of adventitious buds that developed at the proximal, middle and distal zone of aspen sectioned and whole roots after 30 day culture in a half strength MS basal liquid medium. **Distal**: root section adjacent to the root apex, **Middle**: middle section, **Prox.**: root section furthest away from root apex, **Whole**: whole non-sectioned root , **Con.**: control, **T-26** and **T-27**: transgenic line.

The relevance of *rol* gene expression affecting tissue sensitivity to exogenous hormone application is currently being studied.

REFERENCES

Carmi, T. 1994. Bud differentiation and growth from Aspen (*Populus tremula*) root and bud explants in liquid cultures. M.Sc. Thesis. (in Hebrew)

Tzfira, T., Ben-Meir H., Vainstein, A., and Altman, A. (1996). Highly efficient transformation and regeneration of aspen plants through shoot-bud formation in root culture. Plant Cell Reports 15: 566.

ALTERATIONS IN GENE EXPRESSION DURING ADVENTITIOUS ROOT INITIATION IN PEA CUTTINGS

Ann-Caroline Nordström, Marie Bollmark, Helena Forsell, Johanna Wouters, and Lennart Eliasson

Department of Botany, Stockholm University, S-106 91 Stockholm, Sweden.

The formation of adventitious roots in pea (*Pisum sativum* L) cuttings was studied in relation to the endogenous concentration levels of auxin and cytokinin. Now we are engaged in work to identify genes involved in the rooting process and regulated by these two hormones.

The IAA concentration in the rooting zone (as determined by GC-MS) was constant during the whole rooting period, and on the same level as in the seedlings at the time of cutting. That IAA is necessary for rooting is obvious: when the shoot apex (the IAA source) is excised the IAA level in the rooting zone decrease, and no roots are formed. Rooting is restored by exogenous auxin. We conclude that endogenous IAA is a prerequisite for root initiation, and that the level of IAA at the time of cutting suffice for induction of the cellular rooting program (Nordström and Eliasson 1991, 1993). Why then does not root form in the intact seedling? Cytokinins, produced in the root system and transported in the xylem to the shoot, appears to be endogenous inhibitors of root formation, and prevent IAA from activating the rooting program in the intact plant. When the roots are cut off, the cytokinin level (quantified by ELISA and LC-MS) in the rooting zone decreases. Root formation is induced when the cytokinin level has decreased below a certain level. Addition of even low concentrations of cytokinin to the rooting solution inhibits the organized cell divisions leading to root initial formation. When the cytokinin is withdrawn from the solution, root formation is induced, with the same delay as when the roots are cut off (Bollmark et al. 1988, Bollmark et al. 1995). Thus, adventitious root initiation is regulated by the concerted action of auxin and cytokinin.

We are now looking for rooting-related genes, and the effect of auxin and cytokinin on their expression. Using differential display (Mou et al. 1994) on total RNA (Pawlowski et al. 1994) we have found transcripts from seedlings that had disappeared in the cuttings. These may reflect the inactivation of certain genes as a result of the removal of the primary root system. We have also found up-regulated transcripts, that only appeared in the cuttings. Some of these occurred with higher intensities after treatment of the cuttings with IBA. These indicate the activation of new genes in relation to root initiation. The transcripts were verified by northern blot analyses.

REFERENCES

Bollmark, M., Kubát, B., and Eliasson, L., 1988. Variation in endogenous cytokinin content during adventitious root formation in pea cuttings. J. Plant Physiol. 132: 262–265.

Bollmark, M., Chen, H. J., Moritz, T., and Eliasson, L., 1995. Relations between cytokinin level, bud development and apical control in Norway spruce, *Picea abies*. - Physiol. Plant. 95:563–568.

Mou, L., Miller, H., Li, J., Wang, E., and Chalifour, L., 1994. Improvment to the differential display method for gene analysis. Biochemical and biophysical research communications 199:564–569.

Nordström, A.-C., and Eliasson, L., 1991. Levels of endogenous indole-3-acetic acid and indole-3-acetyl aspartic acid during adventitious root formation in pea cuttings. Physiol. Plant. 82: 599–605.

Nordström, A.-C., and Eliasson, L., 1993. Interaction of ethylene with indole-3-acetic acid in regulation of rooting in pea cuttings. Plant Growth Regul. 12:83–90.

Pawlowski, K., Kunze, R., De Vries, S., and Bisseling, T., 1994. Isolation of total, poly(A) and ribosomal RNA from plant tissues. Plant Mol. Biol. Manual D5 1–13.

28

SHOOTBORNE ROOTS — AN ADAPTIVE ORGAN FOR PLANTS OF SAND DUNES

Avinoam Danin

Department of Evolution, Systematics, and Ecology
The Alexander Silberman Institute of Life Sciences
The Hebrew University of Jerusalem
Israel 91904

INTRODUCTION

"Adventitious root" is a term commonly used in the botanical literature. Various kinds of adventitious roots have been reviewed recently with regard to their origin and terminology (Barlow, 1986; 1994; Haissig and Davis, 1994). According to Esau (1977) the term is applied to roots that arise on the aerial parts of plants, on underground stems, on parts of old root, and in places which have been injured. Fahn (1982) adopted a similar definition. Such a definition seems to be too broad (Barlow 1986, Groff and Kaplan 1988, Haissig and Davis 1994, and Barlow 1994). Barlow (1986) suggests the term "shootborne root" for roots developing from shoots, "rootborne shoot" to those shoots which sprout from roots, and reserves "adventitious root" for roots which regrow in place of lost or destroyed parts or organs (Haissig and Davis, 1994). The adventitious roots of grasses which develop from nodes are "nodal roots" (Bell, 1991).

In a recent review Haissig and Davis (1994) have concluded: "We know almost nothing... about the roles of adventitious rooting and roots in the structure and functions of ecosystems of various sorts". While preparing the material for a review on adaptations of plants to desert dunes (Danin, 1996) much adventitious rooting was found in plants of sand dunes of Israel, Sinai, Namibia, Turkey, and California. In searching for a definite indication that adventitious rooting could be considered as an adaptive property I relied upon the general agreement concerning convergence. Convergence is known as the "development of similarities between animals or plants of different groups resulting from adaptation to similar habitats" (Lapedes, 1976). Rost et al. (1984) suggested that convergence often occurs when plants of different origins evolve under similar environmental pressures. In studying morphological adaptations to the dune environment, an attempt was made here to demonstrate convergence in the formation of adventitious roots in dune plants of different floristic regions. Producing shootborne roots by some species in one habitat, and rootborne shoots in another group of species, in another habitat from widely disjunct locations, prove the hypothesis that these traits represent major adaptations to specific dune stresses.

Biology of Root Formation and Development, edited by Altman and Waisel.
Plenum Press, New York, 1997.

EXPERIMENTAL

Geomorphological principles and features of the landscape were used for studying the environment. Morphological and anatomical plant parameters, the occurrence of shootborne roots, and rootborne shoots was recorded in relations to the local environmental conditions in the field. In addition to many occasional observations during phytosociological research in Israel and Sinai, repeated observations of plants in the field followed by microscopical studies in the laboratory were carried out.

RESULTS AND DISCUSSION

Grasses Requiring Sand Accumulation

The common properties of pioneer plants of the desert areas studied and of coastal sands was that many of them belong to the Poaceae, have the ability to elongate, re-emerge and to produce nodal roots shortly after being covered by sand. The roots produce rhizosheaths, which in the seedlings of this group are much more prominent as part of the nodal root than of the seminal root (Danin 1996). Concluding from the studies of Wullstein (1991) and his collaborators, the rhizosheaths have relatively high moisture content which provide a nanohabitat which fits the intensive growth of N_2-reducing bacteria. Hence, the rhizosheaths and their N_2-fixing bacteria may be regarded in the nutrition of perennial grasses as nitrogen-nodules on roots of Fabaceae (Leguminosae). Most of the grass pioneers decease in sites where sand becomes stable. When much of the rhizosphere is exposed for more than a few days, that part of the plant dies. The most common pioneer of mobile desert dunes in Israel, Sinai and NE Africa is *Stipagrostis scoparia;* in Egypt and west of it *S. acutiflora*; in NW Africa *S. pungens;* in Central Asia *S. pennata*; in Namibia *S. sabulicola, S. lutescens* and *S. seelyae. S. amabilis* is confined to mobile dunes of the Kalahari Desert (Leistner 1967); *Swallenia alexandrae* is endemic to one dune in Eureka Valley, near Death Valley, California and *Panicum urvilleanum* is confined to mobile sand of the Mojave Desert, California. Being able to produce nodal roots from stems covered by sand is not a general property of grasses. Even in the common genus of the desert pioneers — *Stipagrostis* — there are many species which are not able to recover from sand cover of more than 10 cm. Such are *S. obtusa* and *S. ciliata* which grow on stable sand sheets and never in sites of sand accretion.

Coastal mobile dunes of non desert areas are common around the Mediterranean, Europe, and N. America. The most common grasses in such habitats, either as a spontaneous or as a naturalized species, are of the genera *Ammophila* and *Elymus*. In addition to their ability to produce nodal roots on vertical stems and recover from being covered with sand, some of them reproduce vegetatively by underground, almost horizontal stems, known as rhizomes or more accurately as rhizodes (Du Rietz 1931, Danin 1996).

Most of the dunes pioneers decease if not covered with sand, possibly because of the decreasing efficiency of conducting systems in aging roots and stems. Studies on growth and development, demonstrating the adaptations of grasses to sand accretion on the coastal dunes of Europe, are reported by Marshall (1965) for *Corynephorus canescens,* and by Crawford (1989) for *Ammophila arenaria.* In each case the root system rapidly senesces and the plants require a fresh deposition of sand around the shoot to encourage the growth of new nodal roots. Death of *Stipagrostis scoparia* on stabilized sand is a common occurrence in Israel and Sinai (Danin 1983, 1996).

Species Adapted to Sand Cover and Produce Shootborne Roots

Plants of this category grow in sites where the main events of sand accretion took place already, and plants grow either on stable sand, or on slightly mobile. *Artemisia monosperma* and *Calligonum comosum* are common psammophytes of sand sheets in Israel, Egypt, Jordan, and other arid Mediterranean countries. They have the ability to produce shootborne roots shortly after being covered by sand, during the rainy season. After producing new roots from the buried shoots they produce new shoots above them and assist the lateral or vertical movement of the plant to a new soil level (Fig. 1). All plants of

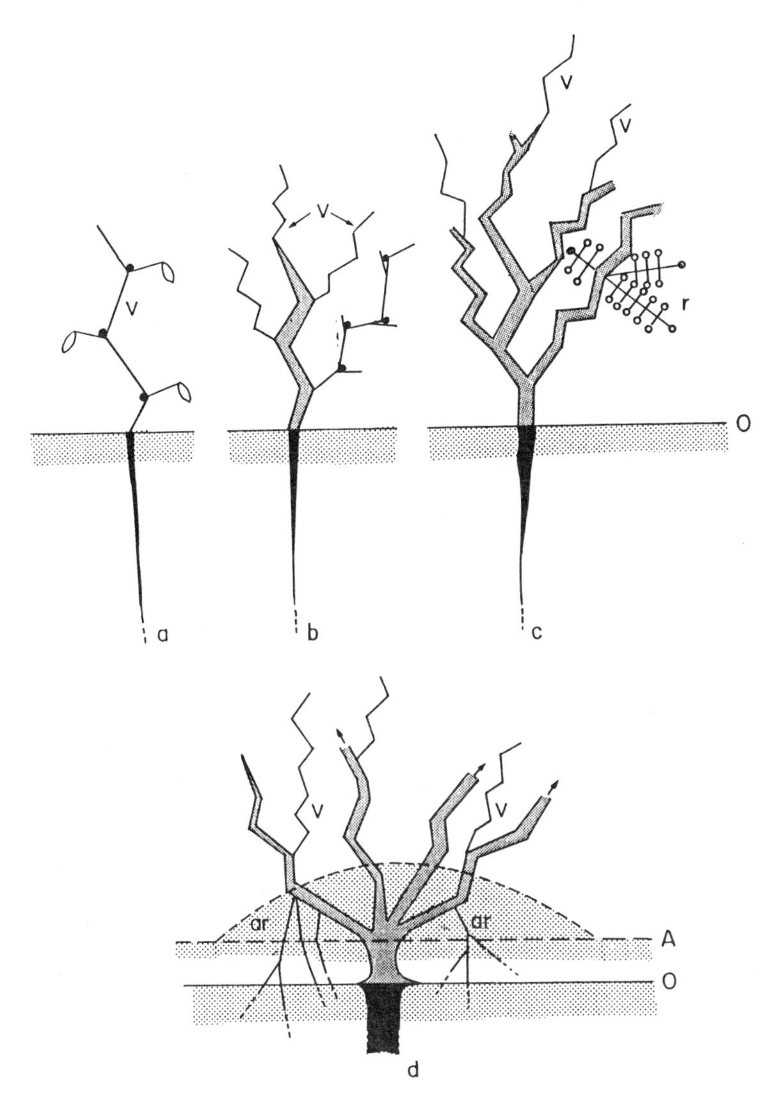

Figure 1. Schematic presentation of structure and development of *Calligonum comosum*. **a** A seedling with leaves; **b** a young plant with a lignified stem and four zigzag-shaped vegetative branches (*v*); **c** vegetative and reproductive branches (*r*); **d** a nebka (*N*) where adventitious roots (*ar*) have developed from the plagiotropic part of old lignified branches; new vegetative branches sprout above the ground close to *ar*. *A*-level marks the new level after sand accretion took place covering the *0* level (From Danin 1996).

this ecomorphological type can be recognized from afar by the accumulated sand mound in which they grow, and known as "nebka" (Walter 1973, Walter and Box 1983a, Bendali et al. 1990). Common species of this group in the desert areas of the east Mediterranean countries are *Ephedra aphylla, E. alata* and the chamaephytic grasses *Pennisetum divisum, Panicum turgidum,* and *Lasiurus scindicus.* Many other species of *Calligonum* growing in sandy areas of Asia function similarly (Danin 1996). *Tetradymia tetrameres, Psorothamnus polyadenius* and *Sarcobatus vermiculatus* cover considerable areas of desert stable dunes in the cold Great Basin deserts of western North America and have the same rooting syndrome. *Haloxylon persicum*, a common psammophyte of extreme sand desert areas in the Asian countries, displayed shootborne roots only occasionally whereas R*etama raetam* which grow in a rather similar habitat do not produce such roots. A few desert trees are capable of producing shootborne roots. These include several species of *Tamarix* and *Prosopis glandulosa,* which tolerate sand accretion by producing roots from buried branches. Shootborne roots of *Tamarix* cf. *ramosissima* growing on alluvial terraces of the Mojave River, near Barstow, California display obvious rhizosheaths (Danin 1996).

Coastal psammophytes may be divided into two categories — those confined to sandy shores, growing in the spray zone of the foredunes and those growing at the boundary of the dunes with the stable land vegetation. *Silene succulenta* and *Otanthus maritimus* are typical plants of this belt in the coasts of Israel, produce shootborne roots and typically make nebkas; *Abronia umbellata* and *Abronia maritima* are typical perennial succulents of foredunes in California and other northwestern American states and similarly have such roots and nebkas (Cooper 1936, Johnson 1977, Danin 1996).

Trees and shrubs of inland woodlands produce adventitious roots when sand invading inland cover their lignified stems. Such Mediterranean semishrubs are *Sarcopoterium spinosum, Coridothymus capitatus*, and *Satureja thymbra* which commonly grow as psammophytes in southern Turkey. Mediterranean trees and shrubs which build considerable nebkas with their rooting lignified branches in the Mediterranean coasts of Turkey are *Pistacia lentiscus, Pistacia terebinthus, Myrtus communis,* and *Rhamnus lycioides*. *Prunus maritima* and *P. serotina* occur in eastern Long Island, New York, in woodlands where they are liable to become covered with sand. *Baccharis pilularis,* which is very common on nonsandy ground of mesic parts of California also forms shootborne roots in stabilizing coastal dunes.

Species Adapted to Sand Deflation and Produce Rootborne Shoots

The plants belonging to this ecomorphological group have a reactive response to sand deflation. When their roots are exposed, they produce adventitious shoots from the exposed roots. In many places the mother plant die whereas its surviving roots produce new plants from the sprouting shoots (Fig. 2). All belong to the Boraginaceae and occur in the Old World. The common species of the East Mediterranean sands are *Moltkiopsis ciliata,* and *Echiochilon fruticosum* in both desert and coastal sands, *Heliotropium digynum* in desert sands and the local endemic *Anchusa negevensis* (Danin 1995, 1996). *Heliotropium arguzioides* of Central Asia (Walter and Box 1983a) appears to be the same.

Sarcobatus vermiculatus is one of the few plants found in sand of N America which has the ability to reproduce from exposed roots. It is basically a phreatophytic halophyte which rarely displays rootborne shoots and only after sand trapped by the plant in nebkas is locally removed by wind erosion of the nebka.

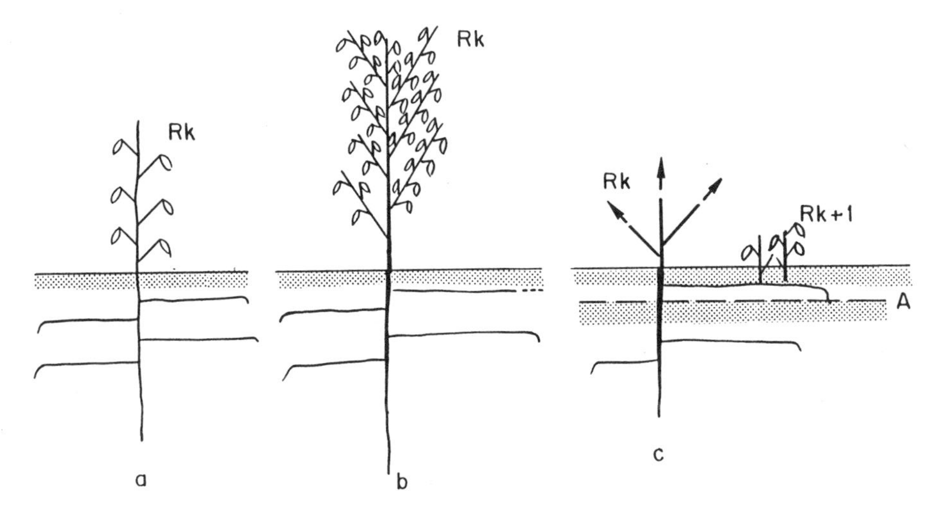

Figure 2. Schematic presentation of structure and development of *Moltkiopsis ciliata*. **a** and **b** growth and development of bimodal root system; **c** sprouting of adventitious shoots from roots after being exposed to level A (From Danin 1996).

Species of Stable Sands that Do Not Produce Shootborne Roots or Rootborne Shoots

Many of the psammophytes growing in stable sand sheets either in desert or in coastal sands do not have the ability to produce shootborne roots or rootborne shoots. This is the majority and they are mentioned here to stress the important adaptive properties of the previous groups. Sites of blowouts in the sandsheets of coastal plain of Israel are recognized not only by the presence of *Moltkiopsis ciliata*, and *Echiochilon fruticosum*, which have rootborne shoots, but also by *Scrophularia hypericifolia* which do not have this property.

Species of Non-Sandy Habitats that Produce Shootborne Roots or Rootborne Shoots

Although only sandy habitats were discussed here as related to a recent review (Danin 1996), there are many more habitats where shootborne roots and rootborne shoots function. Many monocots living in divers kinds of habitats produce mostly shootborne roots. Such are most of the nodal roots of most grasses (Arber 1934), Juncaceae, Typhaceae, and Cyperaceae, especially in waterlogged habitats. Hydrophytic trees such as many species of *Salix*, *Populus*, and *Tamarix* and many species of mangroves easily propagate naturally from detached branches due to their nature of producing shootborne roots. When failing to produce fruits, many *Agave* spp. and *Allium* spp. (Liliaceae) develop seedlings or bulbils on their inflorescence with shootborne roots and reproduce vegetatively. Many gardeners know and use the elementary vegetative reproduction of Cactaceae and cactoid succulents. Detached stems of *Caralluma europaea* var. *negevensis* survived a year of no contact with soil and produced shootborne roots when became in contact with moist soil (Danin 1983). The adaptive value of shootborne root formation in succulents of extreme desert areas is obvious.

Reproducion by rootborne shoots is the rule with *Prosopis farcta* in Israel. Hence large populations are in fact one clone with thousands of ramets. The same is the rule with

Acacia albida and *Lycium europaeum* which do not produce much fruits in most populations in Israel, but easily produce rootborne shoots.

CONCLUSIONS

The ability to produce shootborne roots gives many plants, and perennial grasses in the first place, high biological advantage in sand dunes. Plants are constantly vulnerable there to become covered by mobile sand. When producing roots from stems covered by sand, pioneer plants of sand dunes may grow above the sand cover and do not have to draw resources from aging and deep buried stems and/or roots. Vegetative reproduction which is often associated with difficulties in germination can keep going as long as the plants produce new roots from their shoots.

The ability to reproduce vegetatively from rootborne shoots is a property found in a few species associated with sand deflation and blowouts. However, they are accompanied by plants which do not have this property. Plants able to produce rootborne shoots are not restricted to sand under deflation.

REFERENCES

Arber, A., 1934, The Gramineae. Cambridge University Press, London.

Barlow, P.W., 1986, Adventitious roots of whole plants: their forms, functions, and evolution. In: "New Root Formation in Plants and Cuttings." M.B. Jackson ed., Martinus Nijhoff, Dordrecht, p. 67.

Barlow, P.W. 1994, The origin, diversity and biology of shoot-borne roots. In: "Biology of Adventitious Root Formation." T.D. Davis, and B.E. Haissig eds., Plenum, New York, p. 1.

Bell, A., 1991, Plant morphology. Oxford University Press, Oxford.

Bendali, F., Floret, C., Le Floc'h, E., and Pontanier, R. 1990, The dynamics of vegetation and sand mobility in arid regions of Tunisia. *J. Arid. Environ.* 18:21.

Cooper, W.S., 1958, Coastal sand dunes of Oregon and Washington. *Geol. Soc. Am. Mem.* 72:1–168.

Crawford, R.M.M., 1989, Studies in Plant Survival. Blackwell, Oxford.

Danin, A., 1983, Desert Vegetation of Israel and Sinai. Cana, Jerusalem.

Danin, A., 1995, A new *Anchusa* from Israel. *Edinburgh J. Bot.* 52(3):333.

Danin, A., 1996, Plants of Desert Dunes. In: J.L. Cloudsley-Thompson (ed.) Adaptations of Organisms to the Desert. Springer Verlag, Heidelberg.

Du Rietz, G.E., 1931, Life-forms of terrestrial flowering plants. *Acta Phytogeogr. Suec.* 3:1.

Esau, K., 1977, Anatomy of Seed Plants, 2nd edn. Wiley, New York.

Fahn, A., 1982, Plant Anatomy. 3rd edn. Pergamon Press, Oxford.

Groff, P.A., and Kaplan, D.R., 1988, The relation of root systems to shoot systems in vascular plants. *Bot. Rev.* 54:387.

Haissig, B.E., and Davis, T.D., 1994, A historical evaluation of adventitious rooting research to 1993. In: "Biology of adventitious root formation." T.D. Davis, and Haissig, B.E., eds., Plenum Press, New York, p. 275.

Johnson, A.F., 1977, A survey of the strand and dune vegetation along the Pacific and southern Gulf coasts of Baja California. J. Biogeogr. 7:83.

Lapedes, D.N., ed., 1976. McGraw-Hill dictionary of the life sciences. McGraw-Hill, New York

Leistner, O.A., 1967, The plant ecology of southern Kalahari. *Bot. Surv. S. Afr. Mem.* No. 38.

Rost, T.L., Barbour, M.G., Thornton, R.M., Weier, T.E., and Stocking, C.R., 1984, Botany, 2nd edn. Wiley, New York.

Walter, H., 1973, Vegetation of the Earth. Springer, Berlin, Heidelberg ,New York.

Walter, H., and Box, E.O., 1983a, The Karakum desert, an example of a well-studied eu-biome. In: "Ecosystems of the World, vol 5. Temperate Deserts and Semideserts. "N. West ed., Elsevier, Amsterdam, p. 105.

Walter, H., and Box, E.O., 1983b, The deserts of central Asia. In: "Ecosystems of the World, vol 5. Temperate Deserts and Semideserts. "N., West ed., Elsevier, Amsterdam, p. 193.

Wullstein, L.H., 1991, Variation in N2 fixation (C2H2 reduction) associated with rhizosheaths of Indian Ricegrass (*Stipa hymenoides*). *Am. Midl. Nat.* 126:76.

29

TREE ROOT RESPONSE TO MECHANICAL STRESS

A. Stokes and D. Guitard

Laboratoire de Rhéologie du Bois de Bordeaux
Domaine de l'Hermitage
B.P. 10 33610 Cestas Gazinet
France

ABSTRACT

Until recently, the role of the root system, with regards to tree stability, has been neglected. However, due to the increasing number of storms in Europe over the past years, research into tree anchorage has begun to be investigated.

As the form and strength of a root system influences the stability of a tree, how is it affected by external mechanical factors? Studies carried out during the last thirty years have shown that mechanically stressed woody roots develop large, eccentrically formed root bases. However, recent work has shown that windward and leeward roots of *Picea sitchensis*, grown in wind tunnels, were also found to be more numerous, with highly branched windward roots.

Bending and compression strength tests carried out on the rootwood of several mature tree species showed that the strength of the rootwood was correlated with the shape of the root system. Strength was found to be higher in areas of greatest stress *i.e.* the undersides of lateral roots, where the roots are pushed downwards onto the soil, and parts of the root system which sway the most under wind loading. This increase in strength appears to correspond to a change in the rootwood anatomy, as cells in these regions become more dense with thicker walls and an increased microfibril angle.

It appears that root systems can adapt to imposed mechanical stresses. However, the intrinsic form of a tree root system is a factor which must also be taken into account when considering the stability of a tree. We need to combine this knowledge in order to determine the mechanically optimal form of a tree root system and how it can be manipulated in order to improve tree stability.

INTRODUCTION

The number of storms in Europe over the last years have resulted in huge losses to forest and urban trees (Quine 1991, Fillon 1987, Slodicak 1995) and if predictions that our

Biology of Root Formation and Development, edited by Altman and Waisel.
Plenum Press, New York, 1997.

climate is becoming windier prove to be correct (Gribbin 1990), the loss of forest trees, as well as old and valuable park trees, will increase. In order to attempt to reduce such damage, trees must be selected and planted so that their mechanical stability is maximised. Investigations of tree adaptation to mechanical stress have become increasingly popular over the last few years (see Telewski 1995), as direct practical advice can be given to the forester or arboriculturalist as to how to manage his trees in order to avoid potential mechanical damage (Mattheck and Breloer 1995). However, due to the difficulties of investigating belowground organs, root adaptation to imposed physical stress has been somewhat neglected even though anchorage by the root system is a major component in resisting uprooting. An improved understanding of changes that can occur within the root system, and how these can be manipulated in order to improve stability is needed.

GEOMETRICAL ADAPTATION

To the best of our knowledge, the first experiments to investigate the effects of mechanical action on the growth of tree roots were conducted by M.R. Jacobs in 1939 and 1954. Jacobs attached guy ropes to the stems of young *Pinus radiata* so that only the tops of the plants could sway in the wind. The diameter of the guyed stems was greater than that of the controls above the point of attachment, but smaller below. After two years, the ropes were removed and, in the first high wind, all the stems blew or broke over. Jacobs also found that the guyed trees had thinner woody lateral roots than the free-standing trees. Fayle (1968) repeated this experiment on seven year old *Pinus sylvestris* saplings. After two years, the lateral roots of control trees showed a 75% increase in their annual ring widths, compared to those of the guyed trees. It appears that lateral woody roots can respond to imposed stresses, at least at the root bases. Wilson (1975) analysed the distribution of growth in lateral roots of guyed and free-standing 10 m tall *Pinus strobus*. Roots were significantly thicker in the free-standing trees, but only up to 50 cm from the stem-root base. As there is an increase in radial growth, mechanical stresses must affect the activity of the cambium of such roots. As the amount of stress perceived by the cambium will differ at points around the circumference of the root, it may be expected that the response within the root changes depending on the location of the stress. Such mechanical stimulation to the cambium may explain the often observed eccentricity of woody roots.

Roots that are bent on one axis can become oval or even I-beam in cross-section (Fig. 1) (Cannell and Coutts 1988, Mattheck and Breloer 1992, 1995, Nicoll *et al.* 1995, Nicoll and Ray 1996). In order to quantify the eccentricity of woody root bases subjected to controlled mechanical stress, Stokes *et al.* (1996) measured the vertical and horizontal lateral root diameters of three year old Sitka spruce clones mechanically perturbed in a flexing machine. The vertical root diameters of roots growing along the direction of flexing were significantly larger than the horizontal diameter. The mean cross-sectional area and woody root mass of the same roots were also significantly greater than those of the unflexed plants. This eccentricity increases the roots' resistance to bending in the direction of stress. In a study on 46 year old Sitka spruce (*Picea sitchensis*) growing in a prevailing wind, Nicoll and Ray (1996) found that buttressed parts of roots had greater lateral and vertical secondary thickening above than below the biological centre. Such roots were termed "T-beams" (Fig. 1a). This uneven growth was found to be greater on the lee side of the tree, and largest at 0.5 m from the tree centre in the most shallow-rooted trees. Further from the tree, particularly on the windward side, many roots had increased secondary thickening above and below the root centre, but were "I-beam" in shape (Fig. 1b). T- and

Figure 1. Lateral roots that are bent along one axis grow unevenly and become thicker along the axis of bending. Photographs of cross-sections showing: a) a buttressed "T-beam" shaped root of *Picea sitchensis*, b) an "I-beam" shaped root of *Picea sitchensis*, and c) a root of *Pinus pinaster* showing increased secondary thickening below the biological centre. Photographs 1a and 1b are courtesy of B.C. Nicoll and D. Ray.

I-beams can more efficiently resist vertical flexing due to an increased second moment of area (a function of radius to the fourth power). In Sitka spruce, roots showing only moderate eccentricity were found to be three times more resistant to bending than comparable roots with a circular cross-section (G.J. Lewis, personal communication). As wood is weaker in compression than in tension, more wood is likely to be laid down where the root is placed under compression, as in the case of T-beams, in order to resist rupture. However, in lateral roots of 13 year old Maritime pine (*Pinus pinaster*), the distribution of secondary thickening was found to increase by 26% *below* the biological centre (Fig. 1c) (Berthier *et al.* 1997). Unlike Sitka spruce in maturity, Maritime pine possesses a large tap root, thereby altering the pattern of stress within the root system. As the tree sways in the wind, a concentration of mechanical stress will develop at the bases of the stem and lateral roots, and tap root and lateral roots. As the distribution of secondary growth appears to be linked directly to mechanical stress, the roots may respond to this stress, by laying down more wood on the underside of the root, to increase rigidity in this area.

A more dramatic example of such root eccentricity can be seen in the buttress roots of many temperate and tropical species. The formation of buttress roots has been directly linked to the mechanical environment of the tree (Henwood 1973, Richter 1984, Mattheck 1991, Ennos 1995). Ennos (1995) attached strain gauges around the base of the trunk and the lateral roots of young rain forest trees which were developing buttresses. The young trees were then winched over horizontally and the corresponding deformations (strains) in the tree were measured. It was found that strains were concentrated along places of maximum growth; along the top of developing buttresses and on the trunk above them. Strains were negligible on the sides of the buttresses where little growth occurs. Therefore it appears that the high rates of growth occurring at lateral root bases are stimulated by the mechanical strains set up by wind forces which are concentrated in these regions. Loading forces transmitted down the stem will then be smoothly and quickly transferred into the soil. If buttress roots are not present, these external forces must be dissipated into the soil *via* tap roots or vertically growing roots closer to the trunk (Stokes 1995).

ARCHITECTURAL ADAPTATION

In order to achieve anchorage, plants must transfer loading forces, experienced by the stem, into the ground *via* the roots. The shape of the root system determines how these forces are distributed. If there is a large root surface area, these forces will be quickly dissipated. A large root surface area can be achieved by either larger or more highly branched roots. The former will ensure a greater resistance to bending forces (bending resistance is proportional to the fourth power of radius) whereas the latter will allow a rapid transferral of tensile forces into the soil.

Plant response to mechanical stress has been quantified using various techniques, usually involving mechanical perturbation of the plants by brushing or shaking (see Grace 1977, Telewski 1995). Thigmomorphogenesis, plant growth responses to mechanical stimulation (Jaffe 1973), has been most thoroughly researched using aerial parts of the plant. Typical responses include an increase in stem taper, a reduction in branch length and smaller leaves. The increase in stem taper is usually achieved by a reduction in stem elongation and/or an increase in radial growth (Telewski 1995). The resulting plant may therefore have a "stunted" appearance, thus decreasing the speed-specific drag of the crown (Telewski and Jaffe 1986). Wind loading is reduced on the plant itself, especially if the canopy develops a flag-shaped form, with branches swept onto the more sheltered lee side of the plant (Rees and Grace 1980). If such responses occur at the canopy level, it is likely that comparable changes take place below ground.

Very little work has been carried out on the effects of direct wind action on trees. Only four experiments have included an analysis of changes in root growth of trees grown in wind tunnels. Satoo (1962) observed reductions in both shoot and root dry weight and root length of *Robinia pseudoacacia* seedlings exposed to continuous wind velocities of 3.6 ms^{-1} for four weeks, compared to controls in still conditions. Heiligmann and Schneider (1974) found that shoot and root dry weight were decreased when *Juglans nigra* was grown at similar wind speeds compared with plants grown at wind speeds of 0.1 ms^{-1}. In both experiments, exposure to wind resulted in a general reduction to growth, with no specific effects on roots recorded. However, both these wind tunnel studies used relatively fast continuous windspeeds, unlike field conditions where windspeeds can be low or zero for extended periods. Telewski (1995) states that plants do not respond in the same way to intermittent and continuous wind loading: far greater responses occur when the plants are stressed periodically and have a "gravitropic presentation time" (the minimum stimulation time required to elicit a response).

Taking the gravitropic presentation time into account, Stokes *et al.* (1995a,b) carried out experiments with Sitka spruce and *Larix decidua* (European larch) grown from seed, in wind tunnels for 30 weeks. Wind generators, positioned at ground level, were operated for 6 h during the day and a further 6 h at night. Windspeeds experienced by plants ranged from 0.5 to 2.9 ms^{-1}. The seedlings showed little response of shoot growth to wind loading. However, changes within the root systems were identified. Lateral roots were counted and their orientation relative to the tap root recorded. In both species, there was an increase of almost 60% in the number of large windward roots and of 45% of leeward roots of larch trees, compared with roots growing at right angles to the direction of wind (Fig. 2). An architectural analysis was also carried out on the woody roots of Sitka spruce. It was found that windward lateral roots had a higher incidence of branching than leeward roots. The woody tips of windward roots were also longer with a larger diameter than on leeward roots.

Such changes in architecture also influence the anchorage ability of the plant. Using a method of trenching, cutting and pulling, the most important components of the root sys-

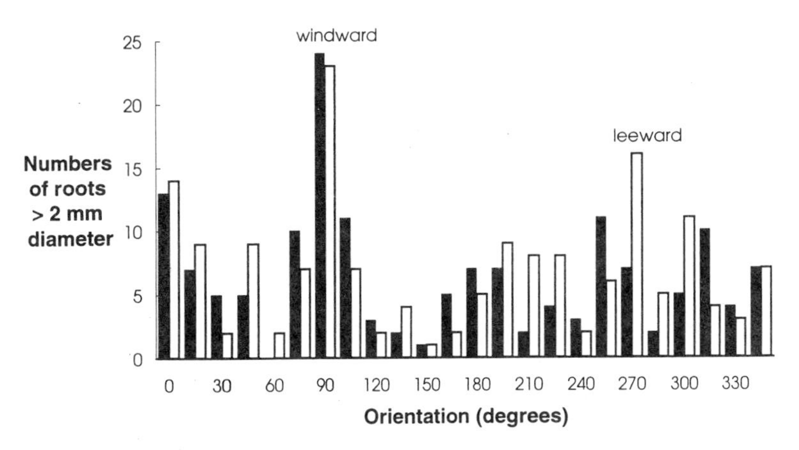

Figure 2. The number of large woody lateral roots in Sitka spruce and larch grown in a wind tunnel are greater on windward and leeward sides. Shaded bars represent Sitka spruce and unshaded bars represent larch (Stokes *et al.* 1995b).

tem with regards to anchorage have been identified in mature *Picea sitchensis* trees (Coutts 1983, 1986). It was found that the "windward" roots, *i.e.* those held in tension during overturning, provided over half the resistance to overturning. Other less important factors contributing to stability included the weight of the soil-root plate, the soil resistance underneath and at the edges of the soil-root plate and the bending of the leeward roots. According to Mohr-Coulomb's law of soil mechanics, soil shear strength increases with the pressure which the shear surfaces exert over each other (Mattheck and Breloer 1995). There are lower shear surfaces on the windward side of a tree due to the lifting of the root-soil plate during tree sway. Soil cohesion is reduced, therefore also decreasing shear stress. Hence, an increase in the surface area of roots may have a reinforcing effect on the soil. A recent study carried out by the authors on windblown Maritime pine has shown that very few of the trees fallen within a waterlogged stand had large lateral windward roots. Most large lateral roots were distributed at right angles to the wind direction, suggesting that an uneven distribution of woody lateral roots around the tree base does decrease tree stability if the trees are loaded in a direction other than the direction of the centre of mass.

ANATOMICAL ADAPTATION

As root wood is of little economic importance, work carried out on root anatomy is relatively scanty. However, in order for a tree to remain upright and hence produce good quality wood, a healthy root system, which can resist mechanical loads is vital. Compared to that of the stem, the anatomy of the lateral root varies considerably, depending on species, site and root (Patel 1965, Fahn 1967, Fayle 1968, Eskilsson 1969). From the changes in anatomical properties (Table 1), it can be expected that root wood is generally weaker than stem wood, however, this is not always the case (see next section). Along a horizontal, lateral root, longitudinal cells generally become wider and longer. Fahn (1967) attributed such a gradient in cell size to the advantageous effect it would have on suction pressure from the stem to the distal parts of the roots of *Retama raetam*, a desert plant. Such a feature would presumably ensure a more efficient flow of water, especially in the horizontal lateral roots which take up water from the soil surface. In the tap roots of

Table 1. Changes in anatomy along a woody lateral root

- At the stem-root base, cells are often more densely packed, with thicker walls than in the stem.
- Pith disappears.
- In hardwoods, the wood changes from ring porous to diffuse porous (light responsive).
- Longitudinal cells become wider and longer along the root.
- Walls of longitudinal cells become thinner along the root.
- Growth rings have less cells.
- Vessel lumina can be more angular than in the stem.
- Number of vessels per unit area is less.
- Rays are wider than in the stem and heterocellular rays become more conspicuous.
- Microfibril angle is often larger than in the stem.
- Pits are larger than in the stem and sometimes form on the tangential walls of cells.
- Tyloses are infrequent.
- Heartwood is infrequent.
- Boundaries between annual rings are less defined.
- Juvenile wood may persist longer than in the stem.
- Young roots ($<$ 16 years old) have little summer wood.
- Cells are less lignified.
- Parenchyma content is higher and fibre content lower than in the stem.

woody plants however, longitudinal cells are shorter and narrower with thicker cell walls than those in the stem (Patel 1965, 1971, Fahn 1967, Fayle 1968). As the tap root is under considerable mechanical stress, due to the weight of the tree, it may be argued that an increase in cell material would better resist the stress imposed upon the tap root. The distal parts of lateral roots are under relatively little mechanical stress, therefore, an increase in strengthening tissue is not necessary to augment rigidity.

The orientation of the cellulose microfibrils (MF's) in the secondary cell wall plays an important role in determining the toughness of wood (Timell 1986). Compression wood forms when the wood is placed under compressive loading and is more flexible than normal wood, thus resisting rupture better. In normal stem wood, the MF angle is between 5–30° to the cell axis, however in compression wood of conifers, this angle is found to be much higher, in the order of 45°. In lateral roots of Maritime pine the author found the MF angle to vary between 30° and 60°. In *Populus* sp., a smaller MF angle of 30–40° has been found in normal root wood (Hathaway and Penny 1975, Boyd 1977). Large MF angles lead to greater creep or visco-elastic relaxation. Such a large MF angle may be a major attribute in preventing root rupture, especially in conifers, as compression wood rarely forms in roots, unless they are exposed to sunlight (Fayle 1968). Tension wood in angiosperms forms in leaning trunks or branches and "pulls" the tree or branch upright by longitudinal shrinkage of the cells. An extra, gelatinous layer forms inside the cell wall, where the MF angle of is 5° to the cell axis. Differentiation of tension wood has been found on the upper side of permanently displaced lateral roots of *Populus* sp. (Sachsse 1974) and gelatinous fibres have been observed in lateral roots of *Fagus, Quercus* and *Robinia* (Patel 1964). Hence, it appears that strain imposed on roots can cause the development of tension wood, as it does in the aerial parts of the tree.

STRENGTH ADAPTATION

The more oblique the root, the greater is the percentage per unit volume of fibre and ray parenchyma content present (Fayle 1968), hence specific gravity (SG) is also altered. Specific gravity is a measure of the wood or the amount of the cell wall material present and is defined by:

$$\text{specific gravity} = \frac{\text{the dry mass of a volume of wood}}{\text{mass of an equal volume of water}}$$

Fayle (1968) also showed that SG measurements taken along the length of an oblique and a vertical root of 130 year old *Tilia americana,* differed greatly between the two roots, at the same difference from the stem. The oblique root had a higher SG and wood anatomy similar to that found in the stem, whereas the horizontal root had a typical lateral root anatomy (Table 1). Oblique and sinker roots will normally be under a greater compression stress than lateral roots, especially if they are sited underneath the centre of the tree, *i.e.* the centre of the tree's gravity, where most of the tree's weight bears down on them. As SG is directly related to mechanical strength, an increase in SG would increase wood strength (Stokes 1995).

The number and distribution of oblique and vertical roots in a root system may influence the strength of the lateral roots (Stokes and Mattheck 1996). Shallow plate root systems consist of slightly tapering lateral roots and few vertical roots, *e.g. Picea* sp. In such systems, lateral roots were found to be stronger further away from the stem than the highly tapered laterals found in heart (*e.g. Larix* sp.) and tap root (*e.g. Pinus sylvestris*) systems, where large vertical and oblique roots were present. External loading forces in plate root systems will be transmitted into the soil further away from the stem due to the lack of vertical branches, therefore a high rate of branching near the stem, or large, rigid tap root, allows a faster dissipation of forces nearer the stem, therefore a high investment in strength further along the root is not necessary.

As a tree sways in the wind, windward lateral roots are lifted up in tension and leeward roots are pushed down onto the hard bearing surface of the soil. Maximum windward root displacement can occur within 50–100 cm of the stem in mature trees and a point of rotation on the leeward side much closer to the stem (Coutts 1983, 1986, Stokes *et al.* 1995c). The strain on the roots in these areas will be much greater than along the rest of the lateral root. Mechanical tests on the compression strength of lateral roots of *Picea abies* showed that wood strength was higher 25–50 cm along the root than the rest of the root or stem at DBH (Stokes and Mattheck 1996). Reidl (1937) found that in *Picea* lateral roots, SG increased rapidly between 10 and 50–80 cm from the stem and then declined. It appears that the changes in SG observed by Reidl may account for the increase in strength observed by Stokes and Mattheck (1996), although a more complete study would be needed to confirm the relationship. It is also interesting to note that the T-beam roots of Sitka spruce (Fig. 1a,b), observed by Nicoll and Ray (1996) and discussed in the first section, were also found to have the most uneven growth 50 cm from the stem centre on the leeward side of the tree. Therefore it appears that even at specific points along a lateral root, responses to mechanical stress can take place in order to improve resistance to rupture. One further example of this phenomenon can be seen in the lateral roots of *Populus* sp. (Stokes and Mattheck 1996). Compression strength was found to increase sharply by 25% on the underside of the roots. As the roots are pushed onto the bearing surface below, the lower sides of lateral roots in shallow root systems, such as poplar, will be placed under enormous compression stress. This action also inhibits radial growth, and in extreme cases, rays even buckle (Fayle 1968). As SG of root wood has been found to increase with decreasing annual ring width, where radial growth is inhibited, cell density must have increased, thereby increasing compression strength.

CONCLUSIONS

Trees with a high resistance to windthrow would be extremely valuable to the forest industry, however, it is necessary to identify those characteristics contributing to tree stability (Table 2). The growth of woody roots, which function for anchorage, can be influenced by environmental factors. Mechanical stresses must affect the activity of the root cambium since there is an increase in radial growth when roots are stimulated. The distribution of resources can also be altered within a root system, as those roots which provide the greatest contribution to anchorage receive more resources. It is not clear, however, what signal is transmitted to the roots, or whether a signal is triggered locally within the root system. The perceptive mechanism by which plants sense movement also remains elusive. It has been found that four genes in *Arabidopsis*, are switched on in response to touch (Braams and Davis 1990). These genes encode calmodulin and related proteins, which bind calcium. Calcium has been implicated in the thigmomorphogenic process in that increased ethylene synthesis is the result of high cytosolic calcium (Pickard 1984). Ethylene production in stems increases under mechanical stress and is thought to be the mediator of increased radial growth and reduced elongation (Telewski 1995).

Table 2. The responses of tree roots to mechanical stress

Dynamic loading (e.g. wind)	Static loading (e.g. tree weight)
Increase in lateral root base diameter (AB)	Increase in cell density at root base
Increase in lateral root biomass (AB)	Increase in cell wall thickness at root base
Development of root eccentricity (AB)	Large microfibril angle
Formation of buttress roots	Higher content of fibres and ray parenchyma in oblique or vertical roots
Reduction in total biomass of root system	Increase in specific gravity in the tap root
Increase in number of lateral roots (AB)	Increase in wood strength, probably due to an increase in specific gravity (MS)
Increased branching of lateral roots loaded in tension	Formation of tension wood in angiosperms (MS)
Increase in length and diameter of lateral roots loaded in tension	Formation of compression wood in gymnosperms (when exposed to light) (MS)
Increase in specific gravity of root wood (MS)	
Increase in root wood strength (MS)	

(AB), along the axis of bending. (MS), at mechanically stressed points along the lateral roots.

The influence of wind on tree growth is highly complex and not yet fully understood. However, adaptation to external loading does occur, enabling a tree to better withstand the potentially dangerous mechanical stresses imposed upon it.

REFERENCES

Berthier S., Stokes A. and Guitard D., 1997. Effects of mechanical stress on wood formation in Maritime pine (*Pinus pinaster*) roots. *in:* Proc. of the Second Workshop of IUFRO W.P. S5.01–04 "Connection between silviculture and wood quality through modelling approaches and simulation software," South Africa 1996. In press.

Braams J., and Davis R.W., 1990. Rain-, wind- and touch-induced expression of calmodulin-related genes in *Arabidopsis*. *Cell* 60:357.

Boyd J.D. 1977. Basic cause of differentiation of tension wood and compression wood. *Aust. For. Res.* 7:121.

Cannell M., and Coutts M.P., 1988. Growing in the wind. *New Scientist* 21:42.

Coutts M.P. 1983. Root architecture and tree stability. *Plant and Soil* 71:171.

Coutts M.P. 1986. Components of tree stability in Sitka spruce on peaty, gley soil. *Forestry* 59:173.

Ennos R., 1995. Development of buttresses in rainforest trees: the influence of mechanical stress. *in:* "Wind and Trees," M.P. Coutts, J. Grace, eds., Cambridge University Press, p. 293.

Eskilsson S., 1969. Fibre properties in the spruce root system. *Cellulose Chem. Technol.* 3:409.

Fahn A. 1967. Plant Anatomy. Pergamon Press.

Fayle D.C.F. 1968. Radial growth in tree roots. Technical report no. 9. Faculty of Forestry, University of Toronto, Canada.

Fillon M., 1987. Les Chablis de Bretagne. *Ann. Meca. Forestière* 42:97.

Grace J., 1977. Plant response to wind. Academic Press, London.

Gribbin J., 1990. Hothouse Earth; the greenhouse effect and Gaia. Black Swan Publications.

Hathaway R.L., and Penny D. 1975. Root strength in some *Populus* and *Salix* clones. *N.Z. J. of Bot.* 13:333.

Heiligmann R., and Schneider G., 1974. Effects of wind and soil moisture on black walnut seedlings. *Forest Science* 20:331.

Henwood K., 1973. A structural model of forces in buttressed tropical rain forest trees. *Biotropica* 5:83.

Jacobs M.R., 1939. A study of the effect of sway on trees. *Aust. Comm. For. Bur. Bull.*, 26.

Jacobs M.R., 1954. The effect of wind sway on the form and development of *Pinus radiata* D. Don. *Aust. J. Bot.* 2:35.

Jaffe M.J., 1973. Thigmomorphogenesis: the response of plant growth and development to mechanical stimulation-with special reference to *Bryonia dioica*. *Planta* 114:143.

Mattheck C., 1991. Trees: the mechanical design. Springer-Verlag, Berlin.

Mattheck C., and Breloer H., 1992. Der Wurzelquerschnitt als Protokoll der Lastgeschichte. *Allg. Forst und Jagd Zeitschrift* 163:142.

Mattheck C., and Breloer H., 1995. The body language of trees. A handbook for failure analysis. HMSO, London.

Nicoll B.C., Easton E.P., Milner A.D., Walker C., and Coutts M.P., 1995. Wind stability factors in tree selection: distribution of biomass within root systems of Sitka spruce clones. *in*: "Wind and Trees," M.P. Coutts, J. Grace, eds., Cambridge University Press, p. 276.

Nicoll B.C. and Ray D., 1996. Adaptive growth of tree root systems in response to wind action and site conditions. *Tree Physio.* 16:899.

Patel R.N., 1964. On the occurrence of gelatinous fibres with special reference to root wood. *J. Inst. Wood. Sci.* 12:67.

Patel R.N., 1965. A comparison of the anatomy of the secondary xylem in roots and stems. *Holzforschung* 19:72.

Patel R.N., 1971. Anatomy of stem and root wood of *Pinus radiata* D. Don. *N.Z. J. For. Sci.* 1:37.

Pickard B.L. 1984. Voltage transients elicited by sudden step-up of auxin. *Plant, Cell and Environ.* 7:171.

Quine C.P., 1991. Recent storm damage to trees and woodlands in southern Britain. Forestry Commission Bulletin 97:83. HMSO, London.

Rees D.J., and Grace J. 1980. The effects of wind on the extension growth of *Pinus contorta* Douglas. *Forestry* 53:145.

Reidl H. 1937. Bau und leistungen des wurzelholzes. Jahrbücher für Wissenschaftliche Botanik. Leipzig: Verlag von Gebrüder Borntrager, 1–75.

Richter W., 1984. A structural approach to the function of buttresses of *Quararibea asterolepis*. *Ecology* 65:1429.

Sachsse H. 1974. Vorkommen und räumliche Verteilung von Richtgewebe im Wurzelholz der *Populus x euramericana* cv. *Robusta. Holz Roh Werkst.* 32:263.

Satoo T., 1962. Wind, transpiration and tree growth. *In:* Kozlowski T.T., ed. Tree growth. New York: The Ronald Press 299–310.

Slodicak M., 1995. Thinning regime in stands of Norway spruce subjected to snow and wind damage. *In:* Coutts M.P., Grace J., eds. Wind and Trees. Cambridge University Press p. 436–47.

Stokes A; 1995. The shape of tree root systems affects root wood strength. Technical Report no 5518, Forschungszentrum Karlsruhe, Germany.

Stokes A. Fitter A.H., and Coutts M.P., 1995a. Responses of young trees to wind: effects on root growth. *in*: "Wind and Trees," M.P. Coutts, J. Grace, eds., Cambridge University Press, p. 264.

Stokes A., Fitter A.H., and Coutts M.P. 1995b. Responses of young trees to wind and shading: effects on root architecture. *J. Exp. Bot.* 46:1139.

Stokes A., Drexhage M., Heinze P., and Guitard D. 1995c. Distribution of strain in the root systems of forest trees. Proc. 3rd Int. Colloque "L'Arbre," Montpellier, France.

Stokes A., and Mattheck C. 1996. Variation of wood strength in tree roots. *J. Exp. Bot.* 47:693.

Stokes A., Nicoll B.C., Coutts M.P. and Fitter A.H., 1997. Responses of young Sitka spruce clones to mechanical perturbation and nutrition: effects on biomass allocation, root development and resistance to bending. *Can. J. For. Res.* 27: 1049.

Telewski F.W., and Jaffe M.J. 1986. Thigmomorphogenesis: field and laboratory studies of *Abies fraseri* in response to wind or mechanical perturbation. *Physiol. Plant.* 66:211.

Telewski F.W. 1990. Structure and function of flexure wood in *Abies fraseri*. *Tree Physiology* 5:113.

Telewski F.W., 1995. Wind-induced physiological and developmental responses in trees. *in*: "Wind and Trees," M.P. Coutts, J. Grace, eds., Cambridge University Press, p. 237.

Timell T.E. 1986. Compression Wood in Gymnosperms 2. Springer-Verlag, Berlin.

Wilson B.F. 1975. Distribution of secondary thickening in tree root systems. *in:* The development and function of roots. eds. J.G. Torrey and D.T. Clarkson. Academic Press, New York.

30

ROOT GROWTH OF COTTON AS INFLUENCED BY CO_2 AND TEMPERATURE

V. R. Reddy

USDA: ARS: BA: NRI: Remote Sensing and Modeling Laboratory
Bldg. 007, Rm. 008, BARC-West
Beltsville, Maryland 20705

Atmospheric carbon dioxide concentration [CO_2] is increasing and it is important to know how this will affect crop growth. Understanding crop response to climate change requires knowledge of how roots respond to changes in the aerial environment. Changes in rate and amount of root growth could affect the root distribution in the soil profile and absorption of water and nutrients. Growth and distribution of cotton (*Gossypium hirsutum* L.) roots were examined at day/night temperatures of 15°/7°C, 20°/12°C, 25°/17°C, 30°/22°C and 35°/27°C and at [CO_2] of 350 and 700 μL. L^{-1} in the shoot chambers. Plants were grown in controlled-environment chambers with a perspex top under nearly natural daylight. Twice a week root observations were made on one 2 m^2 glass side of the soil bin. Root weight was significantly greater in the 700 μL. L^{-1} CO_2 treatment at all depths and at all temperatures. Number of roots increased with increasing temperature up to 25°/17°C but was not affected by the CO_2 treatment. Roots in the 350 μL. L^{-1} [CO_2] treatments were longer (root length per root axis) and penetrated the soil profile faster at the lower temperatures. In the 700 μL. L^{-1} CO_2 treatment, roots were more evenly distributed down the soil profile than in the ambient [CO_2] treatment. Root growth was depressed 63 days after emergence (DAE) in virtually all treatments when fruits (bolls) were developing. The optimum temperature for root growth was also the optimum temperature for shoot growth (30°/22°C). The effect of elevated [CO_2] was to make roots heavier, but there was no evidence that this was translated into a root system with increased length. Roots were shorter in elevated [CO_2], penetrating the soil profile less rapidly but perhaps more thoroughly.

1. INTRODUCTION

Increase in atmospheric carbon dioxide concentration, and associated climatic changes are of interest to agronomists and crop physiologists all over the world. For the last decade there has been enhanced interest in the scientific community on how the projected changes may affect crop production (Kimball, 1983; Post et al., 1990; Newton, 1991).

Biology of Root Formation and Development, edited by Altman and Waisel.
Plenum Press, New York, 1997.

Enhancement of canopy photosynthesis is one important direct effect of raising [CO_2]. Increase in [CO_2] has been shown to increase photosynthesis of cotton (Mauney et al., 1979; DeLucia et al. 1985) and to enhance growth and yield of above ground plant parts (DeLucia et al., 1985). The studies we conducted under controlled environmental conditions in recent years (Reddy et al., 1991a, 1991b, 1992b) established a comprehensive database on the effects of temperature on the above ground parts of cotton plants. The effects of extreme low and high temperatures were investigated and stem elongation, node initiation, leaf expansion, branching, fruiting and square and flower abscission were measured.

To date we have not seen any data on the effects of temperature and [CO_2], and their interactions on cotton root growth. This is partly due to the difficulty involved in measuring root growth non-destructively. The objectives of this study were to evaluate the interactive effects of temperature and [CO_2] on cotton root growth, root initiation rate, depth of root penetration and root dry weight under optimum water and nutrient conditions.

2. MATERIALS AND METHODS

The naturally lighted plant growth chambers used in this study have been described by Reddy et al. (1991a, 1991b). The growth chamber soil bins were filled with a mixture of sand and vermiculite (3:1 by volume) that had incorporated in it slow release micronutrients at the rate of 88 mg L^{-1} of soil prior to filling the bins. Cotton, "Deltapine 50" (DPL 50) seeds were pregerminated in moistened paper towels at 28°/23°C day/night temperatures for 48 hours. The germinated seeds were selected for uniformity and planted in the plant growth chambers.

The temperature and CO_2-controlled chambers were all maintained at 28°/23°C (day/night) until 14 days after emergence (DAE). Temperature and [CO_2] treatments were then imposed, and the air temperatures in the growth chambers were maintained at 15°/7°C, 20°/12°C, 25°/17°C, 30°/22°C and 35°/27°C, averaging 17.8°C, 18.7°C, 22.7°C, 26.6°C and 30.6°C over the duration of the experiment. CO_2 concentrations at the above ground compartments were maintained at 350 and 700 μL. L^{-1} for each temperature. The day time temperature was initiated in these naturally lighted (39° N Latitude) growth chambers at 1h after sunrise and returned to night time temperature 1h after sunset during the experimental period.

Carbon dioxide concentration, air temperature, and irrigation time were controlled by a computer, which also monitored other environmental and plant response variables. The temperatures in the growth chambers were maintained to within ±0.1°C of the set points for at least 95% of the time, using a secondary cooling system and electrical resistance heaters. Continuous circulation of air maintained uniform temperatures throughout the chambers. The chambers were sealed, and the CO_2 concentration was monitored at 10-s intervals and averaged over 15 min. periods. Carbon dioxide was injected from a gas cylinder through a pressure regulator, solenoid valve, needle valve, and flow meter into the chambers as necessary to maintain 350 or 700 $\mu L \cdot L^{-1}$. Graded shade cloths were adjusted around the cabinet edges to plant height to simulate shading effects found in a field crop.

The plants were irrigated with drip irrigation system with one emitter per plant, three times a day. Drippers were calibrated prior to the start of the experiment, and those drippers that emitted more or less than 15% of the set point were replaced. The amount and timing of the irrigations were computer controlled. Insects were controlled as needed during the course of the experiment with insecticide applications.

New root growth, appearing on the glass face of the soil bin, were recorded at least twice each week. The root length was measured and marked with wax pencil. In addition, at each measurement time the number of growing root tips and depth of the deepest root were recorded. At the time of the final destructive harvesting, on 70 DAE, the rooting medium in the soil bins was excavated and the roots were washed out of the medium using a metal screen. This destructive harvesting of the roots was accomplished separately for each 0.1m layers of the root system in each soil bin. The washed roots were dried and weighed.

3. RESULTS AND DISCUSSION

The distribution of root dry weight at 70 DAE is shown in Figure 1. Root dry weight increased with increasing temperature at both [CO_2]. There was about a five-fold increase in root dry weight from 17.8°C to 18.7°C. At the highest temperature, root dry weight was much larger in 700 μL. L^{-1} [CO_2] than it was in 350 μL. L^{-1} CO_2. The high [CO_2] treatment produced higher root weight in all the temperature treatments. The increase caused by additional [CO_2] was 51% at 17.8°C, 16% at 22.7°C, 25% at 26.6°C and 74% at 30.6°C. The largest amount of root dry weight ranging from 60% to 84% of the total root weight was present in the top 20 cm of the soil in all [CO_2] and temperature treatments. As the soil depth increased, the amount of root weight decreased in all [CO_2] and temperature treatments. A similar root distribution pattern was observed in soybean by DelCastillo et al. (1989) and Arya et al. (1975). However, the amount of root dry weight was much lower in soybeans compared to cotton at all the depths observed. The large root dry weight located in the top 20 cm soil is mostly contributed by the large taproot of the plants. At greater soil depths the taproot is present, but its dry weight is relatively small. The effect of [CO_2] on root dry weight accumulation increased with increase in temperature, and at all the soil depths.

The root:shoot ratios at 70 DAE decreased with increasing temperature in both [CO_2] treatments (data not shown). This decrease in root:shoot ratio was primarily caused by the different stages of development of the cotton plants in different temperature treatments. Cotton plants distribute carbohydrates about 50% to roots and 50% to shoots at the time of emergence. However distribution of carbohydrates to roots decreases to less than 10% by flowering time (Hodges et al., 1992). The root:shoot ratios were unaffected by [CO_2] in all temperature treatments.

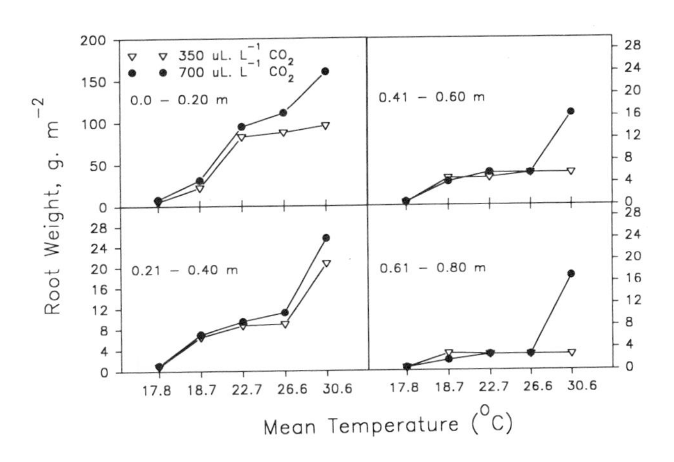

Figure 1. The effects of CO_2 and temperature on root dry weight at different soil depths, when the roots were harvested at 70 DAE.

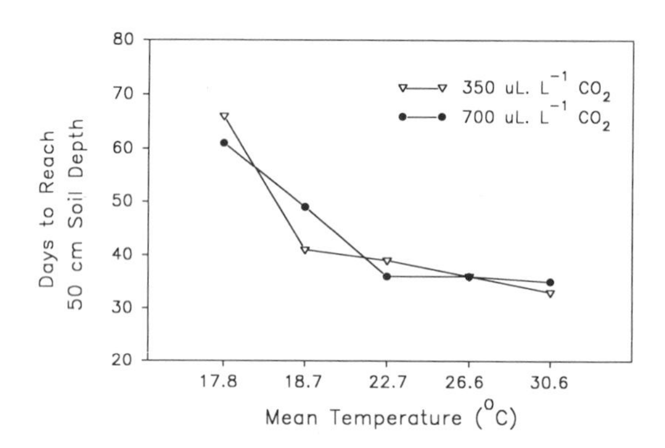

Figure 2. The number of days taken by the roots to reach 50 cm soil depth under different CO_2 and temperature treatments.

3.1. Root Penetration

The time taken for roots to penetrate half the depth (0.5m) of the soil profile was more or less similar at the two [CO_2] (Fig. 2), except at the lowest two temperature treatments. At 17.8°C there was a decrease in the number of days taken for the roots to penetrate to 0.5m depth and at 18.7°C there was a significant increase in the 700 μL. L^{-1} [CO_2] treatment. The time taken for the roots to penetrate to 0.5m soil depth decreased with increase in temperature irrespective of [CO_2]. This increase in root penetration with increase in temperature, probably satisfied the increased demand for water and nutrients because the canopy was much larger at higher temperatures (Reddy et al., 1991a, 1991b). It can also be triggered by the availability of larger amounts of carbon from the larger canopies.

3.2. Maximum Root Elongation Rate and Root Density

The maximum elongation rate of individual root axes, regardless of when it occurred in the life cycle, was higher in 350 μL. L^{-1} CO_2 than in 700 μL. L^{-1} CO_2 except at 17.8°C where the plants in 700 μL. L^{-1} CO_2 had a higher elongation rate. The maximum root elongation rate increased with increase in temperature irrespective of [CO_2] treatment.

The root weight:root length ratio which gives the weight of root per unit length (g. cm^{-1} root) increased with increase in temperature at both [CO_2] levels, although the rate of increase was higher in 700 μL. L^{-1} CO_2 and especially in higher temperature treatments, and the ratio more than doubled in 30.6°C for plants grown 700 μL. L^{-1} CO_2 compared with 350 μL. L^{-1} CO_2. The root weight:root number ratio, which is also an indicator of root density was higher in 700 μL. L^{-1} CO_2 compared to 350 μL. L^{-1} CO_2 except at 17.8°C. The ratio also increased with the increase in temperature at both [CO_2], indicating that the roots were denser at higher temperatures and [CO_2].

3.3. Root Distribution by Soil Depth

The distribution of roots within the soil profile was bimodal with peaks in the top 0.2m and at the 0.5–0.6m soil depth. The CO_2-enriched plants had roots more evenly distributed over the top 0.6m of soil depth. The number of roots increased with increased air temperature at both [CO_2], however the highest number of roots were observed at 22.7°C with 700 μL. L^{-1} CO_2. The [CO_2] appeared to have an influence at the lowest temperature, where there were considerably higher number of roots produced at 700 μL. L^{-1} CO_2. It appears that high [CO_2] produced higher number of roots in the top soil layer at 30.6°C.

4. CONCLUSIONS

The dynamic response of the root systems to temperature and [CO_2] cannot be understood without knowing how those same factors affect shoot and fruit development. Flushes in root growth appeared to be related to shoot activity. Roots flourished when flowers and fruits aborted or when shoot growth declined. High temperatures can limit shoot or fruit growth and favor root development. The greatest root activity occurred at temperatures that are optimal for shoot growth. Proliferation of roots coincides with specific stages in plant development . The location of roots in the soil profile depends on how far they have penetrated into the soil profile when certain plant development stages were reached.

In this study, the elevated [CO_2] made the roots heavier, but there was no evidence that this was translated into a root system with more absorbing power for water and nutrients. There was no increase in the number of root axes. The observed roots were shorter, and they penetrated the soil profile less rapidly. However, root distribution through the soil profile was more uniform under elevated [CO_2].

Temperature increased the numbers of root axes and root length up to 22.7°C. Further increases in root activity depended on how temperature affected the balance between the root and the shoot.

REFERENCES

Arya, L.M., Blake, G.R., and Farrel, D.A. 1975. A field study of soil water depletion patterns in the presence of growing soybean roots: III. Rooting characteristics and root extraction of soil water. *Soil. Sci. Soc. Am. Proc.* 39: 437.

DelCastillo, D., Acock, B., Reddy, V.R., and Acock, M.C. 1989. Elongation and Branching of Roots on Soybean Plants in a Carbon Dioxide-Enriched Aerial Environment. *Agron. J.* 81: 692.

DeLucia, E.H., Sasek, T.W., and Strain, B.R. 1985. Photosynthesis inhibition after long-term exposure to elevated levels of atmospheric carbon dioxide, *Photosynthesis. Res.* 7: 175.

Hodges, H.F., Reddy, K.R., McKinion, J.M., and Reddy, V.R. 1992. Temperature effects on cotton growth and development. Miss. Agrl. Exp. Stn. Bull. 1.

Kimball, B.A. 1983. Carbon dioxide and agricultural yield: an assemblage and analysis of 430 prior observations. *Agron. J.* 75: 779.

Mauney, J.R., Guinn, G.E., Fry, K.E., and Hesketh, J.D. 1979. Correlation of photosynthetic carbon dioxide uptake and carbohydrate accumulation in cotton, soybean, sunflower and sorghum. *Photosynthetica.* 13: 260.

Newton, P.C.D. 1991. Direct effects of increasing carbon dioxide on pasture plants and communities. *New Zealand J. Agric. Res.* 34: 1.

Post, W.M., Tsung-Hung Peng, Enamuel, W.R., King, A.W., Dale, V.H., and DeAngelis, D.L. 1990. The global carbon cycle. *Am. Scientist.* 78: 310.

Reddy, K.R., Reddy, V.R., and Hodges, H.F. 1992b. Temperature effects on early season cotton growth and development. *Agron. J*: 84: 229.

Reddy, V.R., Reddy, K.R., and Baker, D.N. 1991a. Temperature effect on growth and development of cotton during the fruiting period. *Agron. J.* 83: 211.

Reddy, V.R., Baker, D.N., and Hodges, H.F. 1991b. The effects of temperature on cotton canopy growth, photosynthesis and respiration. *Agron. J.* 83: 699.

SAS Institute, Inc. 1987. SAS user's guide: statistics version 5.18, SAS Inst. Inc., Cary, NC.

31

RESPONSE OF ROOT GROWTH TO A COMBINATION OF THREE ENVIRONMENTAL FACTORS

Water Stress Salinity and Soil Compactness

Z. Plaut, Melanie Newman, Evelyn Federman, and A. Grava

Institute of Soils & Water
ARO, Volcani Center
Bet Dagan, Israel

INTRODUCTION

Intensive irrigation in heavy clay soils, especially with brackish water, may induce a saline water table to rise. Since the salts are mostly Na salts, the high sodium content may lead to deterioration of soil structure and reduction of its hydraulic conductance. Problems of soil salinity and lack of aeration in the root zone may be intensified if artificial drainage systems are ineffective or absent. Salinity levels and a low rate of oxygen diffusion may become mostly inhibitory to root growth in deep soil layers. Salinity has been shown to reduce root growth, in studies of different crops (Shalhevet and Bernstein, 1968; Poljakoff-Mayber and Lerner, 1994). The root elongation rate and root length have also been shown to be inversely proportional to the soil impedance (Misra et al., 1986; Bennnie, 1991).

A popular way to overcome the salinity hazard under field conditions is to apply a high rate of irrigation water during the growing season; this will partly serve for leaching. However, this practice increases the salt load of the field, raises the water table and may thus induce soil compaction and lack of aeration.

A preferable way to overcome the problem, at least partly, is to minimize water application, in order to stimulate water use from deep soil layers. A simulation model has shown the advantage of deficiency irrigation under high-water-table conditions, followed by pre-season salt leaching (Bradford and Letey, 1992). The same approach was tested by adopting an irrigation regime which uses less water than the crop requirement and applies the water at long intervals (Cohen et al., 1996). It was considered that this would promote root expansion and water use from deep and wet soil layers. The findings of field trials with cotton, however, did not support this concept. A marked decrease in soil moisture content under deficient irrigation was found only in the top 30 cm, and there was hardly any water use from layers below 60 cm.

Biology of Root Formation and Development, edited by Altman and Waisel.
Plenum Press, New York, 1997.

The above findings contradict what is known about the distribution of water use by cotton, along the soil profile (Ritchie, 1980). It has also been shown that most of the available soil water is depleted at the end of the growing season (Jordan, 1983). The restriction of root development and water use below 60 cm, and the almost total lack of water use below 90 cm in that study could thus be attributed to the effects of salinity, lack of soil aeration or a combination of both factors.

Experiments were conducted to simulate field conditions in soil columns and to distinguish between the effects of salinity and those of soil compaction—which causes the low aeration—on root growth and water use. Water stress, which could be a result of the infrequent irrigation provided under the adopted management regime, might also lead to reduced root growth, and was thus, the third factor to be studied.

EXPERIMENTAL

Clay-soil columns 38 cm high and 10.4 cm in diameter were prepared by filling the lower halves of plastic cylinders (80 cm high). The soil in half of the cylinders was compressed during filling and was then brought to a moisture content of 0.48 by weight (predetermined as being equivalent to a soil moisture tension of 10 kPa). Good-quality or saline water (EC = 1.0 and 8.0 dS/m, respectively) was used to wet the soil. The same two types of water were also used to wet the uncompressed soil columns. The salinized and uncompressed soil columns were treated, prior to cylinder filling, with a synthetic polymer, P-101 (supplied by Hydropolymer Co., Israel) to prevent the clay dispersion and compaction expected to result from adding the Na salt. Soil was then added to the cylinders, without compression, to extend the column up to 75 cm, and the columns were irrigated with good-quality water. The cylinders were placed on top of dry crushed soil taken from the same location, to allow drainage of excess moisture.

Sunflower seeds were planted in the cylinders and seedlings were thinned to one per cylinder shortly after emergence. The soil surface was covered with a layer of small white polystyrene balls to minimize evaporation. Water losses due to transpiration were determined by weighing the cylinders and water was replenished according to the experimental design (Table 1). Two levels of water stress were imposed, by irrigation at two frequencies, which were gradually increased during plant development, from twice and once a week up to daily and three times a week, at full canopy.

In order to avoid possible losses of salts, through leaching with excess irrigation water, the cylinders were placed once a week in buckets containing either salt solution (EC = 8 dS/m) or tap water (EC = 1 dS/m) for 4-h periods. The height of the solution

Table 1. Experimental design and density of soil columns

Treatment[1]	Soil compression	Soil salinity	Irrigation frequency	Soil density ($g.cm^{-3}$)	
				Top	Bottom
Con[1]-FR	–	–	High	1.16	1.16
Con-IF	–	–	Low	1.16	1.16
Sal-FR	–	+	High	1.16	1.20
Sal-IF	–	+	Low	1.16	1.20
Comp-FR	+	–	High	1.16	1.34
Comp-IF	+	–	Low	1.16	1.34
Comp-Sal-FR	+	+	High	1.16	1.33
Comp-Sal-IF	+	+	Low	1.16	1.33

[1]Con = control, Sal = salinity, Comp = compressed, FR = frequent, IF = infrequent.

was set so that capillary rise in the cylinders did not exceed the height of the presalinized soil layers.

Plant water use was determined by weighing, taking into account water addition from the bottom. Changes in soil moisture tension were determined by means of tensiometers located 20 and 60 cm from the top. The effect of the three environmental factors (soil compression, salinity and water stress) on root and canopy development was determined at the end of the growing season, seven weeks after planting. At this time soil columns were divided into seven segments, and soil EC and root distribution were determined in each segment as described previously (Plaut et al., 1996).

RESULTS AND DISCUSSION

The soil EC analyzed at the termination of the experiment shows a distinct difference between the top three soil segments and the bottom segments in the salinized columns (Fig. 1): that at the bottom was approximately 2 dS/m higher under the low-frequency irrigation, probably because of lower water content at the time of sampling.

The low irrigation frequency induced a higher soil moisture tension at both soil depths throughout a significant part of the growing period (Fig. 2). The rise of soil moisture tension was clearly suppressed under salinity, especially in the deeper layer, probably due to a reduced rate of water use from the salinized soil.

Root development in the soil columns was clearly reduced by salinity and water stress (imposed by infrequent irrigation). Soil compression reduced root development only

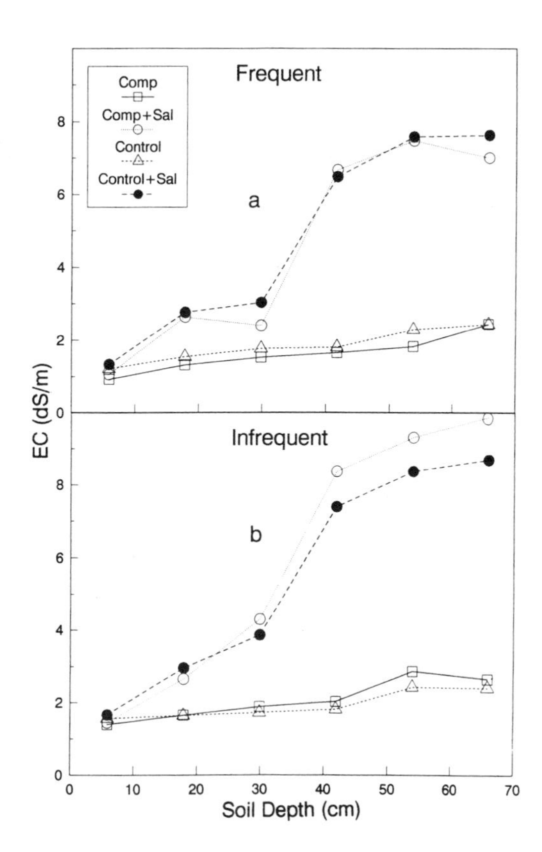

Figure 1. EC of the soil profile under frequent and infrequent irrigation. Comp = compressed soil in bottom half of cylinders, Comp+Sal = compressed and salinized soil, Control = uncompressed and unsalinized soil, Control+Sal = uncompressed and salinized soil.

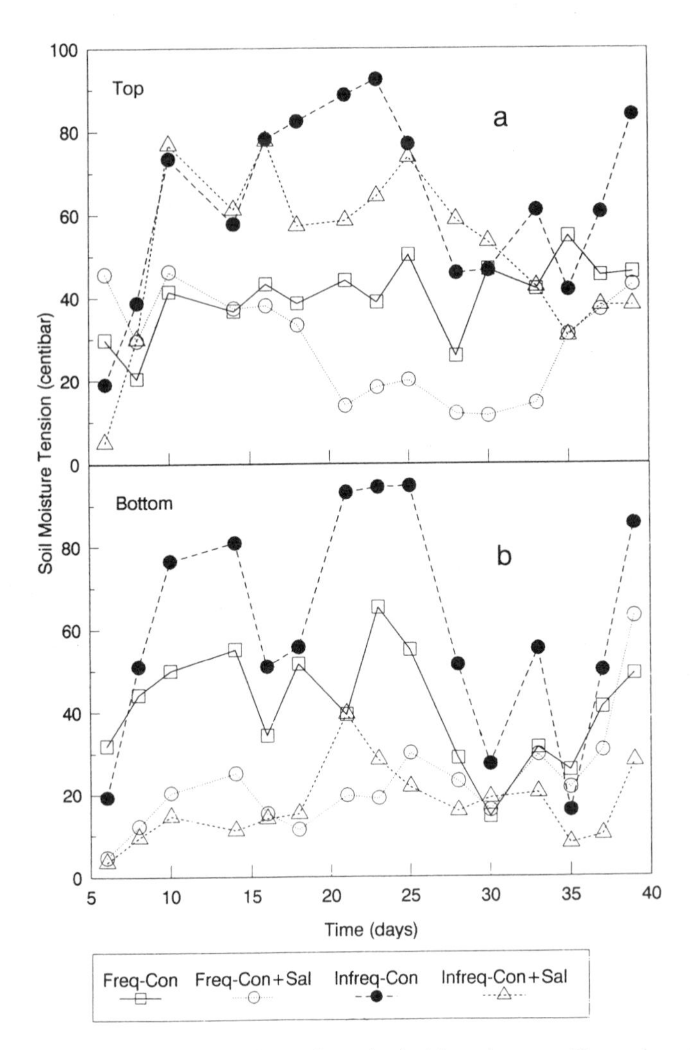

Figure 2. Changes with time of soil water tension as determined with tensiometers. The tensiometers were located at 20 and 60 cm from the surface are designated as top and bottom. Each value is an average of 2 tensiometers and 2–3 measuring days. Freq. and Infreq = High and low frequency of irrigation, respectively. Con = control, uncompressed soil, Sal = salinized.

under salinity (Fig. 3). A clear gradient in root development from the top to the bottom layer was found under all environmental conditions, which is similar to what is known for various species (Klepper, 1987). The distribution of root length density (cm root/cm^3 soil) clearly shows maximal density for the control plants and minimal density for those in compressed and salinized soil (Fig. 3). Under compression, root density in the deep layers was hardly affected in unsalinized soil but was decreased in the salinized soil; especially when there was no water stress (Fig. 3a). Water stress reduced root density by approximately 30–50% in the top two soil layers (apart from the treatment with compressed and unsalinized soil). The effect of water stress on root density gradually diminished toward the lower soil layers, since most of the water was extracted from the top layers, so that there was no stress development in the lower layers.

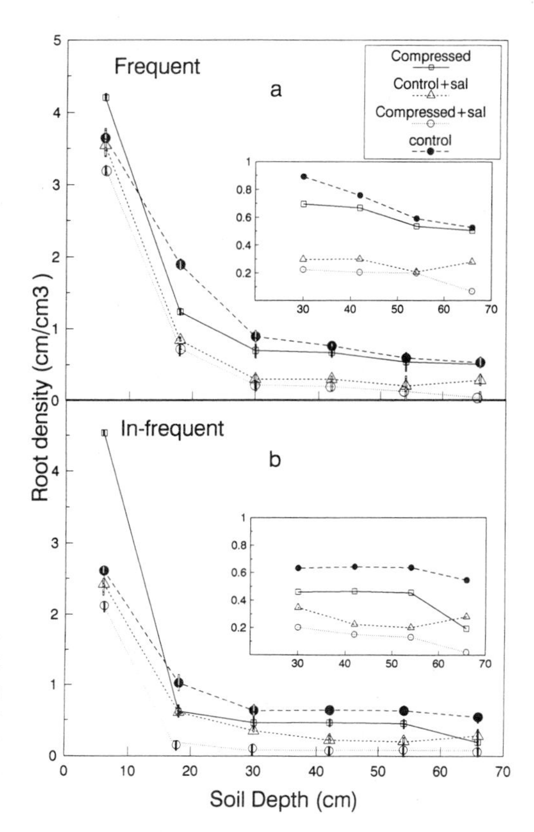

Figure 3. Distribution of root density (cm.cm^{-3} soil) along the soil column under frequent and infrequent irrigation. The distribution was determined among six soil segments, each 12 cm thick, and is shown for the center of each segment. The distribution in the four deepest segments is presented on an extended scale in inserts.

The interaction among the effects of soil compression, salinity and water stress on root development (Fig. 3) was confirmed by combining the separate responses of root dry weight to these three factors (Table 2). Comparison between these calculated responses and the actual decrease in root dry weight, when both salinity and soil compression were effective (Table 2), shows that the actual combined effect exceeded the calculated value.

Table 2. Effect of irrigation frequency, salinity and soil compression on sunflower dry weight production. SE of the means in parentheses

Treatment*			Shoot			Root		
				Relative			Relative	
Irr	Comp.	Sal.	(g/plant)	Meas.	Cal.	(g/plant)	Meas.	Cal.
FR	–	–	75.0 (6.5)	1.00		7.48 (0.95)	1.00	
	–	+	36.8 (4.8)	0.49		5.16 (0.36)	0.70	
	+	–	56.1 (4.7)	0.75		7.01 (0.42)	0.94	
	+	+	29.8 (4.6)	0.40	0.37	2.93 (1.31)	0.39	0.66
IF	–	–	57.4 (3.1)	0.77		5.72 (0.50)	0.76	
	–	+	21.2 (2.3)	0.28	0.38	1.95 (0.24)	0.26	0.53
	+	–	41.7 (2.5)	0.56	0.58	5.88 (0.50)	0.79	0.71
	+	+	22.2 (3.3)	0.30	0.28	2.38 (0.23)	0.32	0.50

*Irr = Irrigation, Comp = Compressed, Sal = Salinity, FR = Frequent, IF = Infrequent, Meas. = Measured, Cal. = Calculate

This applied to root development under both irrigation regimes, frequent and infrequent, but did not apply to shoot growth, especially under frequent irrigation. The combination of soil compression and salinity must, therefore, have had a strong inhibitory effect on root development. Since this inhibition was much less reflected in shoot growth, it seems that it was not necessarily related to conductance of water and ions. The actual combined effect of water stress and salinity also exceeded that calculated on the basis of the effects of those factors acting separately. It should be emphasized that the interaction between water stress and salinity was observed for both shoot and root growth.

The findings on root development imply that if root growth is limited by water stress in one section of the root zone, the possibility of compensatory extra growth in other zones must also be limited. These findings also imply that the possibility of inducing the crop to increase its use of water from deep soil layers, by restricting irrigation, is questionable. Additional, unpublished findings support the conclusion that suppression of water uptake in one region of the root zone by water stress cannot be compensated by additional root growth and water uptake in another region, especially if there are saline conditions in the latter region. This failure could simply be due to a decrease in the number of roots reaching the latter region. Alternatively, it could be attributed to a chemical control, such as ABA accumulation induced by water stress in the roots of the region exposed to water stress and transported within the plant to other growing regions. This possibility awaits further investigation.

REFERENCES

Bennie, A.T.P., 1991, Growth and mechanical impedance. in: "Plant Roots - The Hidden Half," Y. Waisel, Eshel, A., and Kafkafi, U., eds., Marcel Dekker, Inc., New York, NY.

Bradford, S., and Letey, J., 1992, Simulated effects of water table and irrigation scheduling as factors in cotton production. Irrig. Sci. 13: 101–107.

Cohen, M, Plaut, Z., Meiri, A., and Hadas A., 1996, Impact of deficit irrigation on water use from clay soil with shallow ground water table. Agron. J. (In press).

Jordan, W.R., 1983, Cotton. in: "Crop-Water Relations," I.D. Teare, and Peel, M.M., eds. John Wiley & Sons, New York, NY.

Klepper, B. 1987, Origin, branching and distribution of root systems. in: "Root Development and Function," P.J. Gregory, Lake, J.V., and Rose, D.A., eds., Cambridge University Press, Cambridge, UK.

Misra, R.K., Dexter, A.R. and Alston, A.M., 1986, Maximum axial and radial growth pressures of plant roots. Plant and Soil 95: 315–326.

Plaut, Z., Carmi, A., and Grava, A., 1996, Cotton root and shoot responses to subsurface drip irrigation and partial wetting of the upper soil profile. Irrig. Sci. (In press).

Poljakoff-Mayber, A. and Lerner, H.R., 1994, Plants in saline environment. In: "Handbook of Plant and Crop Stress" M. Pessarakli, ed. Marcel Dekker, Inc., New-York, NY.

Ritchie, J.T. 1980, Moving up the yield curve: Advances and obstacles. Am. Soc. Agr. Special Publication No. 39 pp 1–12.

Shalhevet, J., and Bernstein, L., 1968, Effect of vertically heterogeneous soil salinity on plant growth and water uptake. Soil Sci. 106: 85–93.

32

INFLUENCE OF WATER STRESS ON THE MASS OF ROOTS OF BUFFELGRASS SELECTIONS AND ON THEIR DISTRIBUTION

N. F. G. Rethman, P. S. Venter, and J. P. Lindeque

University of Pretoria
Pretoria, Gauteng
South Africa

INTRODUCTION

Approximately two thirds of South Africa is classified as arid or semi-arid. *Cenchrus ciliaris* (buffelgrass) is well adapted to many of these dry areas and as such can play an important role in the reclamation of degraded rangelands and stabilization of abandoned, or marginal, cropland. Over the past six years the University of Pretoria has been involved in a series of trials to evaluate the relative merits of material from Africa, America, Asia and Australia (Venter et al., 1993; Rethman et al., 1995; Venter & Rethman, 1996). One of the characteristics, for which this material has been screened, is that of drought tolerance. In South Africa, where low precipitation is also characterized by high variability, the ability of plants to exploit both light and episodic heavy precipitation is influenced by the distribution and extent of the root system. In this instance use was made of the opportunity—presented by an investigation to assess the influence of water stress on above ground productivity over a two year period (Venter et al., 1994; Venter et al., 1995)—to make a terminal assessment of the mass and vertical distribution of roots of seven buffelgrass selections.

EXPERIMENTAL

To study the influence of water stress on the mass and distribution of roots of seven buffelgrass selections (Lodwar, Makuyuni, Molopo and Tanzania from Africa, T-4464 from America and Biloela and Gayndah from Australia) seedlings were established on a well drained soil with 30% clay and a depth in excess of one metre. Original plans to apply a range of moisture stress treatments, based on the length of the permanently wilted condition, did not prove to be applicable with this species. It was, therefore, decided to apply five levels of irrigation along a water gradient created by replicated dripper lines, where 100%, 80%, 60%, 40% and 20% of the water required to restore field capacity was applied at five points (Venter *et al.*, 1994). After an initial establishment period, with unlimited soil water, treatments were applied over two growing seasons. At the end of this period the root mass and distribution of each selection, at each level of water supply, was destructively measured by removing a 2.5 cm core to a depth of 100 cm directly below the monitor tufts. These cores were divided into 20 cm sections and the roots in each section washed out and oven dried to determine the mass of roots.

RESULTS AND DISCUSSION

Total root mass: Where moisture was not limited (Tables 1 and 2) Tanzania recorded the highest root mass with Gayndah, Lodwar, Biloela, Makuyuni, Molopo and T-4464 producing respectively 43%, 52%, 64%, 68% and 78% less root material. Where, however, the level of water supply was only 20–40% of field capacity the ranking changed somewhat. Tanzania still recorded the highest root mass with Makuyuni, Lodwar, T-4464, Gayndah, Biloela and Molopo producing 25%, 25%, 52%, 63%, 66% and 71% less. When examining the proportionate decline in root mass of individual selections, however, T-4464 and Makuyuni were the least affected, with a decrease in root mass of only 32%. Lodwar recorded a decline of 51% and the others ranged from 68% for Tanzania to 80% for Gayndah.

Vertical distribution of roots: Selections varied widely with respect to the actual mass of roots concentrated in the upper soil horizon (Table 1). Tanzania had the most roots in this layer followed by Gayndah, Biloela, Molopo, Makuyuni, Lodwar and T-4464. The proportion of roots concentrated in the upper 20 cm followed a similar trend, with Lodwar and T-4464 having the smallest proportion of surface roots (31–32%), Tanzania the highest proportion (75%) and Biloela (66%), Molopo (59%), Makuyuni (46%) and Gayndah (45%), being intermediate (Table 1). These rankings were not drastically affected by water stress (Table 3). Reduced water applications did, however, have a differential effect on the "% decline" in surface roots (Table 4). Makuyuni and T-4464, with 19% and 36% "decline," were least affected, Lodwar (62%) was intermediate and the remaining (76–87%) were strongly impacted.

With respect to roots in the deeper soil layers (40–100 cm), which would facilitate the exploitation of a larger soil water "reservoir" in the case of episodic deeper percolation of rain water, the results presented in Tables 5 and 6 have application. With unlimited

Table 1. Vertical distribution of roots of buffelgrass selections with a 100 × 2.5 ϕ cm core with unlimited water, expressed as g dry material and percentage in each horizon

	Horizon					
Selection	0–20 cm	20–40 cm	40–60 cm	60–80 cm	80–100 cm	Total
Molopo	0.38 (59%)	0.14 (22%)	0.03 (5%)	0.04 (6%)	0.05 (8%)	.64
Gayndah	0.51 (45%)	0.22 (19%)	0.16 (15%)	0.15 (8%)	0.09 (7%)	1.13
Biloela	0.47 (66%)	0.08 (11%)	0.07 (10%)	0.06 (8%)	0.03 (4%)	0.70
T-4464	0.14 (32%)	0.12 (28%)	0.10 (23%)	0.06 (14%)	0.01 (3%)	0.44
Lodwar	0.29 (31%)	0.45 (47%)	0.08 (9%)	0.07 (7%)	0.06 (6%)	0.95
Makuyuni	0.32 (46%)	0.19 (27%)	0.08 (12%)	0.07 (10%)	0.03 (4%)	0.69
Tanzania	1.48 (75%)	0.24 (12%)	0.15 (7%)	0.07 (3%)	0.03 (2%)	1.97
Mean	0.51 (52%)	0.21 (24%)	1.10 (12%)	0.07 (8%)	0.04 (5%)	

Table 2. Mean dry weight of root (g) of buffelgrass selections recovered from a 100 × 2.5 ϕ cm core after the application of different levels of irrigation

	Level of water supply				
Selection	100	80	60	40	20
Molopo	0.64 (100)*	0.27 (42)	0.29 (45)	0.13 (20)	0.23 (36)
Gayndah	1.13 (100)	1.14 (101)	0.51 (45)	0.29 (26)	0.16 (14)
Biloela	0.70 (100)	0.68 (97)	0.44 (63)	0.27 (39)	0.15 (21)
T-4464	0.44 (100)	0.25 (57)	0.36 (82)	0.24 (55)	0.36 (82)
Lodwar	0.95 (100)	0.74 (78)	0.78 (82)	0.55 (58)	0.38 (40)
Makuyuni	0.69 (100)	0.48 (70)	0.39 (57)	0.52 (75)	0.41 (59)
Tanzania	1.97 (100)	1.14 (58)	0.77 (39)	0.73 (37)	0.52 (26)
Mean	0.93 (100)	0.67 (72)	0.51 (55)	0.39 (42)	0.32 (34)

*The index of production is given in parentheses after each entry.

Table 3. Influence of water stress on the dry weight (g) and proportion (%) of roots in the upper 20 cm of the soil

	Level of water supply				
Selection	100	80	60	40	20
Molopo	0.38 (59%)	0.11 (39%)	0.17 (58%)	0.05 (35%)	0.08 (34%)
Gayndah	0.51 (45%)	0.49 (43%)	0.24 (46%)	0.10 (35%)	0.07 (43%)
Biloela	0.47 (66%)	0.45 (66%)	0.22 (49%)	0.13 (47%)	0.06 (38%)
T-4464	0.14 (32%)	1.10 (39%)	0.10 (29%)	0.07 (30%)	0.09 (26%)
Lodwar	0.29 (31%)	0.16 (21%)	0.19 (24%)	0.12 (21%)	0.11 (28%)
Makuyuni	0.32 (46%)	0.18 (38%)	0.15 (38%)	0.32 (61%)	0.26 (63%)
Tanzania	1.48 (75%)	0.82 (71%)	0.59 (76%)	0.47 (64%)	0.35 (67%)
Mean	0.51 (51%)	0.33 (45%)	0.24 (46%)	0.18 (42%)	0.15 (43%)

Table 4. The percentage "decline" in root weight in upper 20 cm of the soil profile caused by the reduced rate of water application

	% Reduction in water supply			
Selection	20	40	60	80
Molopo	71%	55%	87%	70%
Gayndah	4%	53%	80%	86%
Biloela	4%	53%	72%	87%
T-4464	29%	29%	50%	36%
Lodwar	45%	34%	59%	62%
Makuyuni	44%	53%	0%	19%
Tanzania	45%	60%	68%	76%
Mean	35%	48%	59%	64%

moisture the ranking of selections on the basis of mass of roots starts with Gayndah and proceeds through Tanzania, Lodwar, Makuyuni, T-4464 and Biloela (indices of 61, 55, 45, 44 and 39 respectively) to Molopo with an index of only 29. The influence of water stress on the "deep root component" is, however, very unclear (Tables 5 and 6). Although the reduction in water supply had an overall negative effect (−37%) on this component, the effect was not as drastic as in the case of surface roots (−52%). Of the selections investigated it was only Makuyuni that exhibited a pattern correlated with water supply!

Table 5. Influence of water stress on the weight and proportion of roots in the deeper soil layers (g in 40–100 cm horizon; % of total root mass)

	Level of water supply				
Selection	100	80	60	40	20
Molopo	0.116 (33%)	0.124 (51%)	0.051 (40%)	0.049 (58%)	0.117 (42%)
Gayndah	0.402 (36%)	0.035 (31%)	0.145 (28%)	0.122 (43%)	0.072 (45%)
Biloela	0.158 (22%)	0.106 (16%)	0.142 (32%)	0.089 (32%)	0.074 (50%)
T-4464	0.176 (40%)	0.099 (39%)	0.0152 (43%)	0.0143 (55%)	0.200 (56%)
Lodwar	0.207 (23%)	0.613 (21%)	0.144 (19%)	0.157 (29%)	0.191 (49%)
Makuyuni	0.182 (26%)	0.176 (37%)	0.175 (46%)	0.121 (23%)	0.079 (19%)
Tanzania	0.244 (12%)	0.110 (10%)	0.100 (13%)	0.122 (17%)	0.102 (20%)
Mean	0.212 (27%)	0.116 (29%)	0.130 (32%)	0.113 (37%)	0.119 (40%)

Table 6. The percentage "change" in root weight in the deeper soil layers (40–100 cm) caused by the reduced rate of water application

	% Reduction in water supply			
Selection	20	40	60	80
Molopo	+7%	–56%	–58%	+15%
Gayndah	–91%	–64%	–70%	–82%
Biloela	–33%	–10%	–44%	–53%
T-4464	–44%	–14%	–24%	+14%
Lodwar	–21%	–30%	–24%	–8%
Makuyuni	–3%	–4%	–34%	–57%
Tanzania	–55%	–59%	–50%	–58%
Mean	–34%	–34%	–43%	–35%

CONCLUSIONS

It would appear that, in terms of the effects on root development, water stress had the least effect on Lodwar. Makuyuni and T-4464 were the next most tolerant followed by Tanzania. Molopo, Gayndah and Biloela, which are widely planted in Southern Africa and Australia, would, however, appear to be the least tolerant.

REFERENCES

Rethman, N.F.G., Meissner, H.H. and Cox, J.R. 1995. Evaluation of *Cenchrus ciliaris* ecotypes in terms of productivity and quality. Proc. Internat. Rangeland Congr. Salt Lake City, U.S.A.

Venter, P.S., and Rethman, N.F.G. 1996. Comparing various parameters to study the drought tolerance of subtropical grasses. Proc. 31st Congr. Grassl. Soc. South. Afr. Nelspruit, S.Afr.

Venter, P.S., Rethman, N.F.G., Wolfson, M.M. and Caetano, J.A. 1993. The use of drought simulating screening technique to evaluate *Cenchrus ciliaris* ecotypes. Proc. 28th Congr. Grassl. Soc. South. Afr. Bloemfontein, S.Afr.

Venter, P.S., Rethman, N.F.G., de Beer, J.M. and Eckard, J.H. 1994. Creation of facilities to study the influence of soil water availability on forage crops. Proc. 29th Congr. Grassl. Soc. South. Afr. Harare, Zimbabwe.

Venter, P.S., Rethman, N.F.G., Joubert, C.J. and Lindeque, J.P. 1995. The use of a water gradient to study the drought tolerance of perennial grasses. Proc. Internat. Rangeland Congr. Salt Lake City, U.S.A.

EFFECTS OF A PLANT GROWTH-PROMOTING RHIZOBACTERIUM (*Pseudomonas putida* GR12-2) ON THE EARLY GROWTH OF CANOLA SEEDLINGS

Bernard R. Glick,[1] Sibdas Ghosh,[2] Changping Liu,[1] and Erwin B. Dumbroff[1,3]

[1]Department of Biology
University of Waterloo
Waterloo, Ontario N2L 3G1
Canada
[2]Department of Biological Sciences
University of Wisconsin-Whitewater
Whitewater, Wisconsin 53190
[3]Kennedy-Leigh Centre for Horticultural Research
Faculty of Agriculture
Hebrew University, Rehovot 76100
Israel

INTRODUCTION

The free-living plant growth-promoting rhizobacterium (PGPR) *Pseudomonas putida* GR12-2 can enhance plant growth by one or more of several mechanisms including the synthesis of siderophores, the production of phytohormones, and the action of 1-aminocyclopropane-1-carboxylic acid (ACC) deaminase which degrades ACC, the immediate precursor of the phytohormone ethylene (1-6). Following the binding of the PGPR to the seed coat or to the root of a developing seedling, the bacterial deaminase sequesters and then degrades ACC leached from germinating seeds thereby increasing the concentration gradient of ACC from the root. This would lower the level of ACC as well as the level of ethylene in plants, hence promoting plant growth. Plants treated with mutant *P. putida* GR12-2/*acd*68, which lacks ACC deaminase activity, produce higher levels of ethylene and have shorter roots than plants treated with wild-type *P. putida* GR12-2. In the present study, we have examined whether the impact of *P. putida* GR12-2 on canola plants is affected by the method used to apply the bacterium, i.e. either as a seed coating prior to planting or by direct application to the soil. We have also extended our investigation to determine whether the bacterium moderates plant response to some of the effects of salt stress or cold night temperatures.

Biology of Root Formation and Development, edited by Altman and Waisel.
Plenum Press, New York, 1997.

EXPERIMENTAL

The PGPR *Pseudomonas putida* GR12-2 (the wild-type bacterium), which promotes canola root elongation under gnotobiotic conditions (4), was provided by Dr. Gerry Brown of Agrium Inc. (Saskatoon, Saskatchewan, Canada). *P. putida* GR12–2/*acd*68 is a mutant of *P. putida* GR12-2 that lacks ACC deaminase activity and does not promote plant growth (1) under gnotobiotic conditions (4,6). Both the wild-type and the mutant bacteria were grown aerobically at room temperature (22°C ± 1°C) on tryptic soybean broth medium.

In all experiments, canola seeds were stored at 4°C prior to their sterilization (3). Plants were grown at a constant temperature of 20°C (unless noted otherwise) under a 12 h photoperiod of mixed fluorescent and incandescent light with an irradiance of 22 $\mu mol/m^2/s$.

To assess the effects of the PGPR on plant growth when applied as a seed coating, equal numbers of sterilized seeds were transferred to a petri dish and then incubated for 1 h at room temperature with 5 mL of 100 mM $MgSO_4$, *P. putida* GR12-2 in 100 mM $MgSO_4$, or *P. putida* GR12-2/acd68 in 100 mM $MgSO_4$. Sixty coated seeds were then planted in 6-inch pots containing soil mixed with 110 g of vermiculite and 461 mL of water. To assess the effects of direct soil application of the PGPR on plant growth, sixty sterilized seeds were planted in 6-inch pots containing soil mixed with 11 mL of bacterial suspension, with an absorbance of 1.2 at 600 nm, plus 110 g of vermiculite and 450 mL of water.

To evaluate the impact of PGPR on plant growth under stress conditions, the bacteria were applied directly to the soil before planting. Salt stress was imposed by adding a 0.5% aqueous solution of NaCl to the soil, in place of distilled water, throughout the test period. Plants were also exposed to low temperatures using a day/night temperature regime of 25°/5°C and a 14-h photoperiod.

RESULTS AND DISCUSSION

To determine the effective use of PGPR in an agricultural setting, we have addressed two important questions. First, whether the way in which the bacterium is added to the plant affects the response of the plant to the bacterium. Second, whether the bacterium provides the plant with some relief from the adverse effects of different kinds of environmental stress.

The wild-type bacterium, irrespective of whether it was added to the seeds or to the soil in which the seeds were planted, promoted root growth significantly when compared to roots from plants in the $MgSO_4$ controls (Fig. 1 A and B). As predicted however, the mutant bacterium, which lacks ACC deaminase activity, did not have any consistent effect on root growth (Fig. 1 A and B). Moreover, root growth in seedlings treated with the wild-type bacterium was apparent somewhat earlier when the seeds were pre-treated than when the bacterium was added directly to the soil. This may reflect the likelihood that, with the pre- treated seeds, ACC deaminase could immediately begin to lower the endogenous level of ACC and hence the level of endogenous ethylene in the emerging seedling. In contrast, when the bacterium was added to the soil, its effect on plant ACC or ethylene levels, and consequently on plant development, would be delayed until it had migrated through the soil and bound to the seed or emerging root.

Comparison of shoot growth of 10-day-old canola plants revealed no apparent differences when seeds were pre-treated with either $MgSO_4$ or the wild-type bacterium (Fig. 2). In contrast, direct application of the bacterium to the soil increased shoot length by

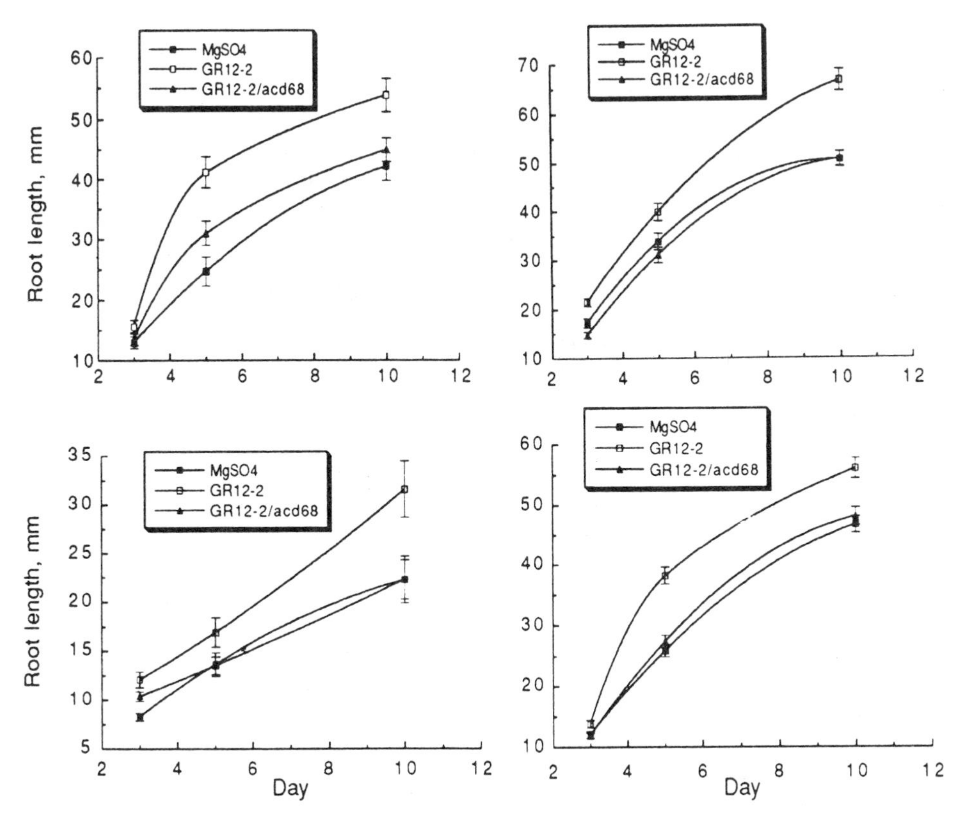

Figure 1. Effects of PGPR on the root growth of canola seedlings. (A) Effects of seed treated with $MgSO_4$, *P. putida* GR12-2 or *P. putida* GR12-2/*acd*68 before planting on the root growth of canola seedlings at a constant temperature (20°C). (B) Effects of soil pre-treated with PGPR on the root growth of canola seedlings exposed to a constant temperature (20°C) following the addition of $MgSO_4$, *P. putida* GR12-2 or *P. putida* GR12-2/*acd*68 directly to the soil. (C) Effects of PGPR on the root growth of canola seedlings exposed to 0.5% NaCl under constant temperature (20°C) following the addition of $MgSO_4$, *P. putida* GR12-2 or *P. putida* GR12-2/*acd*68 directly to the soil. (D) Effects of PGPR on the root growth of canola seedlings exposed to day/night temperatures of 25°/5°C following the addition of $MgSO_4$, *P. putida* GR12-2 or *P. putida* GR12-2/*acd*68 directly to the soil. The error bars indicate 1 ± S.E.M.

approximately 35% compared to the controls (Fig. 2). Moreover, in every instance (Figs. 1 and 2), adding the wild-type bacterium to the soil resulted in significant increases in the lengths of shoots as well as the roots.

When canola seedlings were grown in the presence of both the bacterium and salt, the resultant salt stress caused a decrease in root lengths by approximately half when compared to seedlings grown in the absence of salt (Fig. 1 B and C). Although the bacterium stimulated root elongation in the presence of salt, it did not restore root lengths to the levels observed in the absence of salt. During salt stress, the stimulatory effect of the wild-type bacterium on shoot growth was also retained, with shoot lengths about 38% greater than those from the $MgSO_4$ controls (Fig. 2).

For a bacterial strain to be an effective PGPR in northern climates such as Canada, it must be able to survive long cold winters and then grow at cool temperatures in the spring. In a previous study it was shown that *P. putida* GR12-2 can readily survive freezing temperatures and can grow at temperatures as low as 5°C (7). In addition, this bacterium

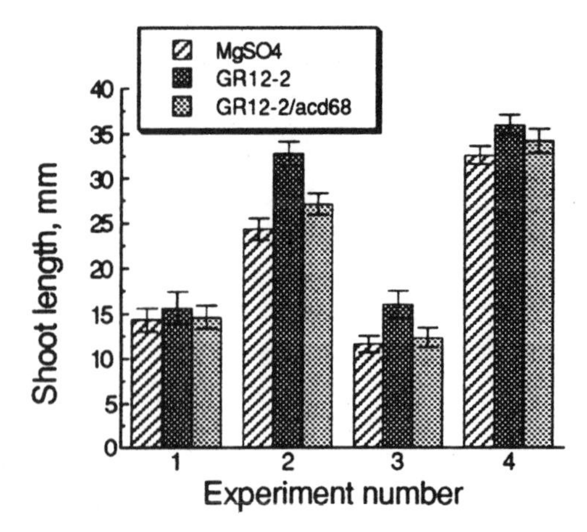

Figure 2. Effects of PGPR on the shoot growth of canola seedlings. (1) Effects of seed treated with $MgSO_4$, *P. putida* GR12-2 or *P. putida* GR12-2/*acd*68 before planting on the shoot growth of canola seedlings at a constant temperature (20°C). (2) Effects of soil pre-treated with PGPR on the shoot growth of canola seedlings exposed to a constant temperature (20°C) following the addition of $MgSO_4$, *P. putida* GR12-2 or *P. putida* GR12-2/*acd*68 directly to the soil. (3) Effects of PGPR on the shoot growth of canola seedlings exposed to 0.5% NaCl under constant temperature (20°C) following the addition of $MgSO_4$, *P. putida* GR12-2 or *P. putida* GR12-2/*acd*68 directly to the soil. (4) Effects of PGPR on the shoot growth of canola seedlings exposed to day/night temperatures of 25°/5°C following the addition of $MgSO_4$, *P. putida* GR12-2 or *P. putida* GR12-2/*acd*68 directly to the soil. The error bars indicate 1 ± S.E.M.

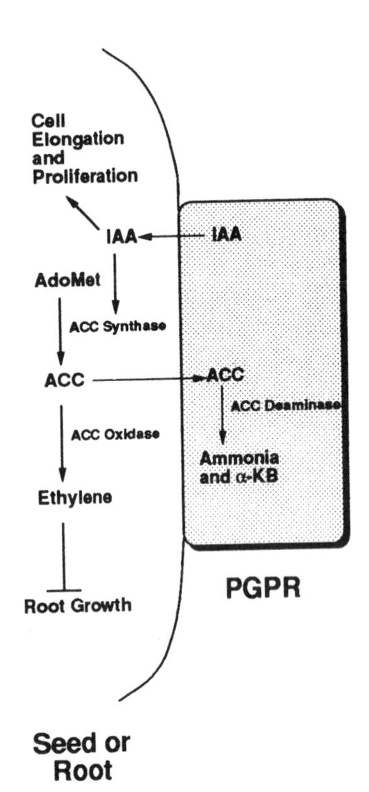

Figure 3. Model of PGPR stimulation of plant growth. Following the binding of the bacterium (PGPR) to either the seed coat or to the root of a developing plant, bacterial ACC deaminase presumably sequesters and degrades leached ACC, the immediate precursor of the phytohormone ethylene, into ammonia. The level of ethylene within the plant is lowered and the promotion of plant growth is enhanced. Abbreviations: IAA, indoleacetic acid; AdoMet, S-adenosyl methionine; ACC, 1-aminocyclopropane-1-carboxylate; α-KB, α-ketoglutarate.

retained its ability to stimulate root elongation of canola seedlings in growth pouches under gnotobiotic conditions at 5°C (7). In the present study, only small decreases in root growth were recorded for the cold-treated, $MgSO_4$ control plants compared to similar plants kept at a constant temperature of 20°C (Fig. 1 B and D). In contrast, under day/night temperatures of 25°/5°C, shoot growth increased compared to plants maintained at a constant temperature of 20°C.

We have previously shown that the number of *P. putida* GR12-2 cells did not decrease during a one-month period in sterilized soil and that it also persisted in non-sterilized soil for the same period of time (6). Based on this and the present results, our observations suggest that adding *P. putida* GR12-2 to the soil at planting could be highly effective in promoting plant growth in the field during early seedling establishment, presumably by degrading ACC and increasing its concentration gradient and rate of loss from the roots with a consequent reduction of endogenous ethylene (Fig. 3).

ACKNOWLEDGMENTS

This research was supported by grants from the Natural Sciences and Engineering Research Council of Canada and Agrium Inc. to B.R.G., from the Natural Sciences and Engineering Research Council of Canada to E.B.D., and from the National Science Foundation (#BIR-9512262), USA and the University of Wisconsin-Whitewater to S.G.

REFERENCES

1. Glick, B.R., Jacobson, C.B., Schwarze, M.M.K., and Pasternak, J.J., 1994, 1-Aminocyclopropane-1-carboxylic acid deaminase mutants of the plant growth promoting rhizobacterium *Pseudomonas putida* GR 12–2 do not stimulate canola root elongation, Can. J. Microbiol. 40:911.
2. Glick, B.R., 1995, The enhancement of plant growth by free-living bacteria, Can. J. Microbiol. 41:109.
3. Glick, B.R., Karaturovic, D.M., and Newell, P.C., 1995, A novel procedure for rapid isolation of plant growth promoting pseudomonads, Can. J. Microbiol. 41:533.
4. Lifshitz, R., Kloepper, J.W., Kozlowski, M., Simonson, C., Carlson, J., Tipping, E.M., and Zaleska, I., 1987, Growth promotion of canola (rapeseed) seedlings by a strain of *Pseudomonas putida* under gnotobiotic conditions, Can. J Microbiol. 33:390.
5. Xie, H., Pasternak, J.J., and Glick, B.R., 1995, Isolation and characterization of mutants of the plant growth-promoting rhizobacterium *Pseudomonas putida* GR12-2 that overproduce indolacetic acid, Curr. Microbiol. 32:67.
6. Tang, W., Pasternak, J.J., and Glick, B.R., 1995, Persistence in soil of the plant growth promoting rhizobacterium *Pseudomonas putida* GR12-2 and genetically manipulated derived strains, Can. J. Microbiol. 41:445.
7. Sun, X., Griffith, M., Pasternak, J.J., and Glick, B.R., 1995, Low temperature growth, freezing survival, and production of antifreeze protein by the plant growth promoting rhizobacterium *Pseudomonas putida* GR12-2, Can. J. Microbiol. 41:776.

ROOT GROWTH OF SEEDLINGS AND TISSUE CULTURED PLANTS OF *EUCALYPTUS*

J. A. McComb,[1] E. Steel,[1] D. T. Bell,[2] P. van der Moezel,[2] I. J. Bennett,[3] B. Dell,[1] and I. Colquhoun[4]

[1]School of Biological and Environmental Sciences
Murdoch University
Perth, Western Australia 6150
[2]Department of Botany
University of Western Australia
Nedlands, Western Australia 6907
[3]Department of Applied Science
Edith Cowan University
Mount Lawley, Western Australia 6050
[4]Alcoa of Australia Ltd.
PO Box 252
Applecross, Western Australia 6153

INTRODUCTION

Plants of dicotyledonous trees raised in tissue culture initially have many adventitious roots rather than the single tap root produced by a seedling. Little is known about the subsequent development of the root morphology of tissue cultured trees. Root architecture affects both survival after planting in the field and the likelihood of windthrow when trees are taller. The seedling tap root of eucalypts does not normally persist as the dominant root. Several lateral roots develop as sinker roots, grow to equal thickness and penetrate to great depths (Dell et al., 1989).

Two species of eucalypts that are being grown in Western Australia as micropropagated clones are *E. camaldulensis* (river red gum) and *E. marginata* (jarrah). *E. camaldulensis* is grown in several countries (Morocco, Spain, Portugal, USA, India, Israel and Burma (Eldridge et al., 1993)) for paper pulp production, but in Australia it is most valued for its ability to grow on secondary salinised soil that is seasonally waterlogged (Marcar, 1995). A selection program was undertaken to identify genotypes highly tolerant of saline waterlogging (van der Moezel et al., 1992). These genotypes were micropropagated (McComb, 1995) and field trials have been established. These have shown that selected clonal lines survive and grow better on saline sites than seedlings of the same provenance (Bell et al., 1994). We were interested to know whether the root architecture

Biology of Root Formation and Development, edited by Altman and Waisel.
Plenum Press, New York, 1997.

of clones and seedlings differed and whether the roots of both types of plants could penetrate clay. This of importance as many soils in the Western Australian wheatbelt are duplex soils of sand over clay.

E. marginata is endemic to Western Australia and is the most important timber tree harvested from native forests. The wood is used for structural purposes and cabinet making and is hard and durable. Under forest conditions it is slow growing.

E. marginata is not normally grown in plantations as on average trees are only 70 cm in diameter after 250 years (Abbott and Loneragan, 1983). The jarrah forest overlies deposits of bauxite which are being mined by Alcoa of Australia and other companies. Alcoa mines 500 ha annually and aims to re-establish a self-sustaining jarrah forest ecosystem when it revegetates bauxite mine sites (Ward and Koch, 1995). One problem is that 14% of the northern jarrah forest is affected by an introduced pathogen, *Phytophthora cinnamomi,* which can kill both jarrah in the native forest and trees on rehabilitated mine sites (Davidson and Shearer, 1989; Colquhoun and Petersen, 1994).

Research was undertaken to select jarrah lines resistant to *P. cinnamomi* by underbark inoculation of seedlings 9–12 months old (Stukely and Crane, 1994). Seedlings with the smallest underbark lesions were cloned and proved resistant to the pathogen in glasshouse and field conditions (McComb et al., 1990; McComb et al., 1992). Micropropagation techniques for jarrah are expensive, and we now wish to study the resistance to *Phytophthora cinnamomi* of the progeny of crosses between clones. Seed orchards of micropropagated trees are placed in the agricultural area of the coastal plain to reduce contamination with pollen from native forests. Although micropropagated plants show around 90% survival in the lateritic soils of the revegetated bauxite mines, on sandy agricultural soils without irrigation, up to the 98% of plants die during the first summer (I. Colquhoun, unpubl.).

We previously investigated the root system of clones (from explants from mature trees) and seedlings in the glasshouse (Bennett et al., 1986). After four months the total root length of the clones was significantly less than that of seedlings, and the bulk of the roots was in the top 10 cm of the pot. The roots of the seedling were more uniformly distributed to the bottom of the 40 cm deep pot. It is possible that the proliferation of fine roots seen in this experiment might also characterise plants in the field, and explain the death of the plants. Consequently we set up an experiment to examine the root development of seedlings and clonal *E. marginata* plants in sandy soil in the field.

MATERIALS AND METHODS

E. camaldulensis seedlings (2 lines) and clones resistant to saline waterlogging (5 lines) were planted when 4 months old (mean height 15 cm) into 200 L drums filled with either sand or with the lower one third filled with river bank clay and the top two thirds with sand. After a further seven months (May to November), the drums were cut away and the roots washed out. It was not possible to extract roots from the clay. Root architecture was described qualitatively, while the parameters measured were shoot height, total root length, root dry weight, the distribution of the root mass in the top 0–20 cm, middle 20–40 cm and >40 cm of the soil profile, and the number of sinker roots present (Bell et al., 1993).

E. marginata seedlings and four clonal lines resistant to *P. cinnamomi,* aged 6–9 months and 20 cm in height, were selected. Clones were grown in 5 cm diam. Winellie pots and seedlings in 4 cm diam jiffy peat pots. Normal planting procedures used by Alcoa for jarrah were used for the planting in an agricultural area in sandy soil on the coastal plain near Waroona (130 km south of Perth, W. Australia). A hole 20×20×20 cm was dug and two fertiliser tablets (Bailey's 100 g DAP with micronutrients) were placed in the hole which was then backfilled to the surface before planting the clones and seedlings. Approximately 25 gm of Osmocote (Langley's Chemicals, Macrocote Brown plus) was applied to the base of each plant. Composted eucalypt bark was spread as mulch around the plants to reduce weed growth, and tree guards (Solartubes) placed around them to prevent frost damage in June to August. There was 2.5 m between plants in the rows, and rows were 2.5 m apart. The design was blocks of five plants(one seedling and one plant of each of the clones) replicated five times for each of two irrigation regimes and three harvest times. Trees were planted after the start of the winter rains in June 1995 and half of each plot was watered to saturation every second day from mid September to March.

Plants were harvested at 3, 6, and 9 months. As it rained between June and August only the "irrigated" plot was excavated in the 3 month harvest. At the three month harvest trenches were dug and soil washed away to reveal the roots. Measurements were made of the greatest horizontal distance of a root from the stem, the greatest root depth and the total root length developed beyond the original root ball (about 4×4×5 cm). At the 6 month harvest, a trench 1 m deep was dug 1 m from the trees. The widest and deepest roots were again recorded. After 9 months in the field plants were treated similarly except that the number of roots 1 m long or longer, and their diameter was recorded. The angles of the major roots (1 m or longer) arising from the root ball were also measured.

Total root length was measured using a Comair root length machine (Commonwealth Aircraft Corporation, Melbourne).

RESULTS

Seedlings of *E. camaldulensis* displayed more variable root architecture within a provenance than was seen within clones. While there was greater uniformity within clones there were consistent differences between them (Table 1). One markedly different clone (A4), which had the shortest shoots had a distinctive root architecture with fewer sinker roots and less fine roots than the other clones. The range of morphology amongst clones was similar to that of the seedlings. All plants had a mean of 6–9 sinker roots of >0.5 cm diameter, and all genotypes except A4 had most root mass in the top 0–20 cm and bottom third of the drums (Table 2). Both seedlings and clonal roots penetrated the clay but we were unable to wash off the clay to quantify the root mass in this layer.

Table 1. Characteristics of *Eucalyptus camaldulensis* grown in deep sand[a]

	Seedlings		Clones				
	Albacatya	Silverton	A4	A6	A20	A28	A34
Shoot height (m)	220 (27)[a]	170 (25)	146 (16)	174 (8)	203 (11)	178 (19)	178 (16)
Total root dry weight (g)	270 (189)	192 (58)	128 (21)	352 (43)	302 (43)	375 (42)	191 (39)
Total root length (m)	4279 (1668)	2312 (1346)	2004 (555)	7810 (1676)	6561 (1599)	5862 (779)	4423 (1017)
No of roots >0.5 cm diam	6.8 (1.9)	9.0 (2.4)	6.0 (2.7)	9.0 (0.8)	7.5 (1.3)	7.3 (0.5)	6.7 (1.7)

[a]Values are means with standard deviations in parenthesis. (Data from Bell et al. 1994)

Table 2. Total root length, and distribution of roots of *Eucalyptus camaldulensis* in the soil profile[a]

Depth (cm)	Albacatya seedlings	Silverton seedlings	Plant source clones				
			A4	A6	A20	A28	A34
0–20	1013 (354)	450 (165)	339 (116)	1754 (702)	842 (292)	1790 (334)	597 (233)
20–40	1456 (730)	533 (539)	302 (171)	2258 (338)	2255 (753)	1576 (628)	756 (291)
>40	1809 (1304)	1530 (795)	1360 (489)	3798 (1244)	3465 (1527)	2497 (294)	3069 (830)

[a]Values are means with standard deviation in parenthesis. (Data from Bell et al. 1994)

The measurements of roots of *E. marginata* plants at 3 months showed that when clones were bulked and compared with seedlings, the seedling roots were larger in all parameters measured ($P \geq 0.05$). This difference was reduced when the plants had been in the field for 6 months (Table 3). It was impossible to recover sufficient of the total mass of roots from plants at 6 or 9 months to make total root length measurements meaningful. At 9 months it was physically impossible to measure mean depth of the longest root as too many roots exceeded 1 m. Overall, the number of roots 1 m or longer, and the diameter of these roots was not significantly different in seedlings and clones (Table 3). Comparisons between seedlings and clones is difficult as only one non-irrigated seedling survived. However in the irrigated plots when data for clones were bulked there was a significant difference ($P \geq 0.05$) between the number of roots 1 m or more long in seedlings and clones (Table 4). There appeared to be architectural differences between the root systems of seedlings and the micropropagated plants. The lateral sinker roots, arising from the base of seedling plants, were at a smaller angle than in micropropagated plants (Table 3). However, the low replicate number and the variation within lines meant that this was not statistically different.

Table 3. Characteristics of *E. marginata* seedlings (S) and four clonal lines (C1–C4) grown in the field[a]

Character	Non-irrigated					Irrigated				
	S	C1	C2	C3	C4	S	C1	C2	C3	C4
3 months										
Mean widest root (cm)						24 (7.0)	18 (13.1)	12 (6.4)	11 (3.9)	12 (7.1)
Mean deepest root (cm)						38 (13.1)	20 (6.9)	22 (5.0)	20 (7.7)	19 (2.6)
Mean total root length (cm)						521 (330)	305 (321)	270 (195)	194 (241)	374 (469)
6 months										
Mean widest root (cm)	32 (20.5)	41 (45.0)	24 –	61 (31.3)	32 (18.6)	44 (10.5)	54 (38.4)	44 (12.4)	47 (16.6)	86 (31.9)
Mean deepest root (cm)	125 (57.3)	79 (13.4)	32 –	95 (37.9)	29 (4.0)	77 (31.9)	72 (19.7)	55 (11.0)	69 (13.9)	64 (19.4)
9 months										
Mean widest root (cm)	18 –	100 –	52 (20.2)	41 (19.5)	49 (45.6)	64 (37.4)	58 (10.4)	48 (16.8)	62 (18.7)	39 (20.0)
Mean angle of roots ≥ 1m depth (°)	0 –	60 –	88 (11.8)	54 (36.8)	63 (38.9)	26 (27.2)	91 (15.8)	36 (42.3)	46 (36.8)	– –
Mean no. roots ≥ 1m depth	0 –	2.0 –	3.0 (2.00)	2.4 (0.89)	1.5 (1.92)	3.8 (4.21)	2.0 (1.00)	1.5 (1.92)	1.4 (1.67)	0.8 (1.79)

[a]Values are means for plants harvested at 3, 6 and 9 months with standard deviation in parentheses.

Table 4. Comparison between *E. marginata* seedlings and clones grown in the field for 9 months with or without irrigation[a]

	Non irrigated		Irrigated	
Character	Seedlings	All clones	Seedlings	All clones
Leaf area (cm^2)	142 (–)	1690 (2037)	5153 (4913)	4784 (3785)
Shoot dry weight (gm)	6 (–)	58 (57)	160 (142)	144 (111)
No. of roots ≥ 1m depth	0 (–)	2.4 (1.4)	3.8 (4.21)	1.4 (1.58)
Mean diam. roots (mm) ≥ 1m depth	0 (–)	3.4 (1.79)	3.8 (1.60)	3.7 (1.99)

[a]Values are means for clones or seedlings with standard deviation in parentheses.

Table 5. Survival of *Eucalyptus marginata* seedlings and micropropagated clones planted in the field in sandy soil, with and without irrigation[a]

	Survival %			
	Non-irrigated		Irrigated	
Months after planting	Seedlings	Clones	Seedlings	Clones
3	100 (15)	88 (60)	100 (15)	97 (60)
6	30 (10)	60 (40)	80 (10)	85 (40)
9	20 (5)	60 (20)	100 (5)	85 (20)

[a]There were initially 150 plants of which 50 were harvested at 3, 6 and 9 months. Values are percentage survival for seedlings and mean percentage survival for clones. The number of plants scored at each time period is in parentheses.

There was higher survival when plants were irrigated, but there was no significant difference between survival of seedlings and clones (Table 5). Irrigated *E. marginata* plants had significantly more leaf area and shoot mass than non-irrigated plants (Table 4).

DISCUSSION

There was no consistent difference in root architecture between seedlings and clones of *E. camaldulensis*. Initially the seedling tap root was quite distinct from the fibrous laterals of the micropropagated plants. By 11 months of age the tap root was not usually the dominant root, and both types of plants had developed numerous sinker roots of equal thickness. Although some of the clones could be distinguished on the basis of root morphology, no features could be attributed to tissue culture as there was a similar range of variation in seedlings.

We were unable to harvest complete root systems of *E. marginata* plants after 6 or 9 months growth in the field which meant that parameters that may have been valuable, such as total root length and mean depth of the longest root, were not available. The variability between root systems within a clone was greater in *E. marginata* than in *E camaldulensis* and the replicate size was insufficient to be certain whether or not the clones and the seedlings differed consistently. However, it could be seen that root architecture of the clones included laterals arising from the crown of the plant at a wider angle than for seedlings. Without irrigation, both seedlings and clones had smaller shoots and showed less survival than irrigated lines. Compared with 9 month old seedlings, clones had fewer sinker roots 1m or longer, but the diameter of these roots was similar in seedlings and clones. It is clear that to establish clonal seed orchards on the sandy soils of the coastal plain, irrigation over the first summer will be essential. The poor survival of the clones without irrigation is not due to differences in root architecture, imposed from their origin through micropropagation, as seedlings had a similar low level of survival.

CONCLUSIONS

Micropropagated *Eucalyptus camaldulensis* plants showed a high survival rate after transfer to the field whereas those of *E. marginata* had low survival. We investigated root growth and architecture of seedlings and micropropagated plants of *E. camaldulensis* and *E. marginata* to determine whether differences in root growth might explain these observations.

Roots of 11 month old plants of *E. camaldulensis,* grown in 200 L drums, showed the same arrangement of sinker roots and of fine roots in both micropropagated plants and seedlings. The morphology of the root systems was consistent within clones and the variation between clones was similar to that between seedlings. Seedlings and clones of *E. marginata,* grown for 9 months in the field, survived poorly if they were not irrigated. In comparison to seedlings, the clones had lateral roots arising at a wider angle from the plant axis, and less sinker roots. The difference between the survival of *E. camaldulensis* and *E.marginata* in the field appears more related to the general vigour of plant growth than to the root form of micropropagated plants.

REFERENCES

Abbott, I. and Loneragan, O., 1983, Growth rate of jarrah (*Eucalyptus marginata*) in relation to site quality in cut over forest, Western Australia, *Aust. For.* 46:91.

Bell, D.T. van der Moezel, P.G., Bennett, I.J., McComb, J.A., Wilkins, C. F., Marshall, S. C. and Morgan, A., 1993, Comparisons of growth of *Eucalyptus camaldulensis* from seeds and tissue culture: root, shoot and leaf morphology of 9 month old plants grown in deep sand and sand over clay, *For. Ecol. Manag.* 57: 125.

Bell, D.T., McComb, J.A., van der Moezel, P. G., Bennett, I.J. and Kabay, E. D., 1994, Comparison of selected and cloned plantlets against seedlings for rehabilitation of saline and waterlogged discharge zones in Australian agricultural catchments, *Aust. For.* 57: 69.

Bennett, I. J., Tonkin, C., Wroth, M., Davison, E.M. and McComb, J.A., 1983, A comparison of growth of seedlings and micropropagated *Eucalyptus marginata* (jarrah). 1. Early growth to 2 years, *For. Ecol. Manag.* 14:1.

Colquhoun, I. J. and Petersen, A.E., 1994, The impact of plant disease on mining, *J. Roy. Soc. W.A.* 77:151.

Davison, E.M. and Shearer, B.L., 1989, *Phytophthora* species in indigenous forests of Australia, *New Zealand Journal of Forestry Science*. 19:277.

Dell, B. and Malajczuk, N., 1989, Jarrah dieback - A disease caused by *Phytophthora cinnamomi, in*: "The Jarrah Forest a Complex Mediterranean Ecosystem", B. Dell, J. Havel, and N. Malajczuk, eds., Kluwer; Dordrecht, p. 67.

Eldridge, K., Davidson, J., Harwood, C. and van Wyk, G., 1993, "Eucalypt Domestication and Breeding," Claredon Press, Oxford.

Marcar, N., Crawford, D., Leppert, P., Jovanovic, T., Floyd, R., Farrow, R., 1995, "Trees for Saltland: a guide to selecting native species for Australia," CSIRO, Canberra.

McComb, J.A., 1995, Clonal propagation of eucalypts, *in*: "Plant Tissue Culture Manual" Supplement C8, K. Lindsay, ed., Kluwer, Dordrecht, p.1.

McComb, J.A., Bennett, I.J., Stukely, M. and Crane, C., 1990, Selection and propagation of jarrah for dieback resistance. *Comb. Proc. Intern. Pl. Prop. Soc.*, 40:86.

McComb, J.A., Bennett, I.J., Stukely, M. and Crane, C., 1992, Selection, propagation, laboratory and field testing of jarrah (*Eucalyptus marginata*) resistant to dieback (*Phytophthora cinnamomi*) *in*: "IUFRO/AFOCEL Conference on Mass Propagation Technology for Genetically Improved Fast Growing Tree Species" Bordeaux, France, Sept, 1992. Vol.2: 451.

Stukely, M. and Crane, C., 1994, Genetically based resistance of *Eucalyptus marginata* to *Phytophthora cinnamomi*, *Phytopathology* 84: 650.

Van der Moezel, P.G., Bell, D.T., Bennett, I.J., Strawbridge, M. and McComb, J.A., 1992, The development of salt tolerant clonal trees in Australia. *Comb. Proc. Intern. Pl. Prop. Soc.* 40: 73.

Ward, S.C. and Koch, J.M., 1995, Early growth of jarrah (*Eucalyptus marginata* Donn ex Smith) on rehabilitated minesites in south-west Australia, *Australian Forestry.* 58: 65.

ECTOMYCORRHIZAL FUNGUS INOCULATION AND NUTRIENT SUPPLY EFFECTS ON THE MORPHOLOGY OF CORK-OAK ROOTS

C. Noronha and M. A. Martins-Loução

Departamento de Biologia Vegetal, Faculdade de Ciencias de Lisboa, Campo Grande, Bloco C2, Piso 4, 1700 Lisboa, Portugal.

INTRODUCTION

Cork-oak ectomycorrhizal roots occur naturally under field conditions, but micropropagation eliminates them. Induction of mycorrhization plays an important role on acclimatization of micropropagated cork-oak plantlets, enhancing growth and survival (Romano and Martins-Loução, 1994). The objective of this study was to compare the effects of ectomycorrhizal fungi inoculation on *in vitro* root system morphology of corkoak tissue-cultured plantlets. *Pisolithus tinctorius* and *Hebeloma crustuliniforme,* were chosen since they are commonly observed in association with this species, under natural conditions. GD culture medium (Gresshoff and Doy, 1972), adequated for cork-oak micropropagation, was also compared to modified Melin-Norkans medium (MMN) (Marx, 1969), used on ectomycorrhizal fungi growth and generally employed in pure culture synthesis of mycorrhizae.

EXPERIMENTAL

Cork-oak *(Quercus suber* L.) plantlets were obtained according to Romano et al. (1992). Rooting was induced on perlite moistened with GD nutrient solution supplemented with glucose (40 g l^{-1}). After 1 month, rooted plantlets were maintained in GD medium, or transferred to perlite moistened with MMN nutrient solution, with 5 g l^{-1} of glucose. Ectomycorrhizal fungus inoculum consisted on *H. crustuliniforme* or *P. tinctorius* mycelial liquid MMN cultures. Plantlets were induced to mycorrhizae for 3 months with 1 ml of fungus inoculum. Effects of mycorrhyzae on root system morphology was evaluated in terms of root length, number of roots, number of lateral short roots and mycorrhization degree along one centimeter-long root sections. Mycorrhization was confirmed by scanning electron microscopy.

RESULTS AND DISCUSSION

The number of roots was not affected by mycorrhizal induction, since they have been formed in the same root medium. However, root length decreased in inoculated plantlets, effect that was more pronounced in GD nutrient solution. Number of lateral short roots was also significantly increased in inoculated plantlets. Root branching and mycorrhization degree increased along upper root sections, similarly to what is observed in other mycorrhized plants. GD was more effective than MMN for mycorrhization degree. Apparently, this can be explained by an enriched carbon concentration in GD medium. No differences were found between the fungi assayed in all evaluated morphological parameters.

CONCLUSIONS

GD proved to be more adequated to root system development and mycorrhization than MMN nutrient solution. This system, by permitting in vitro rooting and mycorrhization to occur in sequence without transfer, provides a simple and efficient protocol.

ACKNOWLEDGMENTS

C. Noronha gratefully acknowledges JNICT for financial support.

REFERENCES

Gresshoff, P., and Doy, C. 1972, Development and differentiation of haploid *Lycopersicon esculentutn* (Tomato), *Planta* 170:161.

Marx, D., 1969, The influence of ectotrophic mycorrhizal fungi on the resistance of pine roots to pathogenic infections. I. Antagonism of mycorrhizal fungi to root pathogenic fungi and soil bacteria, *Phytopothology* 59:153.

Romano, A., Noronha, C., and Martins-Loução, M.A., 1992, Influence of growth regulators on shoot proliferation in *Quercus suber* L., *Ann. Bot.* 70:531.

MORPHOLOGICAL AND MOLECULAR RESPONSES OF *Prosopis alba* ROOT SEEDLINGS INDUCED BY SALINE CONDITIONS

G. Muñoz,[1] R. Pring,[2] E. Couve,[3] and P. Barlow [2]

[1]Catholic University of Valparaiso, Chile. [2]Long Ashton Research Station, Bristol, England. [3]University of Valparaiso, Chile.

Plant productivity is greatly affected by environmental factors. Salinity of soils is one of the main abiotic stress factors that lead to severe crop loss every year especially in arid and semiarid areas (Greenway and Munns, 1980). Although these environments may contain a variety of salts, sodium chloride is the most commonly considered source of salinity (Flowers et al., 1977). The primary effect of salt is experienced by the roots, i.e. by the most sensitive part of the plant to adverse factors which are present in the soil. Root elongation is critical to the ability of plants to survive salinity (Ashraf et al., 1986). The mechanism underlying salt stress tolerance involves morphological and biochemical responses related to the maintenance of cell viability. The analysis of ultrastructural and molecular changes in roots from *Prosopis alba*, a native tree of the Atacama desert in northern Chile, could provide information useful to understanding the design of strategies for the improvement of agriculture in saline zones.

Prosopis alba seeds were germinated in distilled water, 300 mM NaCl and pure sea water and grown in hydroponics at 28±1°C. The percentage seed germination and root length were measured. 5-day-old seedlings were obtained for further analysis. In vivo protein phosphorylation was analysed for seedlings incubated with 0.1 mCi/ml of $H_3{}^{32}PO_4$; protein extracts from roots were analysed by 10% polyacrylamide gel electrophoresis as previously described (Laemmli, 1970). Phosphoproteins were visualized by autoradiography and the protein profile by Coomassie Blue staining. Incorporation of ^{3}H-tyrosine was studied by light autoradiography in root tips from seedlings incubated for 1 h with ^{3}H-Tyr 50 Ci/mmol. X-ray microanalysis was performed on freeze-dried samples with a Link AN10000 analyser and a thin window detector interfaced with the SEM. Root exocellular acid phosphatase was determined essentially as described earlier (McLachlan, 1980). For transmission electron microscopy, root tips of 5-day-old seedlings were processed using standard methods.

RESULTS AND DISCUSSION

P. alba seedlings, growing under salt stress incorporate sodium and chloride in their roots and hypocotyls. A notoriously greater incorporation was observed in the cotyledons. A delay and a reduction in germination percentage and root length, in comparison with the results obtained in distilled water, were observed. Nevertheless, ultrastructural analysis of longitudinal sections shows, in central cells of root caps, the presence of enlarged amyloplasts, mitochondria, and wider cell walls. Such morphological characteristics are related with a salt tolerance response. Moreover, turgid cells were observed, revealing that *P. alba* has an efficient system of osmotic adjustment. Also, the results obtained by light autoradiography show a greater incorporation of ^{3}H-Tyr into root tips of *P. alba* seedlings growing in sea water, an observation which could be related with a salt tolerance response as described in halophilic bacteria (Lobyreva et al., 1994). Other responses to salt stress were the appearance of three predominant polypeptides of 50, 34 and 25 kDa and a phosphoprotein of 37 kDa. Moreover, an increase in the root exocellular acid phosphatase activity was obtained in reponse to salt stress. Our observations supports the conclusion that *Prosopis alba* presents morphological and molecular responses related with salt tolerance, important to the survival of plant species in natural environments, where drought and salinity are the prevailing conditions.

Acknowledgements. We are indebted to Dirección General de Investigación y Post Grado, Universidad Católica de Valparaíso, Chile.

REFERENCES

Ashraf, M., McNeilly, T. and Bradshaw, A.D. 1986.The potential for evolution of salt (NaCl) tolerance in seven grass species. New Phytol 103:299–309

Flowers, T.J., Troke, P.F. and Yeo, A.R. 1977. The mechanism of salt tolerance in halophytes. Annu. Rev. Plant Physiol. 28:89–121.

Greenway, H. and Munns, R. 1980. Mechanism of salt tolerance in nonhalophytes. Annu. Rev. Plant Physiol. 31:149–190.

Laemmli, U.K. 1970. Cleavage of structural proteins during the assembly of the head of bacteriophage T4. Nature 277:680 -685.

Lobyreva, L.B., Kokoeva, M.V. and Plakunov, V.K. 1994. Physiological role of tyrosine transport systems in *Halobacterium salinarium*. Arch. Microbiol. 162:126–130.

McLachlan, K.D. 1980. Acid phosphatase activity of intact roots and phosphorus nutrition in plants. I. assay conditions and phosphatase activity. Aust. J. Agric. Res. 31:429–440.

ROOT GROWTH AND WATER UPTAKE BY FLOWERING MAIZE PLANTS, UNDER DROUGHT CONDITIONS

Timotej Ješko, Ján Navara, and Katarína Dekánková

Slovak Academy of Sciences, Institute of Botany, Dúbravska cesta 14, SK-842 23 Bratislava, Slovak Republic.

Water stress seems to be the most important factor for plants under the changing climate, particularly during their generative phase. Several mechanisms have developed in plants to minimize drought damages, e.g. drought tolerance, and drought avoidance (Levitt, 1972; Ješko, 1996). Drought avoidance and particularly that type is expressed by developing of "compensating roots" in favourable soil layers, is most common for maize plants which grow under the middle European conditions.

The effects of long term drought after anthesis, on root growth and water uptake (WU) were studied in *Zea mays* L., hybrid CE 330 and TO 360 plants. WU by the seminal and the vegetative nodal roots were measured separately using vertically-divided compensation lysimeters (Navara et al., 1993). WU from three different soil layers were measured using horizontally-divided lysimeters with plaster blocks. Six to sixteen lysimeters were used for WU and evaporation loses measurements and up to 36 lysimeters were used for growing plants for destructive measurements in growth analysis. The lysimeters, of a total depth 1.2 m and 22 cm width, were filled with a loamy sand soil. They were protected from rain by a transparent cover, so the periods of drought could be reasonably simulated. Micrometeorological data were registered.

Plants of the vertically-divided lysimeters, showed the growth of "compensating roots", mainly on seminal roots. WU by these roots have increased 3- to 6-fold during drought periods. Results from horizontally-divided lysimeters showed that under the conditions of unlimited water supply 51.4% of WU of the total plant water uptake, during the season, was taken up by the roots in the soil layer 0–33 cm, 42.6% in the 33–66 cm, and only 6.6% in the 66–99 cm. The distribution of the roots' dry matter had a similar trend. After the application of drought condition of the whole rooting profile, the distribution of the root dry matter in the 66–99 cm layer, increased evidently, from 11.4% to 22.4%. The WU increased from 6.6% to 20.6%. But this was insufficient to recover the plant dry matter and grain production, which were only about half of control in this variant. A similar situation in dry matter distribution also occured in the experiments where soil drying was induced in the layers of 0–33 cm and 33–66 cm, but an unlimited water supply was given to the layer of 66–99 cm. This has simulated the underground water source. The dry matter distribution in layer 66–99 cm was increased from 11.4% to 24.4%, while the WU increased substantialy from 6.6% to 45.3%. The total dry matter production in these plants was about the same as in control plants and grain production per cob was even significantly greater.

The development of "compensating roots" in deep and wet soil layers, manifested the action of the drought avoidance mechanism with a consequent improvement of grain yield.

REFERENCES

Ješko, T., 1996, Plant strategies in overcoming drought conditions, *in*: "Proc. 7th Days of Plant Physiology", Biologia 51:57.

Levitt, J., 1972, Responses of plants to environmental stresses, Academic, New York.

Navara, J., Ješko, T., Ziegler, W., and Duchoslav, S., 1993, Water uptake by maize (*Zea mays* L.) root system. Biologia, 48:113.

DENSITY DEPENDENT HABITAT SELECTION IN PLANTS

Gersani Mordechai, Zvika Abramsky, and Omer Falik

Department of Life Science, Ben-Gurion University of the Negev, Beer-Sheva, 84105 Israel.

INTRODUCTION

The theory of density-dependent habitat selection was developed by Fretwell and Lucas (1970) and by Fretwell (1972). For a single species, Fretwell deduced that, in the absence of strong contest competition, individuals of a species are better off settling in habitats so that no individual has a higher per capita population growth rate than any other. This is accomplished by individuals distributing themselves unequally in different habitats: more productive habitats will have more individuals than less productive ones. Although the theory was developed for mobile animals we think that it may also fit plants. Furthermore, we suggest that the theory of density-dependent habitat selection can also describe the processes occurring at the level of one individual plant, in which its root biomass is analogous to an animal population.

MATERIALS AND METHODS

Pea seedling *Pisum sativum* L. were forced to form a split-root system, that contained two equal roots. The roots of one split-root plant, the "fence-sitter", were planted in two adjacent pots (Fig. 1). In one pot either 1, 2, 3, or 5 competitor plants were planted. These competitor plants also had split roots, though they were planted in only one pot. Control experiments received a similar treatment of 0, 1, 2, 3, or 4 competitor plants for both pots of a pair. The pots were saturated every two days with Hoagland nutrient solution, and once a week with distilled water.

We harvested the plants after seventy days. We separated the roots of the fence-sitter from the vermiculite and from the roots of the competitor plants, and estimated the dry weight of pods, shoots, and roots.

RESULTS AND DISCUSSION

The fitness, expressed by the number and by the dry weight of seeds of the fence sitter, was constant, regardless of the density of the competitor plants, while the fitness of the competing plants decreased with density (Fig 2). The total size of the root system of the fence sitter, in both pots, was not influenced by the number of competitors. However, a shift of the fence sitter root system from the pot with competition to one that is free of competition was proportional to the number of competitors. Each competitor had the same influence on the distribution of the root system of the fence sitter (Fig 3). The success of the fence sitter plant to reach a constant fitness, can be explained by the phenotypic plas-

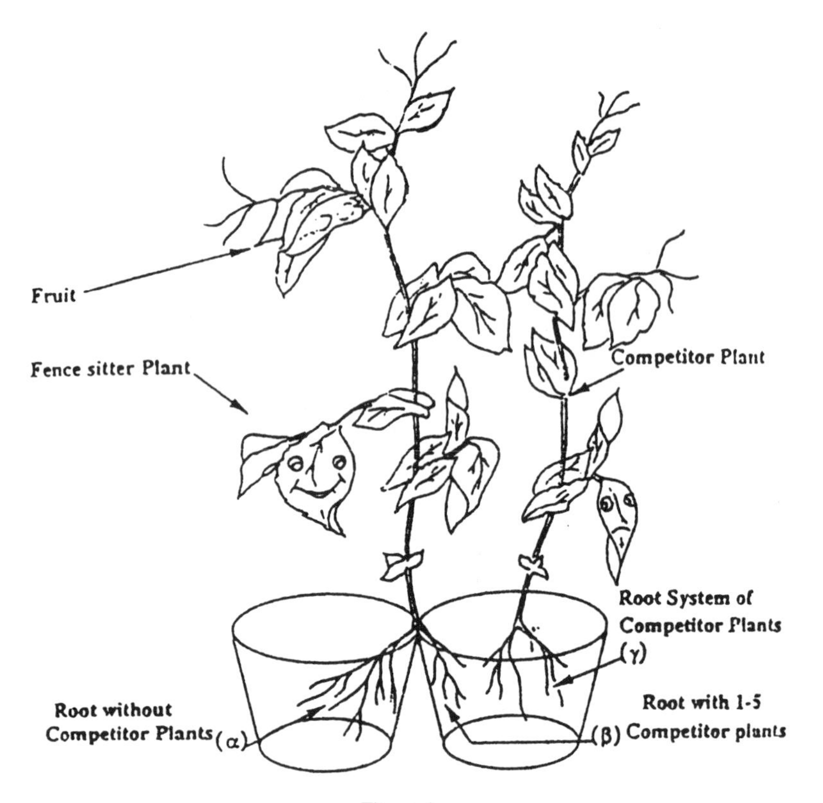

Figure 1.

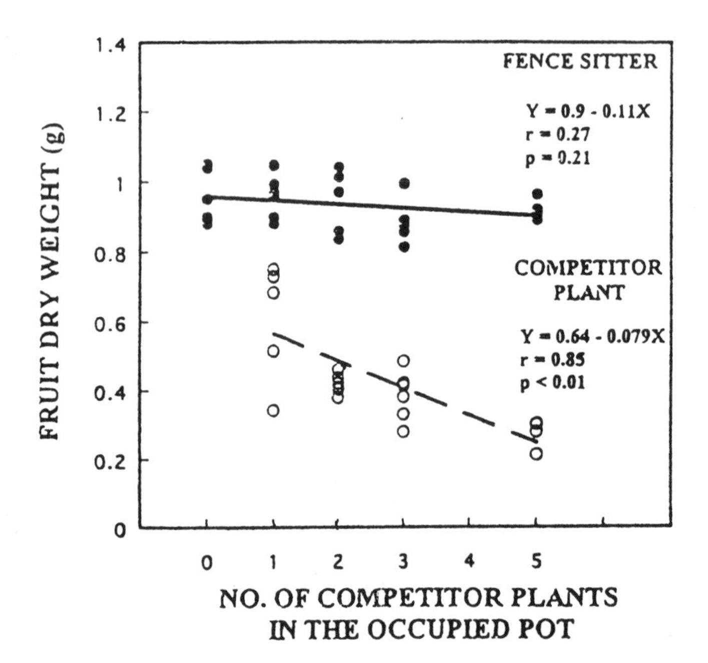

Figure 2.

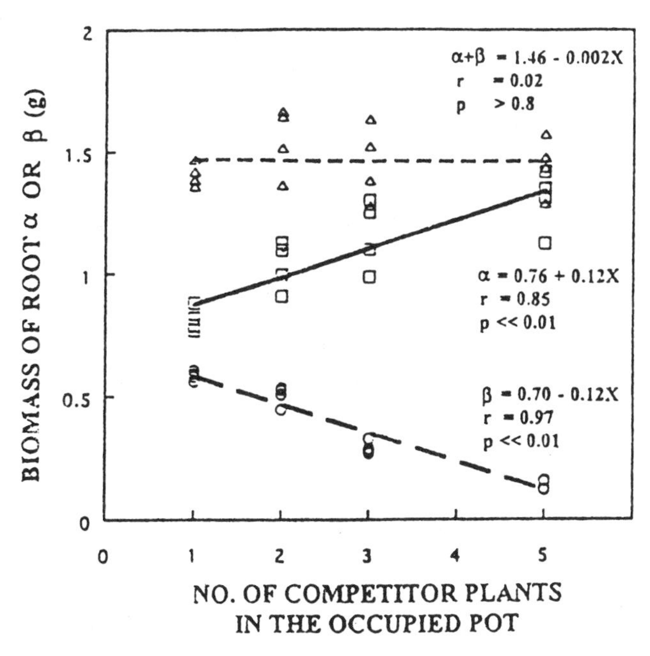

Figure 3.

ticity of the plant to respond to the nutrients' availability in each patch and to invest in each root accordingly. The fence sitter plant avoided competition, demonstrating a density dependent habitat selection, by shifting its roots to better sites—the pots without competitors. The distribution of the roots of the fence-sitter fits Fretwell's model (1972), and supports the hypothesis that the approach of models of habitat selection developed to describe distribution of mobile organisms are probably also capable of predicting the distribution of roots of an individual plant in a heterogeneous environment (Fig 4).

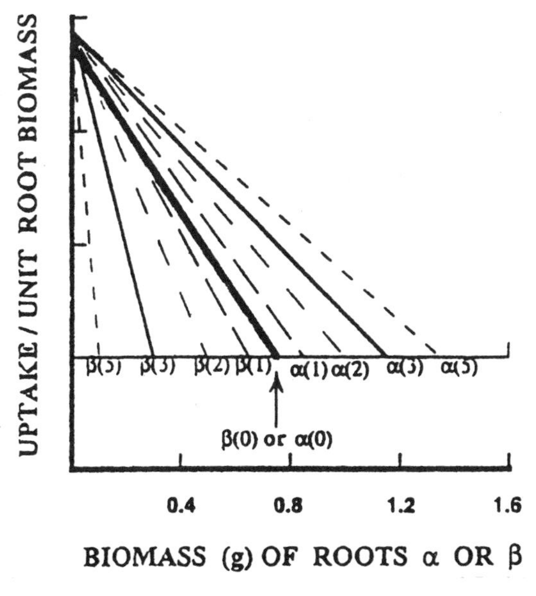

Figure 4.

REFERENCES

Fretwell, S.D. and Lucas, H.L., Jr (1970) *On territorial behavior and other factors influencing habitat distribution in birds*. I. Theoretical development. Acta Bioth. 19.16–36.
Fretwell, S.D. (1972) *Population in a seasonal environment.*, Princeton University Press, Princeton, N. J.

EFFECT OF RAPE SEED CAKE COMPOST ON ROOT GROWTH AND NUTRIENT UPTAKE OF THE KOREAN LAWN PLANT (ZOYSIA KOREANA STEUD)

Choi Byung-Ju[1] and Hoon Park[2]

[1]College of Industry and Science, University of Kongju, Yesan, Korea. [2]Korea Ginseng & Tobacco Res. Inst., Taejeon, Korea.

INTRODUCTION

Lawns of the Korean grass grow everywhere in Korea and has long been used mainly for covering ancestor's tombs. Recently the number of golf courses is rapidly increasing and heavy fertilization of the lawns which might cause environmental pollution is controversial. Peat (Kim et al., 1992) and the poultry dung (Ham et al., 1993) were very effective for lawn growth of the Korean lawn (Choi et al., 1994). In the present investigation we analysed the relationships between growth and mineral content and RSCC application.

EXPERIMENTAL

PVC pipes (20cm 60cm Long) were filled with a low fertility sandy loam soil. Korean lawn was grown for 3 months with single or combined treatments of fertilizer (N–P–K of the commercial fertilizer 21-17-17 at a level of 40g/m2), RSCC (600g/m2), $MgSO_4 \cdot H_2O$ (40g/m2) and granular active carbon (400g/m2). Seeds were sawn and watered with 1l per day, for one month. Then, watering was continuated every 3 days, for 2 months until harvest. Minerals were analysed by $HClO_4$–H_2SO_4–H_2O digestion.

RESULTS AND DISCUSSION

Root yields were significantly higher for the RSCC and for the RSCC+NPK plot. Areal shoot growth was linearly correlated with root growth. Nutrient content (me/100g.dw) was in the order of N>K>Ca>Mg>P in the root, but N>K>Mg>Ca>P in the shoot. The shoot/root ratio of nutrients was in order of P>K>Mg>N>Ca. RSCC increased the nitrogen and phosphorus content of the root and the nitrogen, potassium and magnesium content of the shoot. RSCC caused a decrease in calcium content of the shoot and of potassium in the root. Root yield showed a positive correlation with N (p=0.05), with P (p=0.05), and with Mg content (not significant) and a negative correlation with K (p=0.05) and Ca in the root. Root yield showed a positive correlation with the shoot/root ratio for K (p=0.01), Mg (p=0.1) and N. It was negative with the values of P (p=0.05). These trends were same with shoot yield. N content of the root was positively correlated with P (p=0.05) Ca and Mg. It was negatively correlated with K that had positive correlation with Ca (p=0.1). P had positive with Mg and negative with K and Ca. The shoot/root ratio of N was positively correlated with that of Mg (p=0.01) and K (p=0.05), but negatively correlated with zinc (p=0.1), P, Mn and Ca. That of P had positive with that of Zn (p=0.05) and Fe (p=0.1) Mn, and negative with that of K (p=0.05), Mg and Cu.

CONCLUSION

The positive effect of RSCC seems to be due to the continuous supply of N to the roots that stimulates translocation of Mg and K to the shoots. In addition, the continous supply of P, by the RSCC, to roots inhibits translocation of Fe to the shoots.

REFERENCES

Choi, B. J., Shim, J. S. and Park, H. 1994. Effect of organic material, active carbon and magnesium on the growth of Zoysia koreana. Kor. Turfgrass Sci., 8(3):193–199

Ham, S. K., Lee, J. J. and Kim, I. S. 1993. Effect of organic fertilizer on the growth of Korean lawn. Kor. Turfgrass Sci., 7(1):61–66.

Kim, D. C., Shim, J. S., Chang, H. J. and Chung, W. I. 1992. Effect of peat and chemical fertilizer on the growth of Korean lawn. Kor. Turfgrass Sci., 6(2):83–88.

EFFECTS OF CADMIUM, LEAD AND ZINC ON ROOT MERISTEM, ROOT HAIR FORMATION, XYLOGENESIS AND DEVELOPMENT OF LATERAL ROOT PRIMORDIA IN *Ocimum sanctum* L. AND *Festuca rubra* L. CV. MERLIN

Nina M. Cadiz[1] and Michael Stuart Davies[2]

[1]Plant Biology Division, University of the Philippines Los Baños, College, Laguna 4031. [2]Department of Pure and Applied Biology, University of Wales, Cardiff, United Kingdom CF1 3TL.

INTRODUCTION

Toxic ion has received considerable attention not only because of its occurrence in the environment but also because of its role in reducing crop productivity. Roots are very sensitive to elevated levels of ions, thus their growth and development in a metal-stressed environment becomes an important factor in assessing the capability of the plants to populate a wide range of land. *Ocimum sanctum*, a dicot herb which is commonly colonizing open and waste places in the Philippines, and *Festuca rubra* cv. Merlin, a Pb/Zn tolerant grass cultivar from Wales, were used to evaluate the overall effects of three heavy metals;

namely, cadmium (Cd), lead (Pb) and zinc (Zn) on the growth and differentiation of roots. In addition, the present study also determined some root traits that can be used as possible markers of metal tolerance in plants.

EXPERIMENTAL

Seeds of *Ocimum sanctum* and *Festuca rubra* cv. Merlin were germinated on circular polystyrene rafts with fibreglass matting which were floated on aerated -P Rorison nutrient solution. The seedlings were first established for 7 d before they were treated for 7 d with various concentrations of Cd, Pb and Zn (0, 2.5, 5 and 10 μM) supplied in a background of 0.1 strength Rorison's -P nutrient solution. The experimental set-up was maintained in a growth chamber at 23±0.5°C with white light illumination at a fluence rate of 227 μmole m^{-2} s^{-1} PAR for 12 h d^{-1}.

RESULTS AND DISCUSSION

A 7 d exposure to Cd, Pb and Zn, generally reduced the length of the root meristem (i.e. delineated by the densely stained region). In *Festuca*, Cd, Pb and Zn reduced elongation by 73%, 77% and 83% of the control value, respectively. The proportional reduction in *Ocimum* was smaller. In contrast, the meristem width of *Festuca* was increased by the three metals up to the concentration of 2.5 μM, but was little affected in *Ocimum*. The pattern of change in meristem volume closely paralleled that of meristem width in *Festuca*. In *Ocimum*, increasing metal concentrations tended to shorten the distance from the proximal boundary of the meristem to the first lignified xylem element and to the firts root hair. In *Festuca*, root hairs were noted to form farther from the proximal boundary of the meristem. This can be regarded as a mechanism which enables this species to tolerate a give stress by decreasing the surface area for absorption. The metals in general stimulated the formation of lateral root primordia in both species but Zn caused the greatest induction.This is thought to operate like the apical dominance where its removal results in the growth of lateral branches. In the present study, the metal-induced reduction of the meristem size is regarded as a kind of "chemical decapitation" releasing the dominance effect of the root apex, thus increasing the number of lateral root primordia.

CONCLUSION

The overall response of plants to the metals, but more particularly so of *Festuca*, can be regarded as indication of tolerance mechanism of this plant to metal stress. However, it is clear from our results that different species respond in different ways.

REFERENCES

Cadiz, NM, CC de Guzman & MS Davies. 1996. Tolerance strategies of plants to heavy metals: cellular changes, accumulation pattern and intracellular localization Cd, Pb and Zn in *Festuca rubra* cv L. Merlin (red fescue) and *Ocimum sanctum* L. (holy basil). Phil Jour of Crop Science vol 21 (Supp. No. 1).

Powell, MJ, MS Davies & D Francis.1988. Effect of zinc on meristem size and proximity of root hair and xylem elements to the root tip in a zinc-tolerant and non-tolerant cultivar of *Festuca rubra* L. Annals of Botany. 61:723–726.

GROWTH RESPONSE OF *Phaseolus vulgaris* L. TO HIGH TEMPERATURE AND HYPOXIA IN THE ROOT ZONE

L. Incrocci,[1] A. Pardossi,[2] F. Tognoni,[2] and G. Serra[1]

[1]Scuola Superiore di Studi Universitari e Perfezionamento S. Anna, Pisa, Italy.
[2]Dipartimento di Biologia delle Piante Agrarie, University of Pisa, Italy.

In the Mediterranean regions, during spring and summer, soilless-grown greenhouse plants may undergo severe stress due to high root zone temperature which can be coupled with a reduction in the oxygen content of the nutrient solution. We report here the first results of a study initiated to separate the effects of heat and hypoxia stress on hydroponically-grown bean seedlings.

Bean seedlings (*Phaseolus vulgaris* L. cv. Borlotto) were differently treated on the basis of air/root temperature (25/25, 35/25, 25/35 and 35/35°C) and oxygen content (7.5–8 or 1.5–2.0 mg l^{-1}, achieved by insufflating compressed air or nitrogen into the nutrient solution). The nutrient solution was controlled daily for pH (6.5–7.5) and EC (1.6–1.8 mS cm^{-1}) and replaced totally every 4 days. For each treatment, fresh and dry weight of roots, leaves and stem, leaf area and leaf number were measured after 8 days of treatment.

Bean plants appeared to tolerate better an increase in air temperature than in root temperature. Increasing air temperature from 25 to 35°C caused a marked reduction of leaf area (LA) and root (RDW) and total dry weight (TDW) . Growth inhibition by the same increase in the root temperature was much strong (Table 1).

The high root temperature decreased the relative growth rate (RGR) but did not affect the net assimilation rate (NAR). Root cooling to 25°C reduced the growth depression by high air temperature (35°C).

Hypoxia caused a strong inhibition of plant growth, but this effect was less evident when the plants were placed at sopra-optimal temperature.

Work is in progress to investigate the physiological effects of hypoxia and high temperature on root zone with special emphasis to leaf water relation, gas exchange and assimilate partitioning.

Table 1. Root and total dry matter, leaf area, root/shoot ratio and growth index of bean plants grown in hydroponics and subjected to different experimental treatments. The data are expressed as % of plant grown in the aerated solution at 25/25°C.

	Air/root temperature							
	25/25°C		35/25°C		25/35°C		35/35°C	
	+AIR	$+N_2$	+AIR	$+N_2$	+AIR	$+N_2$	+AIR	$+N_2$
RDW	100.0	21.4	58.4	13.1	23.9	13.3	18.0	6.2
TDW	100.0	39.4	58.4	27.4	48.7	37.1	38.2	20.1
LA	100.0	20.7	46.0	15.7	21.7	29.2	16.0	11.2
R/S	100.0	48.1	103.7	44.4	44.4	29.6	40.7	25.9
RGR	100.0	41.1	67.5	24.9	53.8	36.5	43.1	10.1
NAR	100.0	69.4	100.9	49.5	96.5	46.0	105.2	17.1
LAR	100.0	62.5	73.2	56.5	57.1	82.7	44.6	54.8
SLA	100.0	83.1	79.0	72.3	76.4	98.2	59.6	72.8

COLD-INDUCED ALTERATIONS IN PROTEIN COMPOSITION OF MAIZE ROOTS

Otília Gašparíková and Ladislav Tamas

Institute of Botany, Slovak Academy of Sciences, Dúbravská 14, SK-842 23 Bratislava, Slovak Republic.

There is mounting evidence to indicate that most cold tolerant plant species synthesize new sets of proteins that are correlated with the increase in cold hardiness (Howarth and Ougham, 1993). Some of the surveys suggests the possibe role of protein synthesis also in chilling tolerance in plant species native to tropical or subtropical habitats (Xin and Li, 1993). However, the data are scarce. Therefore, the objective of this study was to identify soluble and microsomal membrane proteins that are specifically induced by chilling and examine activity and isozymic composition of such enzymes in maize roots. Primary roots of two genotypes, chilling-sensitive Penjalinan and chilling-tolerant Z7, were used. Seedlings were grown on Hoagland's nutrient solution, in a growth chamber, at 24°C, 70% relative humidity and 12 h photoperiod, until the third leaf stage. Then plants were subjected to a 5 days stress period at 6°C, followed by a 4 days recovery at 24°C. Control plants were maintained at 24°C for the whole period. Protein extraction, SDS-PAGE and detection were carried out as described previously (Gasparikova et al., 1996).

The exposure of plants to a low temperature treatment (6°C) caused the changes to be predominantly in membrane-associated proteins. A group of peripheral membrane proteins, with an apparent mass of 10, 11, 12, 18 and 27 kDa was induced in both chilling sensitive and chilling tolerant genotypes. By contrast, levels of 15, 16, 21, 23, 25, 29, 30 and 36,5 kDa proteins increased specifically in roots of chilling-sensitive plants. These results together with the changes in abundance of the integral membrane proteins suggest that root cell membranes undergo molecular reorganization under the chilling.

The changes in the soluble proteins were not so strong. Whereas almost no apparent changes could be detected in patterns of the soluble proteins in Z7 genotype, an increase of 33, 40, 55 and 90 kDa proteins appeared in the sensitive Penjalinan.

The increase in soluble and in membrane protein contents paralleled increases in activities of cytosolic and microsomal membrane acid phosphatase, non-specific esterase, and exoprotease, but not of peroxidase. No new isoenzymes were synthesized neither for acid phosphatase nor for the non-specific esterase. Some quantitative isozymic differences were observed in response to chilling. These were completely repared during recovery stage especially with Z7 genotype. Comparing the synthesis of cold-shock proteins in maize genotypes which differ in chilling tolerance showed no relationship between the synthesis of cold-shock proteins and chilling tolerance.

REFERENCES

Gašparíková, O., Èiamporová, M., Tamas, L., Trgiòová, I., and Luxová, M., 1996, Cold-induced changes in protein patterns and ultrastructure of root cells of maize seedlings, Biologia (Bratislava) 51:449.

Howarth, C.J., and Ougham, H.J., 1993, Tansley Review No. 51 Gene expression under temperature stress, New Phytol. 125:1.

Xin, Z.G., and Li, P.H., 1993, Alteration of gene expression associated with abscisic acid-induced chilling tolerance in maize suspension-cultured cells. Plant Physiol. 101:277.

TRANSCRIPTION OF Δ^1-PYRROLINE-5-CARBOXYLATE SYNTHASE IN ALFALFA ROOTS DURING SALT STRESS

Idit Ginzberg,[1,2] Yoram Kapulnik,[2] and Aviah Zilberstein[1]

[1]Department of Botany, Tel-Aviv University, Tel-Aviv 69978, Israel. [2]Institute of Field and Garden Crops, The Volcani Center, P.O. Box 6, Bet-Dagan 50250, Israel.

Soil salinity is an important determinant of plant growth and survival. In legumes, saline conditions impose a significant limitation of productivity related to the combined effects on the host growth and the root nodulation by *Rhizobia*. Salt stress conditions initiate various biochemical and physiological changes in plants including a dramatic increase in intracellular concentration of proline. Proline is synthesized from glutamate via two successive reductions. In *Vigna aconitifolia* and *Arabidopsis thaliana* the reduction of glutamate to its semialdehyde intermediate was demonstrated to be catalyzed by a single bifunctional enzyme Δ^1-pyrroline-5-carboxylate synthase (P5CS). To isolate alfalfa-*P5CS*, a cDNA library prepared from salt stressed alfalfa (*Medicago sativa* L.) roots was screened using *Arabidopsis P5CS* cDNA clone as a probe. Two clones were identified, *MsP5CS-1* (2.6Kb) and *MsP5CS-2* (1.25Kb), with homology of only 74% in the nucleotide sequence and 75% in the deduced amino acid sequence. *MsP5CS-1* and *MsP5CS-2* show 82% and 75% homology to *Vigna* cDNA, respectively. Both show 72% homology to the *Arabidopsis* cDNA clone. Southern blot analysis revealed the presence of two different genes. The transcription pattern of the two genes in alfalfa roots was studied in 6 d-old hydroponically grown seedlings that were exposed to 100mM NaCl for 72 hours, and compared to control seedlings. Roots samples were taken periodically, total RNA was extracted and the level of the transcripts was estimated by RT-PCR with primers specific to each gene. The results show an increase in the transcriptional level of the two genes, characterized by two dominant peaks at 6 and 48 hours. The induction of *MsP5CS-2* transcription was higher than that of *MsP5CS-1*. It is suggested that the primary increase induced by the salt stress is feed-back inhibited by the accumulating proline. The second induction might be due to adaptive processes caused by a long exposure to NaCl.

DEVELOPMENT AND INITIATION OF THE ROOT SYSTEM OF OLD BLACK SPRUCE (*Picea mariana* (MILL.) B.S.P.) AND BALSAM FIR (*Abies balsamea* (L.) MILL.) TREES

Cornelia Krause and Mylaine Fournier

Consortium de Recherche sur la Forêt Boréale Commerciale, Université du Québec à Chicoutimi, 555 blvd. de l'Université, Chicoutimi, Québec, Canada G7H 2B1.

Little knowlegde is attributed to below ground growth of mature trees (Hardley and Smith 1987). Some work has been done to classify the distribution of the roots and to group the tree species into three main classes (horizontal root, tap root and heart root). Other studies were concerned with the length of the root branches and the calculation of below ground biomass. In this study, the root installation data were examined and whenever possible, related to age at the root/shoot interface. The root/shoot interface represents the zone between the roots and the trunk and gives the exact age of the trees (Telewski 1993).

The study sites were located in the boreal zone of eastern Canada, in the province of Quebec (48°50′ N, 70°88′ and 49°40 N, 72°60′ W). Three balsam fir sites and two black spruce sites were chosen. Insect outbreaks and forest fires are mainly responsable for the forest succession in the region. A total of thirty balsam fir and twenty black spruce tree stems and root systems were sampled. The entire root systems of the chosen trees were excavated up to a root diameter of 2 cm and cut into 2 cm thin sections successively. The stem and root sections were cross-dated with pointer years, in order to obtain an exact calendar year of the establishment of each section. This cross-dating method was less successful in certain sites, due to two main reasons: 1) the heart of the stumps was rotten and the piths were located outside of the stump wood, and 2) a stage of horizontal juvenile growth during the establishment period.

The results show that all dated root branches have an adventitous caracter and that black spruce and balsam fir regenerate their adventive root system more than once in their lifetime. The establishment of the root system occurred over a time range of 8 to 120 years later than the trees' germination period. The difference between the root age and the actual age of the trees varied with the trees' age; if a tree is old the difference between both ages is large. This situation is the opposite in younger trees. It was also demonstrated that the root branches and ramifications grew in considerable lengths during a defined period of several years, whereas during many years diameter growth only was observed.

In conclusion, the dendrochronical study shows the possibility of analyzing through time old growth conifer trees. The datation of the different sections in the stump allows to establish the years of installation of the trees, whereas for the roots, a section closest to the ground level is dated. All trees analyzed had an adventitve root system only. The dendrochronological analysis allowed us to reconstruct the growth of woody roots (over a diameter of 2cm) in time and space. Moreover, it is obvious that for a defined period of years the roots grow in length and then suddenly stops for many years, before restarting its growth activity later on.

DATING BLACK SPRUCE (*Picea mariana*) WITH THE AWARENESS OF ITS ADVENTITIOUS ROOT SYSTEM

Annie Desrochers and Réjean Gagnon

Laboratoire d'Écologie Végétale, Université du Québec à Chicoutimi, 555 boul., Université, Chicoutimi, Québec, G7H 2B1 Canada.

INTRODUCTION

Black spruce (*Picea mariana* [Mill.] B.S.P.) is one of the most important species in Canada for the Pulp and Paper Industry. Despite its importance, we still have some difficulties in dating the mature trees; black spruce develops an adventitious root system, which in turn conceals the base of the stem and the root collar in the stump wood. The usual methods for dating mature trees, which consist in taking a cross section or an increment core at the base of the stem, do not consider this phenomenon, and may lead to misinterpretations about the species dynamics. The age under-estimation increases with every year since the fire, with the development of the adventitious root system and the elevation of the ground level with time. The objective of this study was to experience a new dendrochronological dating method, which permits to locate the juvenile growth rings and the root collar under the ground level, and compare the results with a conventional dating method.

EXPERIMENTAL

Forty-four black spruce stumps were excavated and cut into 2 cm transverse sections. Each of these were anatomically examined to locate the root collar, and then cross-dated in order to retrace the juvenile growth rings present between the root collar and the ground level.

RESULTS AND DISCUSSION

The stump cross-dating method permitted to retrace 3 to 19 years more than with the conventional dating method (ground level cross section). The buried base of the stem was often incompletely included in the stump, preventing to locate the collar and find the actual age of the trees. Cross-dating difficulties, for causes of decay, compression wood, absent rings and abundance of resin, also prevented the determination of the total ages. It was observed that the trees experienced a horizontal juvenile growth period, probably related the development of their adventitious root system; the curvature of the stem permitted to have the base of the stem in contact with the organic matter, factor that greatly promotes the development of adventitious roots (Aubin 1996). A reverse taper phenomenon was observed in all examined trees, the last rings being absent from the sections near the root collar. This phenomenon could be due to the location of this part of the stem, under the nutritive exchange between the roots and the crown (Knight 1961), and to the mechanical stress caused by wind that no longer affects the buried part of the stem and stays higher at the ground level (Heikkinen 1994).

CONCLUSIONS

Dating trees that form adventitious roots on the stem is more difficult since they obscure the location of the root collar. The stump cross-dating method allowed to greatly diminish the age under-estimation brought by usual dating methods.

REFERENCES

Aubin, N. 1996. Influence du contenu en eau du substrat et de la profondeur de plantation sur la formation de racines adventives caulinaires, la croissance et l'allocation glucidique de semis d'épinette noire (*Picea mariana* (Mill.) B.S.P.). Master's thesis, Université du Québec à Chicoutimi, Quebec, Canada.

Heikkinen, O. 1994. Using dendrochronology for the dating of land surfaces. *In* Dating in exposed and surface contexts. *Edited by* C. Beck. University of New mexico Press, Albuquerque, pp. 213–235.

Knight, R.C. 1961. Taper and secondary thickening in stems and roots of the apple. East Malling Research Station, an. rep. 1960, pp. 65–71.

USE OF GROWING DEGREE DAYS TO TIME ROOT DEVELOPMENTAL EVENTS IN THE GRAMINEAE

Betty Klepper, R. W. Rickman, S. E. Waldman, and D. A. Ball

USDA, Agricultural Research Service, Oregon State University Columbia Basin Agricultural Research Center, PO Box 370, Pendleton, Oregon 97801.

Root development can be related to shoot development in cereals such as winter wheat (*Triticum aestivum* L.). Individual root axes can be seen to arise from specific nodes and can be identified using a naming system that relates the axis to a node and to a direction of growth with respect to that node. A phyllochron system can then be used to relate the appearance of specific seminal and crown root axes to main stem leaf number. The general pattern of root axis appearance for grasses is therefore predictable based on a growing degree day (heat unit) calculation and specific root axes are observable at specific points in shoot development. This pattern can be generalized from wheat to other grasses for use by modellers and morphologists.

Eight different grass species were examined in field plots over a two-year period in order to generalize these patterns. Included were winter wheat, barley (*Hordeum vulgare* L.), triticale (X *Triticosecale* Whittmack), bulbous bluegrass (*Poa bulbosa* L.), downy brome (*Bromus tectorum* L.), jointed goatgrass (*Aegilops cylindrica* Host.), Italian ryegrass (*Lolium multiflorum* Lam.), and wild oat (*Avena fatua* L.).

Main stem leaf production rates were slower for wheat, triticale, and jointed goatgrass than for the other species. The requirement ranged from 123 growing degree days (GDD) per leaf for jointed goatgrass to 58 GDD per leaf for downy brome. Thus the time from emergence to development of roots at the first foliar node would be about half as long for downy brome as for jointed goatgrass. Thus one factor influencing the appearance time of crown roots is the rate of leaf development as expressed by the length of the phyllochron.

A second factor can be related to two patterns of seedling morphology. In one, the crown is set at the coleoptilar node and in the other it is set at the first foliar node. These two patterns produce two types of plant in terms of root axis developmental pattern. In one, the coleoptilar axes are produced at the level of the seed and in the other, they are produced at the level of the crown. Thus, the second type, illustrated by downy brome, wild oat, and Italian ryegrass has earlier production of crown roots relative to shoot development than the first type, which includes wheat, barley, triticale, bulbous bluegrass, and jointed goatgrass.

These results allow plant root experts to understand the patterns of root axis appearance for members of the grass family and will promote the development of plant rooting models which are based on heat unit concepts for easy integration of root and shoot comparative morphology. The results are also of interest to weed scientists who must use differential developmental information in the formulation of strategies for weed control.

REFERENCES

Ball, D. A., B. Klepper, and D. J. Rydrych. 1995. Comparative above-ground development rates for several annual grass weeds and cereal grains. Weed Sci. 43:410–416.

Dotray, P. A., and F. L. Young. 1993. Characterization of root and shoot development of jointed goatgrass (*Aegilops cylindrica*). Weed Sci. 41:353–361.

35

RELATIONS BETWEEN EARLY ROOT GROWTH AND FLOWER YIELD OF CLONAL ROSE ROOTSTOCKS

D. P. de Vries and L. A. M. Dubois

Centre for Plant Breeding and Reproduction Research (CPRO-DLO)
6700 AA Wageningen
The Netherlands

1. INTRODUCTION

Rootstocks for roses fall into two categories: (i) seedling stocks and (ii) clonal stocks, including the own roots of the scion variety. In North-Western Europe, seedling stocks are used for both glasshouse- and outdoor-grown roses, while clonal stocks are exclusively applied in the glasshouse. This report refers to clonal stocks only.

Comparing research in clonal rose rootstocks with e.g. those of fruit, rose stocks lag behind more than 80 years. Owing to that, rose growers, horticultural advisers and scientists urgently need a series of clonal stocks, of which the induced vigour can be reliably forecasted.

An obvious way to clarify the situation is to test combination plants of numerous stocks and scion varieties over a range of years. However, such experiments are endlessly repeated, ill-published and usually raise more new questions than answering old ones (De Vries, 1993). Another, very expensive possibilty is to qualify and quantify cytokinins in the bleeding sap of plants on various stocks, and predict vigour by correlating rootstock vigour and cytokinin concentration (Verstappen, 1992).

In our experiment a different approach to predict rootstock vigour was investigated. As a basis, an important parameter of vigour: flower yield per year per plant, was used. Yield had been quantified previously in an experiment with eight "Sonia"-rootstock combinations, including the own "Sonia" roots (Table 1). Owing to the absence of scion-rootstock interaction as to shoot yield (De Vries, 1993), only this scion variety was examined. Our hypothesis was: "the shoot yield of the scion variety, as induced by a series of rootstock genotypes is associated with some root or plant characteristic that can be assessed in an early stage of plant development". In other words: is there a characteristic of early growth of rootstock genotypes that is correlated with its induced shoot yield?

The present study deals with the growth and development of roots and shoots of "Sonia" combination plants on 8 clonal stocks over a 12-day period, in relation to known rootstock vigour.

Biology of Root Formation and Development, edited by Altman and Waisel.
Plenum Press, New York, 1997.

Table 1. The number of flowers harvested per plant over a 12-month period of the rose cv "Sonia," grafted onto 8 rootstock genotypes, grown in rockwool in the glasshouse[a]

Multic	55.8a	Moneyway	52.0a
Sonia	38.0b	Marleen	37.0b
Manettii	36.8b	Indica major	34.0bc
Kuiper	32.9bc	In-IVT-4	27.3c

[a]Figures followed by the same letter do not differ significantly at $p = .001$.

2. MATERIALS AND METHODS

Glasshouse-grown "Sonia" shoots, bearing a just-coloured flower bud, were used. Scions with a subtending 7-leaf were cut from the middle of a shoot. Leaf area of the scions was standardized at approximately 100 cm^2. The internodes of eight glasshouse-grown clonal stocks, standardized for diameter and length, were used in cutting-grafting with "Sonia" (Van De Pol and Breukelaar, 1982). Basal parts of thus constructed combination plants were dipped in 0.5% IBA in talcum, before striking in flats containing a sand:vermiculite:perlite=1:1:1 (v:v:v) mixture. Flats were placed in the nursery greenhouse under "double glass" (25°C, light intensity 70 $\mu mol\ m^{-2}\ sec^{-1}$ (Philips HPI-T, 400 W lamps)). When adventitious roots were 1 cm long, plants were transferred to a growth chamber of the CPRO-DLO phytotron. Combination plants were grown for 12 days in aerated nutrient solution (De Kreij and Kreuzer, 1989) at a constant temperature of 20°C, light intensity 135 $\mu mol\ m^{-2}\ sec^{-1}$ (Philips HPI-T, 400 W lamps). Undesired, unequal sprouting of the scion, which affects carbohydrates distribution, was prevented by removing the axillary bud of scions.

Observations. Daily: number and length of 1st order roots of each plant. *Each 2nd day*: (destructive) measurements of length of 1st and 2nd order roots, number of 2nd order roots per 1st order root (manually), leaf area (Licor area meter); dry weights of leaves, roots, stems, of 3 plants/stock; *Day 12*: diameter of 1st order roots (using binoculars, at 1 mm from root origin), root porosity pyknometer method (Van Noordwijk and Brouwer, 1993).

Experimental. 24 Plants per scion-rootstock combination (8 stocks × 24 = 192 plants), were equally divided over 24 blocks of 8 plants; combination plants were randomized within each block. Statistical analysis was carried out via Epistat. Linear, exponential or logarithmic analyses were used to fit curves. Statistical significance was estimated at 1% or 5% level. RGR was calculated from $(\log W_2 - \log W_1):(T_2 - T_1)$, expressed as $[mg\ g^{-1}]\ d^{-1}$ (Hunt, 1982).

3. RESULTS AND DISCUSSION

Combination plants thrived so well under the conditions described, that observations on root growth and development were concluded after 12 days.

The number of 1st order roots per stock (Fig. 1, top-left) steadily increased in the 12-day period. It is notable that stocks with an initial high number of roots ("Multic", "Sonia") kept forming new roots at a higher level than those with few roots (indica Major, "Kuiper", In-IVT-4).

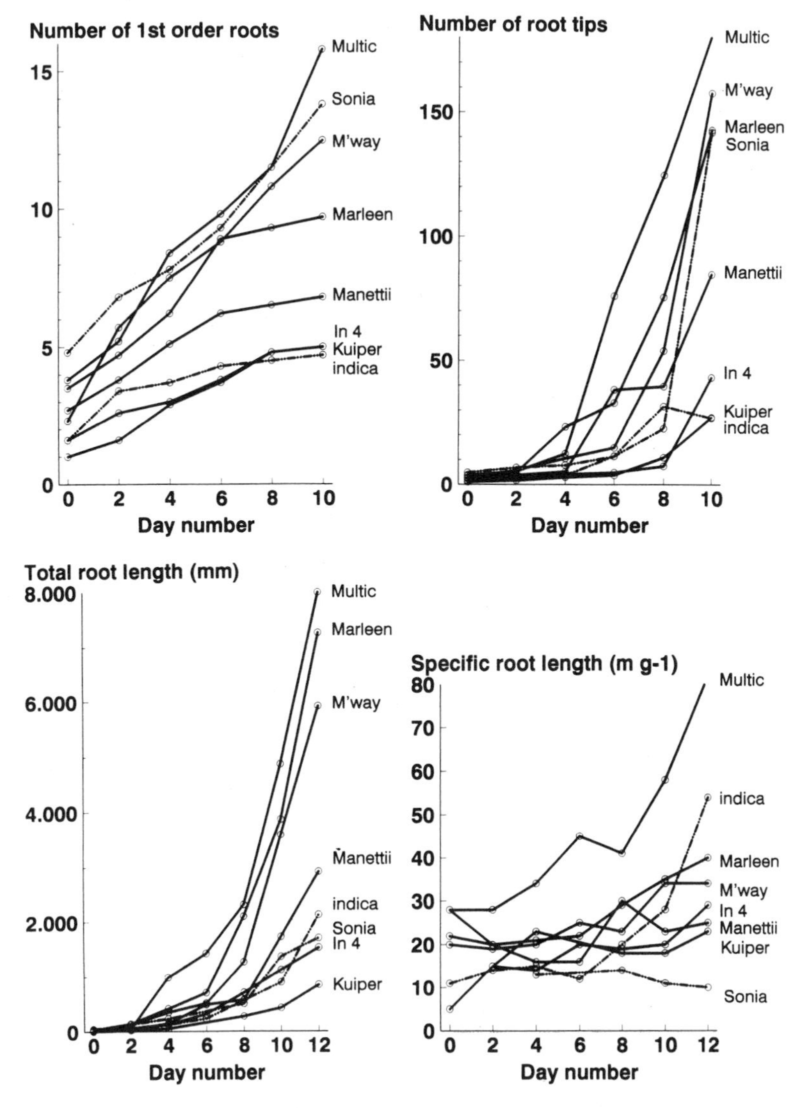

Figure 1.

The number of root tips per rootstock (Fig. 1. top-right) steeply increased from the 2nd day on. Significant differences (day 10) indicated that "Multic", "Moneyway", "Marleen" and "Sonia", had more tips than indica Major, "Kuiper" or In-IVT-4.

The longest root of stocks (Fig. 1, bottom-left), which is always a 1st order one, grew almost linearly. The slopes of the curves for "Multic" and "Moneyway", however, were much steeper than those of "Sonia" or In-IVT-4, resulting in significantly differing lengths after 12 days.

Total root length of stocks (Fig. 1, bottom-right) demonstrates dramatic root growth. Hence, "Multic" and "Marleen" reached lengths between 7000 and 8000 mm, whereas In-IVT-4 and "Kuiper" stalled at 800 and 1500 mm. Significant differences in length occurred between all stocks at day 10. The dry weigths of shoots, leaves and roots of all combination plants showed a continuous increase during 12 days. Differences in *total dry weight* per plant (sum of root, shoot and leaf weight) after 12 days were significant for all stocks; highest weight had plants on "Sonia" and "Marleen", lowest those on indica Major and "Kuiper" stocks. As a consequence of minimal root development at the start, *S/R (dry) weight ratio* (Fig. 2, top-left) of combination plants was initially very high, but as root weight increased more than shoot weight, ratio was about 1 after 12 days. Usually, in roses, after a phase in which fast root growth gives rise to S/R ratios < 1, ratio will return to > 1 as plants grow older.

Specific root length (SRL) (Fig. 2, top-right) is an indirect measure of *mean* root diameter (De Willigen and Van Noordwijk, 1987), thus increasing SRL indicates thinner roots. In all stocks, besides "Sonia", SRL increased as roots grew older. Increase may be explained from the fact that in most stocks the root mass after 12 days mainly consisted of young and thin 2nd order roots. In spite of that, "Multic" or indica Major have basically thinner roots than "Manettii", "Kuiper" or "Sonia".

Root porosity (the percentage of air present in a fresh root mass) would have significance for the growth in wet artificial substrates. Relative low porosity had "Marleen"

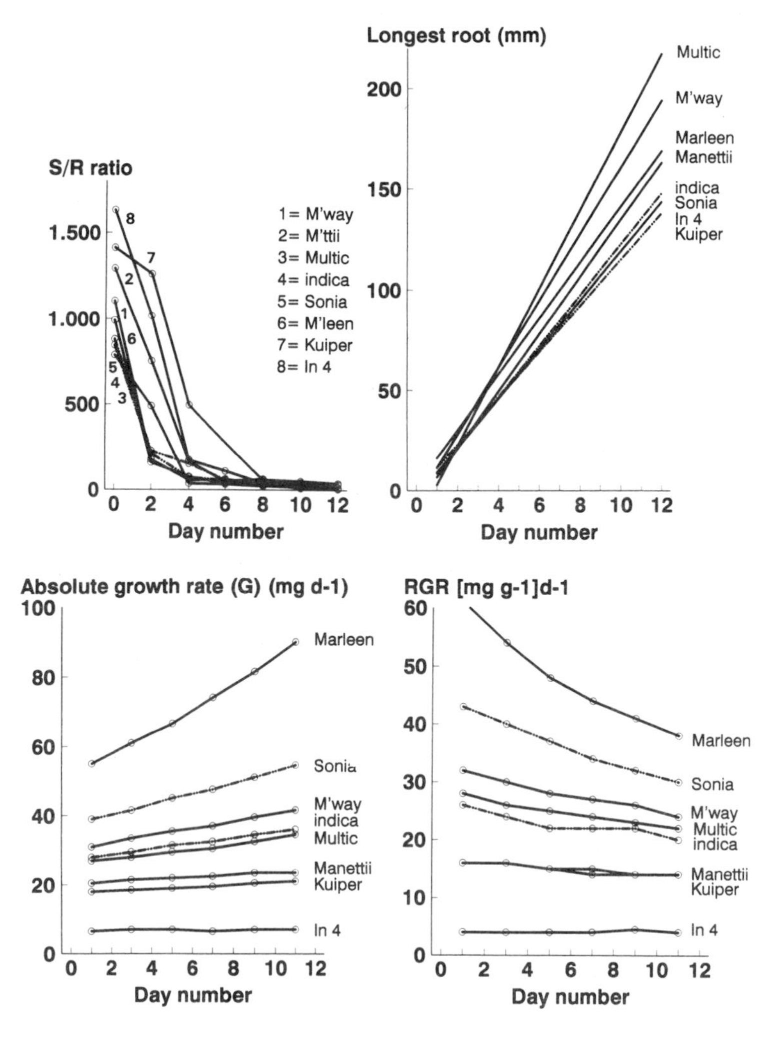

Figure 2.

(4.4%), "Multic" and 83382–95 (6.8%), "Manettii" and "Sonia" (6.9%) and "Natal Briar" (7.7%); high porosity occurred in indica Major and "Moneyway" (12.4%), In-IVT-9 (14.0%) and In-IVT-10 (14.4%). Porosity of these stocks generally appeared to be unrelated to the length of longest root (Fig. 1, bottom-left) or total root length (Fig. 1, bottom-right).

Root diameter was measured in mm: "Multic"(0.89), "Marleen"(0.98), "Manettii" and "Natal Briar"(1.20), indica Major (1.22), 83382–95 (1.26), "Moneyway"(1.30), "Sonia"(1.38), In-IVT-10 (1.47), In-IVT-9 (1.68). It is notable that both In-IVT stocks, which are selections from *R.canina* "Inermis" have particularly thick (1st order) roots. Only as to "Multic", diameter agrees with SRL (Fig. 2, top-right), but in the other stocks this relation was obviously absent. As indicated by correlation ($r = .72^{++}$) root porosity was higher as root diameter increased.

The mean absolute growth rate (G) of stocks (Fig. 2, bottom-left) shows marked differences. The daily increase of some genotypes followed an exponential growth pattern ("Marleen", "Sonia"), but increase of In-IVT-4, "Kuiper" or "Manettii", was almost constant over the 12-d-period. For the stocks with a high absolute growth rate, the *relative growth rate* (RGR) (Fig. 2, bottom-right) steeply decreased, whereas in the "slow" growing ones RGR was almost constant. Maximum RGR was already attained in the first period.

Correlations between flower yield and the above mentioned parameters as well as some additional ones, are presented in Table 2. Unexpectedly, parameters like total dry weight, absolute growth rate or root dry weight, that seemingly express vigour, were not associated with yield. Each of the five parameters that were significantly correlated, however, seem to be important in predicting rootstock vigour. If, in other series of known stocks "the length of the longest root" remains indicative of induced vigour, a very simple parameter can be utilized, which should be vindicated.

The significantly correlated parameters are not completely independent and even may have common physiological bases. Hence SRL, indicating mean root diameter in an indirect way, is likely to be determined by the presence of many fine (2nd order) roots, which shows an indirect relation with the number of root tips. Likewise, the number of 1st order roots is related with the number of root tips, which is the sum of 1st and 2nd order root numbers. Total root length, being the sum of 1st and 2nd order root lengths, is simi-

Table 2. Correlations between the shoot yield of "Sonia" combination plants on 8 rootstock genotypes grown in the glasshouse for 12 months, and parameters of root growth and development of young combination plants on the same rootstocks grown in aerated nutrient solution in the phytotron[a]

Character	Correlation
Total dry weight	– 0.05 n.s.
RGR_{max}	+ 0.21 n.s.
Absolute growth rate (12 d)	+ 0.22 n.s.
Root (d_w) (12 d)	+ 0.31 n.s.
Root porosity (10 d)	– 0.32 n.s.
Root diameter (10 d)	– 0.45 n.s.
Shoot/root (d_w) ratio (12 d)	– 0.47 n.s.
Specific root length (12 d)	+ 0.77 $^{+}$
Number of root tips (12 d)	+ 0.83 $^{++}$
Number of primary roots (12 d)	+ 0.83 $^{++}$
Total root length (12 d)	+ 0.83 $^{++}$
Length of longest root (12 d)	+ 0.89 $^{++}$

[a]r is significant at p = .05 ($^{+}$) or p = .001 ($^{++}$), n.s. = not significant.

larly not independent of the length of the longest root, because long 1st order roots generally bear more 2nd order roots than short 1st order ones (data not presented).

Our results confirm that the induction of vigour by rose rootstocks is not only determined by (root)biomass production in a certain time. Thus, the absolute growth rate of the most vigorous stocks "Multic" and "Moneyway" is on a moderate level only (Fig. 2, bottom-left), while highest growth rates occur among the moderately yielding "Marleen" and "Sonia". Of the stocks presently studied, apparently those with many thin roots, and consequently many root tips, induce highest shoot yields. Root tips are known sources of cytokinins which are generally held responsible for the rate of bud-break of scion varieties (Lockard and Schneider, 1981). Because degree of bud-break, which is another parameter of vigour in cut roses, is supposed to be affected by endogenous cytokinin (De Vries, 1993), it is not unlikely that vigour induction of stocks is ultimately controlled by cytokinin production in their root tips.

4. ABSTRACT

Our hypothesis that the induced vigour of clonal rose rootstocks is related to early root growth and development of combination plants was investigated. 3-w-old "Sonia" combination plants, cutting-grafted on 8 clonal rootstocks, were grown in the phytotron in aerated nutrient solution for 12 days. Significant correlations ($p \leq .001$) with yield occurred for: specific root length, number of root tips, number of primary roots, total root length, length of longest root. Induced vigour of stocks is discussed.

REFERENCES

De Kreij, C. and Kreuzer, A.D.H., 1989. Voedingsoplossingen voor de teelt van rozen in kunstmatige substraten. No.5: Voedingsoplossingen glastuinbouw, Exp.Sta.Naaldwijk, The Netherlands, 40p.

De Vries, D.P., 1993. The vigour of glasshouse roses. Scion- rootstock relationships, effects of phenotypic and genotypic variation. Thesis Agric.Univ.Wageningen, 169 p.

De Willigen, P. and Van Noordwijk, M., 1987. Roots, plant production and nutrient use efficiency. Thesis Agric.Univ.Wageningen, 282 p.

Hunt, R., 1982. Plant growth curves. Arnold, London, 248 p.

Van De Pol, P.A. and Breukelaar, A., 1982. Stenting of roses. A method of quick propagation by simultaneously cutting and grafting. Scientia Hort.17: 187–196.

Van Noordwijk, M. and Brouwer, G., 1993. Gas-filled root porosity in response to temporary low oxygen supply in different growth stages. Plant and soil 152: 187–199.

Verstappen, F.W.A., 1992. Selecteren onderstammen in ver verschiet. Cytokinineproductie maat voor groeikracht roos. Vakbl.Bloemisterij 20: 49.

IN VITRO ROOTING OF FRUIT TREES BY *Agrobacterium rhizogenes*

The Case of Almond and Apple

C. Damiano, E. Caboni, S. Monticelli, P. Lauri, and S. Nicolini

Fruit Trees Research Institute
00040 Ciampino Aeroporto
Rome, Italy

ABSTRACT

Several almond cultivars are strongly recalcitrant to rooting even when cultivated *in vitro*. So far, transformation of leaf tissue with *Agrobacterium tumefaciens*, carrying the plasmid pBinGUSint, has been reported. In analogy to Ti-plasmid, Ri-plasmid of *A. rhizogenes* can also be transferred to tissues of almond. In this case, basal part of microcuttings is infected and roots are expected to be formed from transformed tissue at site of the infection. Many factors influence the rooting response: the genetic background of the cultivar and its susceptibility to the bacterial strains, concentration of the bacterial suspension, the presence of auxin in the medium, etc. The cultivar Supernova shows a very strong sensitivity to the bacterial infection, and the rooting percentage on hormone free medium reached about 60% versus 0% of the control. When localized infection with *Agrobacterium rhizogenes*, combined or not with IAA, was applied to the base of the apple rootstock MM106 microcuttings, which is an easy-to-root genotype, the response was rather inhibitory: without infection, IAA treatment induced most of the explant (95%) to root, and 22.5% of explants rooted even in the hormone free medium; after the infection, the rooting is strongly inhibited both with or without IAA treatment. Evidence of transgenic nature of roots obtained by infection is indicated by hybridization between total DNA and the fragment of pBin19::*Eco*15 containing *rol* genes.

1. INTRODUCTION

Many species are difficult to propagate because of their inability to give an acceptable rooting percentage, either when cuttings or microcuttings are used. The almond (*Prunus dulcis*, Mill.) is also among recalcitrant to rooting, in both cuttings or micro-

Biology of Root Formation and Development, edited by Altman and Waisel.
Plenum Press, New York, 1997.

propagation (Hartman *et al.*, 1990; Tabachnik and Kester, 1977; Rugini and Verma, 1983). The *in vivo* rooting is sometimes improved when soft-woody cuttings are used under fog in glasshouses. A small increase of the *in vitro* rooting percentage is obtained by an appropriate hormone choice and by an induction of a slightly etiolated micro-explants (Caboni and Damiano, 1994). New possibilities seem to arise with the application of transformation techniques. In a previous paper some preliminary results have shown good response to localized infection with *Agrobacterium rhizogenes* 1855, for inducing *in vitro* rooting in the cv Supernova (Damiano *et al.*, 1995) and in the walnut species (Caboni *et al.*, 1996). Infections with different strains of *A. rhizogenes* improved number and quality of roots in olive, pistachio, apple and almond seedlings (Rugini and Mariotti, 1991) and in *Pinus* and *Larix* (McAfee *et al.*, 1993) and bacterial stimulation of adventitious rooting of *Pinus elliottii* Engelm. was found recently (Burns and Schwarz, 1996). Strong increase of rooting was also found when explants of *Pinus nigra* were infected with different strains of *A. rhizogenes*; in this case the transgenic nature of the roots was confirmed by an opine test (Milhalievic *et al.*, 1996). The examples given above are encouraging for extending the experiments of "localized transformation" to other cultivars and species. The results obtained with some difficult to root almond varieties, and with apple rootstock (an easy-to-root among woody species) are presented in the present work.

2. EXPERIMENTAL

2.1. Plant Material

Microcuttings (2.0 cm in length) of difficult-to-root almond (*Prunus dulcis* Mill.), cultivars Fascionello, Ferragnes, Tuono, two clones of cv Supernova, and microcuttings of easy-to-root apple rootstock MM106 (*Malus pumila*, Mill.), were used in the present research. Shoots were grown on MS medium (Murashige and Skoog, 1962) containing 3% sucrose, 0.65% agar (Riedel de Haen). The hormonal composition of the multiplication media were: for cv Supernova, IAA 0.85 μM and BAP 3.33 μM; for cv Tuono and Ferragnes, IAA 0.57 μM and BAP 1.85 μM; for MM106, IAA 0.57 μM and 0.89 μM BAP; for cv Fascionello IAA 0.57 μM and BAP 4.07 μM and in the latter the MS macro-nutrients were also modified (NH_4NO_3 8.24 mM, KNO_3 16.43 mM). The pH was adjusted to 5.6 prior autoclaving. Explants were maintained under a cool white light (Osram L40 white fluorescent), 45 μM $m^{-2}s^{-1}$ at a 16hrs photoperiod, 21 ± 1°C.

2.2. Bacterial Suspension and Inoculation Procedure

A. rhizogenes (wild type, strain 1855 NCPPB) was grown in YMB medium (Hooykaas *et al.*, 1977) at 28°C. The localized infection was done, as described previously (Damiano *et al.*, 1995), with the bacterial suspension, after centrifugation at 2500xg for 10′ at 4°C and adjustment to an optical density of 0.6 at 600 nm with liquid MS medium. MM106 micro cuttings were transferred to MS (Murashige and Skoog, 1962) medium supplemented with 3 μM IAA, or without auxin, for a co-cultivation period of 24hrs, whereas almond cultivars were transferred on BN (Bourgin and Nitsch, 1967) medium supplemented with 3% sucrose, 0.7% bacto-agar (DIFCO) without hormones. Apple microcuttings were always maintained in light, while almond explants were maintained for 12 days in dark after the infection. The culture room condition were the same reported for the multiplication phase.

Number of rooted explants were recorded 21 days and 30 days after the beginning of the rooting treatment, for MM106 and almond cultivars, respectively. Thirty microcuttings per treatments were used for each experiment; data shown are the means of three separate experiments ± S.E (percentage of explants forming shoots were previously transformed with the formula arcsin √ %). The χ^2 test was used to determine significant differences among rooting behaviour in different cultivars and clones of almond.

2.3. Molecular Analysis

2.3.1. MM106. Total DNA was isolated from 0.5 g of roots and from the same amount of tissue of the basal part of unrooted microcuttings, according to a modified Dellaporta protocol (Damiano *et al.*, 1995). Plant material was washed once in NaOH 200 mM + SDS 1% for few minutes, and three-four times in sterilized water, to avoid bacterial DNA contamination in plant total DNA.

The DNA was then restricted with *Eco*RI (Boehringer) and fractionated in 1% agarose (Sigma) gel electrophoresis in Tris-Borate-EDTA buffer (Sigma) (Sambrook *et al.*, 1989) containing 1 μg ml^{-1} of ethidium bromide (Sigma). Electrophoresis was performed at room temperature for 16hrs at 1.5 volts cm^{-1}.

The DNA was then transferred to a nylon membrane (Zeta-Probe Biorad) and hybridized with fragments of Ri plasmid random priming labelled with d-CTP^{32}P (Amersham), using U.S.B. kit, and hybridizations were performed in 30% formamide (Sigma) for 16hrs at 42°C. At the end of hybridization periods, filters were washed 30 min twice in 2x SSC + 0.1% SDS at room temperature and 30′ twice in 0.1x SSC + 0.1% SDS (Sambrook *et al.*, 1989; Sigma products) at 65°C. Filters were then exposed on X-ray films (Hyperfilm Amersham) at -80°C for 3 days.

*Eco*RI-*Eco*RI fragment (4374 bp), containing *rol* genes (Cardarelli *et al.*, 1987) and coming from pBin19::*Eco*15, was used to detect the presence of T_L-DNA in the *Eco*RI digest of total DNA extracted from explants. To avoid bacterial DNA contamination in the DNA extracted from transformed microcuttings, *Eco*RI-*Eco*RI fragment of pMP162 plasmid (Pomponi *et al.*, 1983), containing *vir* genes, was used as control.

2.3.2. Supernova. Total DNA was extracted from roots (about 1.5 g f.wt., corresponding to the whole content of a vessel) of treated microcuttings, after washing as described above. The extraction and purification methods used were a modification of the procedure previously described (Dellaporta *et al.*, 1983; Ziegenhagen *et al.*, 1993): diethyldithiocarbamic acid (50 mM) (Na salt; Sigma) was added to the extraction buffer to inhibit phenoloxidase activity and polyvinylpolypyrrolidone (20% w/w) (Sigma) was added to the roots during grinding to reduce polyphenol oxidation (Howland *et al.*, 1991; Damiano *et al.*, 1995). The DNA restriction was performed as for apple. Fragments were labelled with ^{32}P, using Quick Prime random priming kit (Pharmacia). The hybridizations were performed as in apple. Filter was then exposed on X-ray films (Kodak Biomax MS) at −80°C for 36hrs.

3. RESULTS AND DISCUSSION

Apple microcuttings maintained in the hormone free medium in absence of *A. rhizogenes* treatment, showed 22.5% root formation, but only 5.7% rooting was obtained when the basal part of microcuttings was infected. Microcuttings treated only with IAA showed

Table 1. Effect of localized infection with *Agrobacterium rhizogenes*, IAA (3μM) and/or cefotaxime on rooting of MM106 apple rootstock microcutting (data collected 21 days after transferring to the rooting medium)

			Rooting (%)
+	*A. rhizogenes*	HFM + CX	5.7 ± 0.5
–	*A. rhizogenes*	HFM + CX	22.5 ± 0.7
–	*A. rhizogenes*	HFM – CX	22.7 ± 0.8
+	*A. rhizogenes*	IAA + CX	24.9 ± 1.7
–	*A. rhizogenes*	IAA + CX	95.1 ± 1.4
–	*A. rhizogenes*	IAA – CX	96.5 ± 2.0

HFM = Hormone Free Medium, CX = Cefotaxime. Results are the mean of three separate experiments (n = 30) ± S.E (Percentage scores were previous transformed with the formula arcsin √ %).

high rooting (96.5%), while combining the *Agrobacterium* infection and the IAA treatment decreased the rooting percentage (24.9%). The addition of cefotaxime to the rooting medium affected the rooting response neither in combination with IAA nor in the hormone free medium (Table 1).

The hybridization between total DNA extracted from roots or basal parts of microcuttings and the fragment containing *rol* genes, coming from pBin19::*Eco*15, provided molecular evidence of the transgenic nature of the roots and of the callus that were induced by IAA + *A. rhizogenes* (Fig. 1). The absence of hybridization with the fragment containing *vir* genes, coming from pMP162, showed that there was no bacterial contamination in the explants used for DNA extraction (data not shown).

Localized infection with *A. rhizogenes* of microcuttings of the apple rootstock MM106, an easy-to-root genotype, strongly inhibited rooting, probably producing extensive modification of the endogenous hormone balance and/or hormone sensitivity (Spanò *et al.*, 1988; Estruch *et al.*, 1991; Nilsson *et al.*, 1993).

Unlike apple rootstock, all four cultivars of almond did not root on control hormone-free media. This confirms the data obtained previously with Supernova (Damiano *et al.*, 1995). A different rooting ability among cultivars was also evident when localized infection was applied (Table 2). The transgenic nature of the roots was confirmed by Southern blot (Fig. 2). The contamination of bacterial DNA was excluded, as in apple, by the absence of hybridization with pMP162 (data not shown). Since the DNA was extracted from

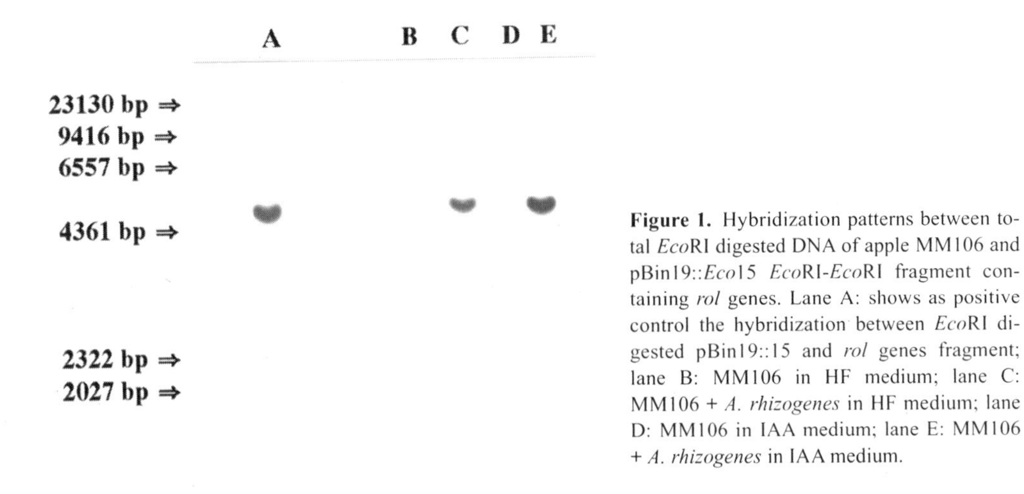

Figure 1. Hybridization patterns between total *Eco*RI digested DNA of apple MM106 and pBin19::*Eco*15 *Eco*RI-*Eco*RI fragment containing *rol* genes. Lane A: shows as positive control the hybridization between *Eco*RI digested pBin19::15 and *rol* genes fragment; lane B: MM106 in HF medium; lane C: MM106 + *A. rhizogenes* in HF medium; lane D: MM106 in IAA medium; lane E: MM106 + *A. rhizogenes* in IAA medium.

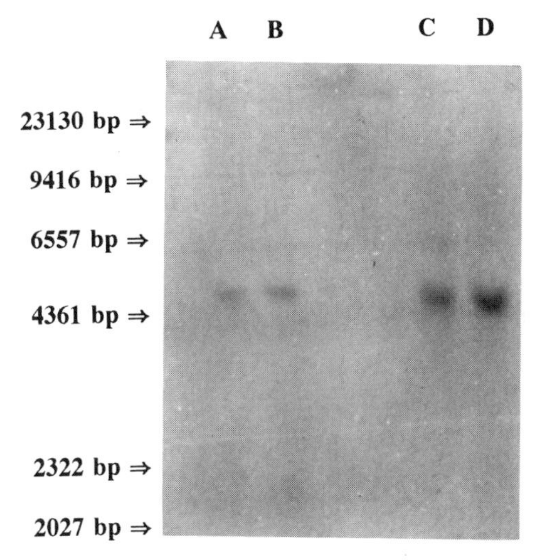

Figure 2. Hybridization pattern between total DNA extracted from roots of Supernova and *Eco*RI fragment from pBin19::*Eco*15 plasmid (4374 bp). Lanes A,B,C and D: hybridization between rol genes and roots DNA of groups of infected cuttings cultured in BN medium.

all the roots of rooted explants, it was not possible to determine whether all roots were transgenic. The root systems were clearly non-geotropic in all the cultivars (Fig. 3), confirming the behaviour induced by *Agrobacterium* infection (Tepfer, 1984; Spanò *et al.*, 1987). Roots were very thick and originated from cambium; they emerged within 10–13 days (Fig. 4). Rooted explants developed a large root mass (Fig. 3, left) and were successfully transferred to greenhouse where they will be further evaluated (Fig. 5).

4. CONCLUSIONS

In conclusion, the basal part infection of microcuttings was confirmed to be very useful in inducing rooting in recalcitrant species such as almond, and can be possibly used in various difficult-to-root varieties. Otherwise, localized infection may inhibit rooting in easy-to-root species. The DNA transfer was confirmed both by Southern analysis and type of root system development, but it is still to be clarified whether all the roots are transgenic, or there is a root induction signal to other cells after the differentiation of the first transgenic ones.

Table 2. Rooting percentage (± S.E.) and roots per microcuttings (± S.E.) in almond explants infected with *A. rhizogenes* and not infected, 30 days after infection, in each cultivar tested[a]

	Infected				Not infected			
Cultivar	rooting %	S.E.	roots/explants (mean)	S.E.	rooting %	S.E.	roots/explants (mean)	S.E.
Fascionello	34.73b	±3.95	2.70	±0.54	0	0	0	0
Ferragnes	38.85c	±9.16	3.68	±1.16	0	0	0	0
Supernova								
Clone 1	58.26a	±3.07	3.04	±0.49	0	0	0	0
Clone 2	54.13ac	±5.28	2.50	±0.34	0	0	0	0
Tuono	60.00ac	±18.39	2.30	±0.68	0	0	0	0

[a]Rooting percentage values followed by the same letter(s) are not significantly different (χ^2, P=0.05).

Figure 3. Rooting in treated microcuttings of almond. A-B: non geotropic behaviour in cultivar Tuono; C: large mass of root primordia in cultivar Fascionello.

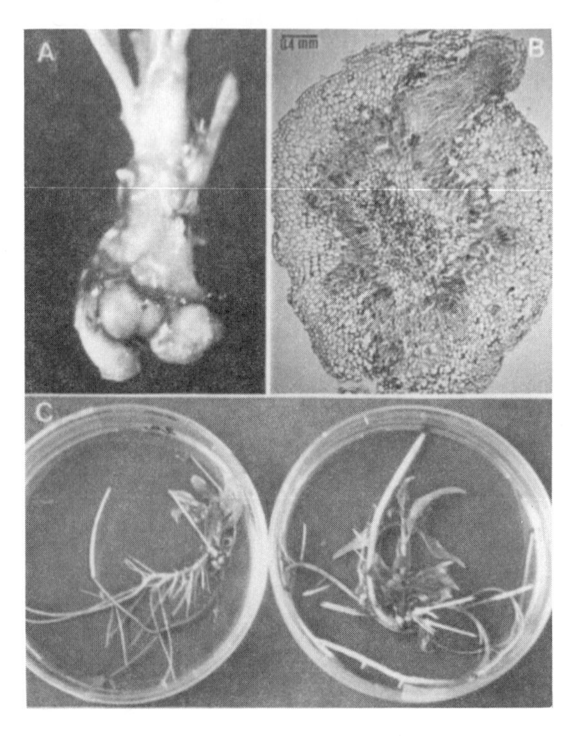

Figure 4. Rooting in treated Supernova microcuttings. A: 10 days after infection; B: cross section of basal part of microcutting with an emerging root, 10 days after infection; C: further growth.

Figure 5. Supernova: different phases of acclimatization. A: 1 month after transfer to the greenhouse: single rooted plant (left), control (right), B: acclimatization of explants under fog respectively; C-D: further development (3 and 5 months after acclimatization, respectively).

REFERENCES

Bourgin, J.P. and Nitsch, J.P., 1967. Production of haploid *Nicotiana* from excised stamens. *Ann. Physiol. Veg.*, 9:377–382.

Burns, J.A. and Schwarz, O.J., 1996. Bacterial stimulation of adventitious rooting on *in vitro* cultured slash pine (*Pinus elliottii* Engelm.) seedling explants. *Plant Cell Report*, 15:405–408.

Caboni, E. and Damiano, C., 1994. Rooting in two almond genotypes. *Plant Science*, 96:163–165.

Caboni, E., Lauri, P., Tonelli, M., Falasca, G., Damiano, C., 1996. Rooting induction by *Agrobacterium rhizogenes* in walnut. *Plant Science*, in press.

Cardarelli, M., Mariotti, D., Pomponi, M., Spanò, L., Capone, I., Costantino, P., 1987. *Agrobacterium rhizogenes* T-DNA gene capable of inducing hairy root phenotype. *Mol. Gen. Genet.*, 209:475–480.

Damiano, C., Archilletti, T., Caboni, E., Lauri, P., Falasca, G., Mariott, D., Ferraiolo, G., 1995. *Agrobacterium* mediated transformation of almond: in vitro rooting through localized infection of *A. rhizogenes* w.t.. *Acta Horticulturae*, 392:161–169.

Dellaporta, S.L., Wood, J., Hicks, J.B., 1983. A plant DNA minipreparation version II. *Plant Mol. Biol. Rep.*, 1:19–21.

Estruch, J.J., Scheli, J., Spana, A., 1991. The protein encoded by the *rol* B, plant oncogene hydrolyses indole-glucosides. *EMBO J.*, 10:3125–3128.

Hartmann, H.T., Kester, D.E. and Davies, F.D., 1990. Propagation methods and rootstock for the important fruit and nut species, in: Hartmann, H.D., Kester, D.E. and Davies, eds, Plant propagators, Principles and practices, 5th ed..Prentice Hall Inc., Englewood, N.J., pp.527–565.

Hooykaas, P.J.J., Klapwjik, P.M., Nuti, M.P., Schilperoort, R.A. and Rorsch, A., 1977. Transfer of the *A.tumefaciens* Ti plasmid to avirulent *Agrobacteria* and *Rhizobium ex planta*. *J. Gen. Microbiol.*, 98:477–484.

Howland, D.E., Oliver, R.P. and Davy, A.J., 1991. A method of extraction of DNA from Birch. *Plant Mol. Biol. Rep.*, vol.9 (4), 340–344.

McAfee, B.J., White, E.E., Pelcher, L.E., and Lapp, M.S., 1993. Root induction in pine (*Pinus*) and larch (*Larix*) spp. using *Agrobacterium rhizogenes*. *Plant, Cell, Tissue and Organ Culture*, 34:53–62

Mihalievic, S., Stipkovic, S., Jelaska, S., 1996. Increase of root induction in *Pinus nigra* explants using agrobacteria. *Plant Cell Reports*, 15:610–614.

Murashige, T. and Skoog, F., 1962. A revised medium for rapid growth and bioassays with tobacco tissue cultures. *Physiol. Plant.*, 15:473–497.

Nilsson, O., Crozien, A., Schmulling, T., Studberg, G., Olsson, O., 1993. Indole-3- acetic acid homeostasis in transgenic tobacco plants expressing the *Agrobacterium rhizogenes rol* B gene. *The Plant Journal*, 3:681–689.

Pomponi, M., Spanò, L., Sabbatini, M.G. and Costantino, P., 1983. Restriction endonuclease mapping of the root-inducing plasmid of *Agrobacterium rhizogenes* 1855. *Plasmid*, 10:119–129.

Rugini, E. and Mariotti, D., 1991. *Agrobacterium rhizogenes* T-DNA genes and rooting in woody species. *Acta Horticulturae*, 300:301–308

Rugini, E. and Verma, D.C., 1983. Micropropagation of difficult-to-propagate almond (*Prunus amygdalus* Batsch) cultivar. *Plant Sci. Lett.*, 28:273–281.

Sambrook, J., Fritsch, E.F., Maniatis, T., 1989. Molecular Cloning: A Laboratory Manual. Cold Spring Harbour Laboratory Press, NY.

Spanò, L., Mariotti, D., Cardarelli, M., Branca, C., Costantino, P., 1988. Morphogenesis and auxin sensitivity of transgenic tobacco with different complements of R_1 T-DNA. *Plant Physiol.*, 87:479–483.

Spanò, L., Mariotti, D., Pezzotti, M., Damiani, F., Arcioni, S., 1987. Hairy root transformation in alfalfa (*Medicago sativa* L.). *Theor. Appl. Genet.*, 73:523–530. Tabachnik, L. and Kester, D.E., 1977. Shoot culture for almond and almond-peach hybrid clones *in vitro*. *Hortscience*, 12:545–547.

Tepfer, D., 1984. Transformation of several spp. of higher plants by *Agrobacterium rhizogenes*: sexual transmission of the transformed genotype and phenotype. *Cell*, 37:959–967.

Ziegenhagen, B., Guillemaut, P. and Scholz, F., 1993. A procedure for mini- preparation of genomic DNA from Needles of Silver Fir *(Abies alba)*. *Biol. Rep.*, vol. 11(2):117–121.

FACTORS CONTROLLING ADVENTITIOUS ROOT FORMATION ON STEM EXPLANTS OF ROSE (*Rosa hybrida* "Motrea") *IN VITRO*

R. L. M. Pierik

Department of Horticulture
Agricultural University
6708 PM Wageningen, The Netherlands

INTRODUCTION

The formation of adventitious roots is an essential process in propagation systems for rose. For that reason, we decided to study adventitious root formation on excised stem explants in a well defined culture medium in vitro, under controlled physical growth conditions. This system helped us to gain a better insight into the plant, nutritional and physical factors determining the formation of adventitious roots.

MATERIALS AND METHODS

In almost all experiments, plants of *Rosa hybrida* "Motrea" were grown in a heated greenhouse so that plant material was available year-round. Homogeneous material of the same diameter and developmental stage was selected among young axillary shoots from buds positioned directly below the flower buds.

After removal of the leaves and shoot tip, the upper two elongating internodes were surface-sterilized as follows: a few seconds in alcohol 70% (v/v), 20 minutes in 1% NaOCl (with a few drops of Tween 20), and rinsed three times (for 3, 5, and 15 minutes, respectively) in sterile tap water.

Sterilized cylindrical stem segments, 3 mm in length without the bud, were cut aseptically and transplanted on the culture media with their basal ends up. Explants were always at random divided over all treatments. Each treatment consisted of 4 Petri dishes (each dish was an experimental unit) with 6 stem explants each. Explants were grown at regular distances from each other. Petri dishes (diameter 6 cm) contained 15 ml of medium, and were sealed with Parafilm.

Biology of Root Formation and Development, edited by Altman and Waisel.
Plenum Press, New York, 1997.

The basic culture medium consisted of MS-macroelements (half strength), MS-microelements (full strength, exept Fe), NaFeEDTA 37.5 mg/l, glucose 45 g/l, IBA 2.0 mg/l, Vitamin B_1 0.4 mg/l, méso-inositol 100 mg/l, agar 7 g/l (MC 29 from Lab M, Amersham, England), pyrex-distilled water, at a pH 5.8 before autoclaving.

Unless otherwise stated, stem segments were incubated in a growth chamber at 23°C in complete darkness. Occasionally, 16 hours photoperiod was employed provided (fluorescent tubes, Pope, FDT/58W, 84HF, 8–10 Wm^{-2}). All experiments had one variable factor.

After 4 weeks rooting was evaluated by determining the % R and the MNR in all explants, except infected ones. Only those roots that were visible as root initials or as elongated roots were counted.

A number of rooting experiments were done, to eliminate limiting factors for the regeneration of adventitious roots. The standard composition of the medium and the physical growth conditions were established after numerous preliminary trials.

RESULTS

Swelling of the tissue occurred immediately after isolation and callus was rapidly formed. This was followed by root initiation and root formation, mainly at the basal side of the explants above the medium. This indicated that polarity of root regeneration was maintained, even when the explants were placed upside down on the medium. Occasionally, some roots were formed at the upper part of the explant (in the medium). The first macroscopically visible adventitious roots were detected after 6–7 days. Four weeks after the start of the experiments, the increase in the percentage of rooting and the mean number of roots almost ceased.

Comparison of 7 MS-macroelement concentrations (Fig. 1) showed that rooting increased by raising the concentration of the macroelements from 0 to 1.0 strength, with a clear optimal response at 1.0 strength, whereas rooting decreased from 1.0 to 1.5 strength.

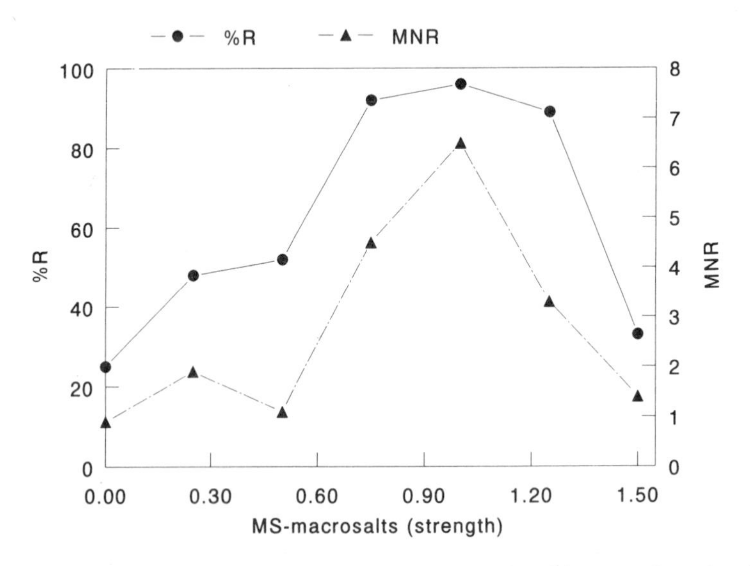

Figure 1. The effect of the strength of the MS-macroelements on adventitious root formation. MNR, mean number of roots per explant; %R percentage of adventitious root formation.

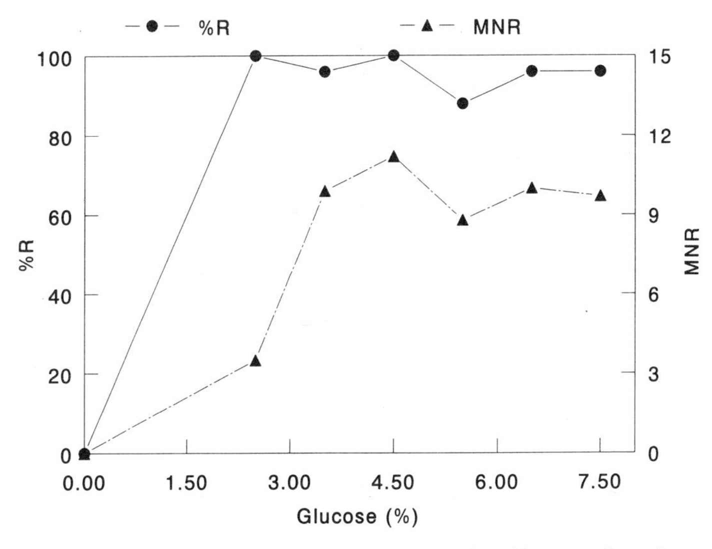

Figure 2. The effect of glucose concentration on adventitious root formation.

No rooting occurred without sugar, and all explants died. Rooting increased by increasing the glucose concentration from 0 to 4.5% (Fig. 2) and by increasing the sucrose concentration from 3.5 to 8.5% (Fig. 3), reaching optima respectively at 4.5 and 8.5%. Comparison of Figures 2 and 3 shows that glucose was much more efficient than sucrose to induce rooting.

The effect of the agar brand (Fig. 4) was remarkable. Agar 1 was very bad for rooting, whereas agar 4 was clearly the best, and rooting on agars 2 and 3 was intermediate. This experiment shows that the choice of the agar brand is extremely important.

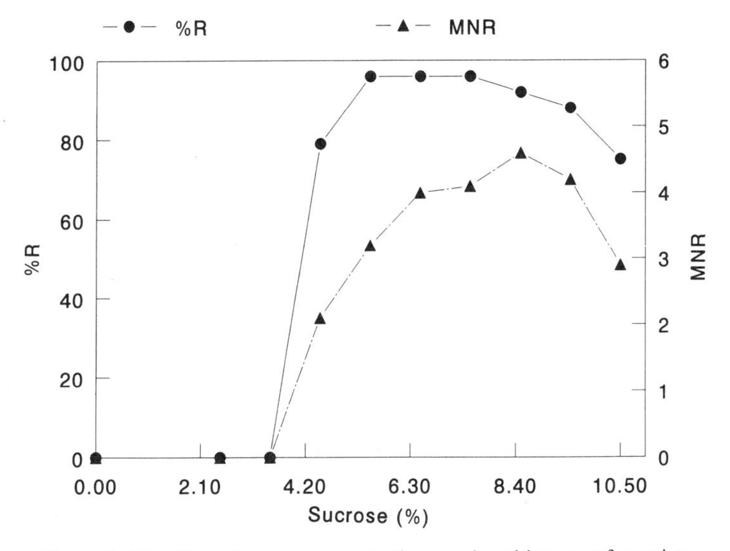

Figure 3. The effect of sucrose concentration on adventitious root formation.

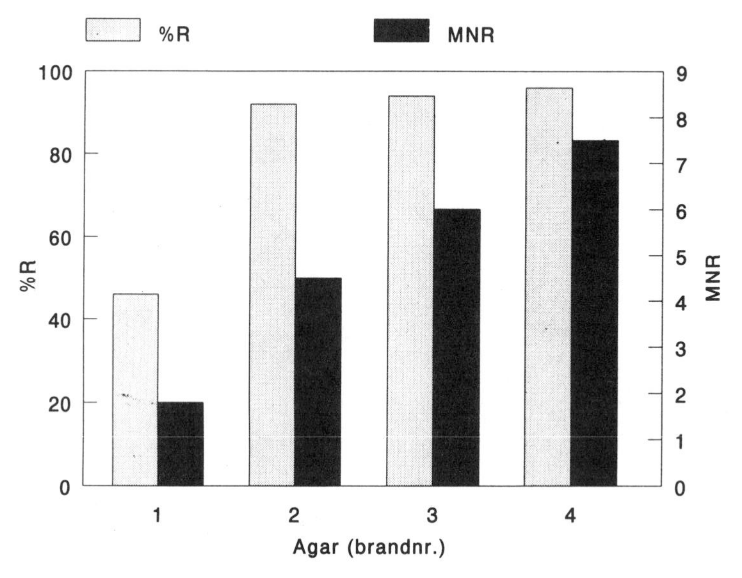

Figure 4. The effect of agar brand on adventitious root formation. Agar 1, Becton Dickinson purified; agar 2, Becton Dickinson Grade A; agar 3, Difco Bacto agar; agar 4, MC 29 from Lab M, Amersham, England.

No rooting occurred on an auxin-free medium, indicating that there is an absolute auxin requirement. Fig. 5 shows that rooting was very stable and optimal at the IBA range from 0.25–1.00 mg/l, whereas rooting decreased in the range from 1.0–5.0 mg/l IBA. Since auxin was essential for rooting, the length of the period during which auxin was applied, was varied (Fig. 6). Explants were initially grown on medium containing 2 mg/l IBA, and transferred to an auxin-free medium after various periods. At the range of 0–24 hours initial auxin supply, there was an increase in rooting, but no further change in rooting was observed at longer auxin treatments.

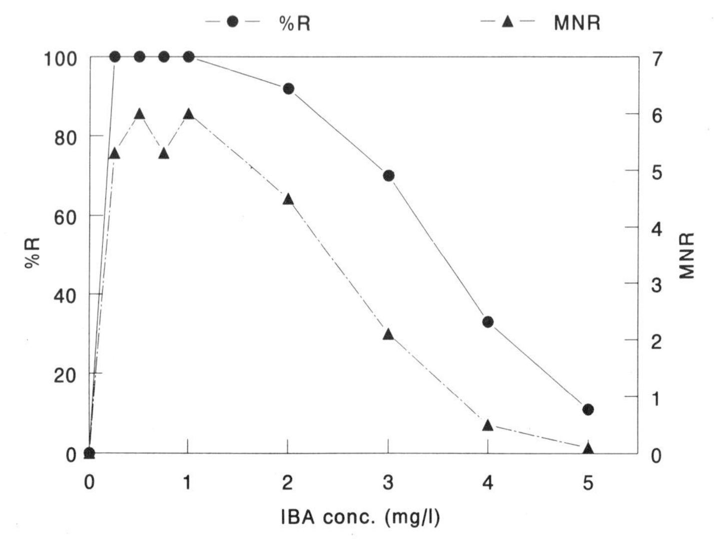

Figure 5. The effect of IBA concentration on adventitious root formation.

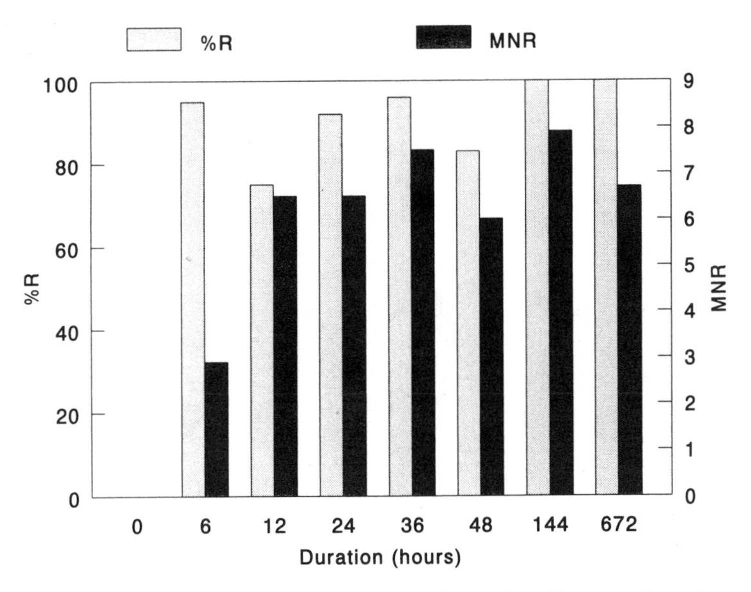

Figure 6. The effect of duration of the IBA supply on adventitious root formation.

The influence of light/darkness on rooting was examined by giving an initial dark period (0, 7, 14, 21 and 28 days), followed by a light period (28, 21, 14, 7 and 0 days, respectively). Fig. 7 shows that continuous darkness resulted in better rooting than continuous light. However, when the initial dark period was 7 days, rooting was decreased compared with the other treatments. Further increase of the length of the dark period, from 7 to 28 days increased rooting. The influence of light/darkness was also examined by applying an initial period of light (0, 7, 14, 21 and 28 days), followed by a dark period (28, 21, 14, 7 and 0 days, respectively). In this experiment (Fig. 8), an initial light period of 7

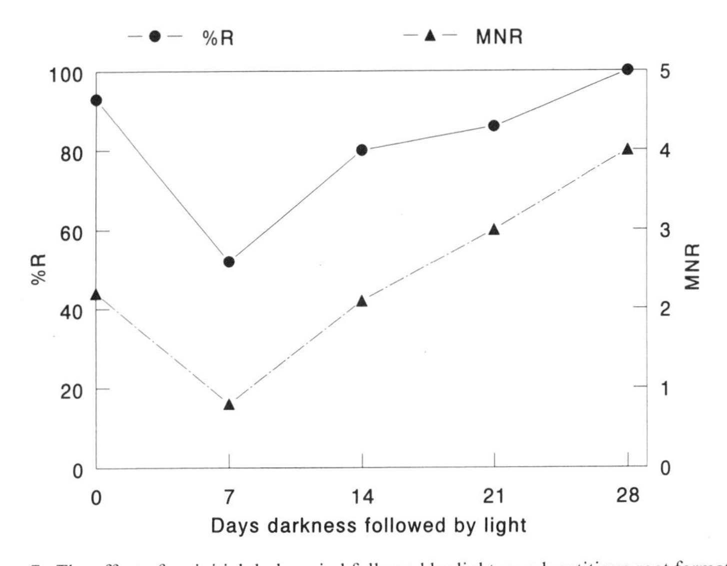

Figure 7. The effect of an initial dark period followed by light on adventitious root formation.

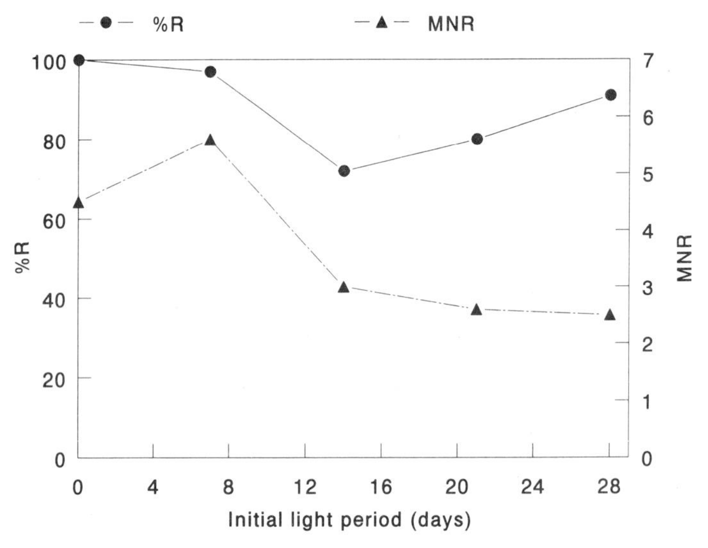

Figure 8. The effect of an initial light followed by a dark period on adventitious root formation.

days, followed by darkness, appeared to be optimal for rooting in comparison with the other treatments. This suggests a promoting effect of light in the first week.

The role of temperature (Fig. 9) on rooting was examined in the range 21 to 25°C. Rooting increased from 21 to 23°C, but slightly decreased from 23 to 25°C.

Fig. 10 shows a clear genotypic effect on adventitious rooting. Our standard cultivar "Motrea" rooted much better than "Madelon", and "Sonia". However, it is possible that the optimal rooting conditions for "Motrea" (as used in this research) were not optimal for the cultivars "Madelon" and "Sonia".

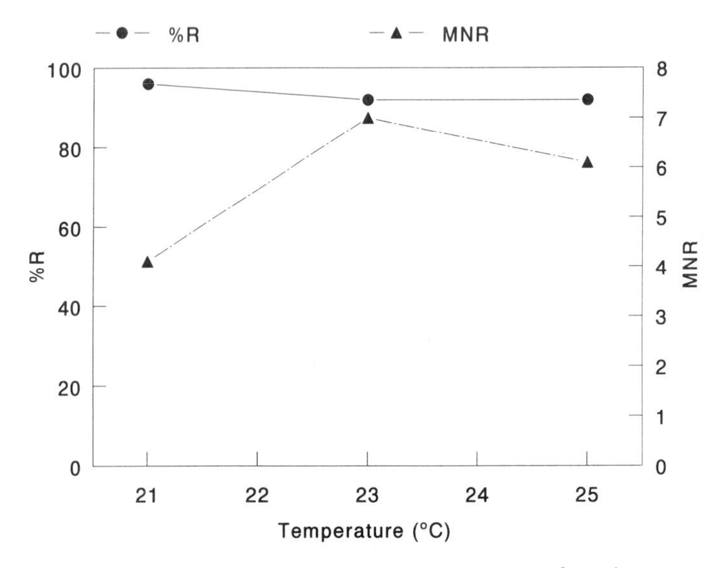

Figure 9. The effect of temperature on adventitious root formation.

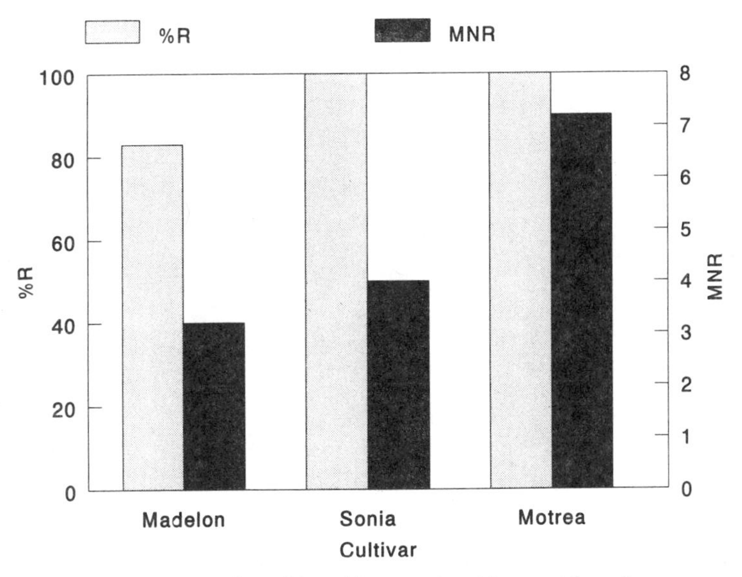

Figure 10. The effect of the cultivar on adventitious root formation.

Comparison of the Figures 11 and 12 clearly shows that placing the segments upside-down (Fig. 11) was much better for rooting than positioning the explants with basal ends down (Fig. 12). Rooting of explants with basal ends down increased by increasing explant length from 3 to 7 mm (Fig. 12), but this did not occur in explants with basal ends up (Fig. 11).

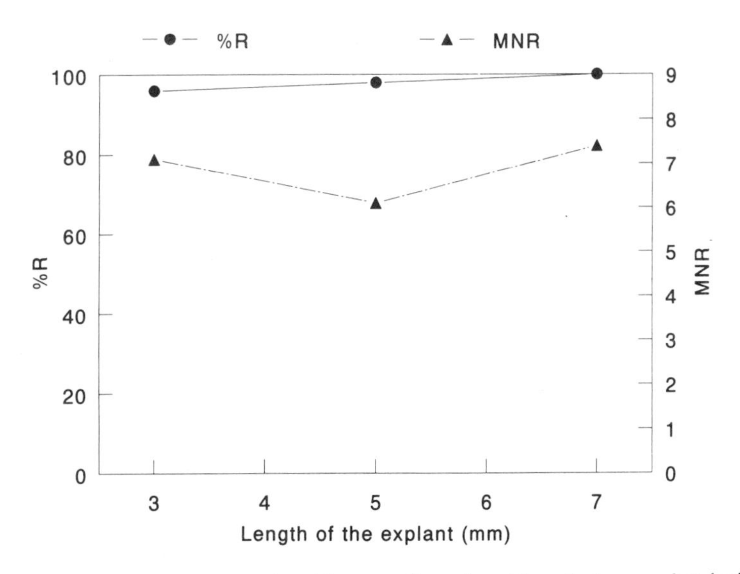

Figure 11. The effect of explant length on adventitious root formation. All explants were placed with their basal ends up.

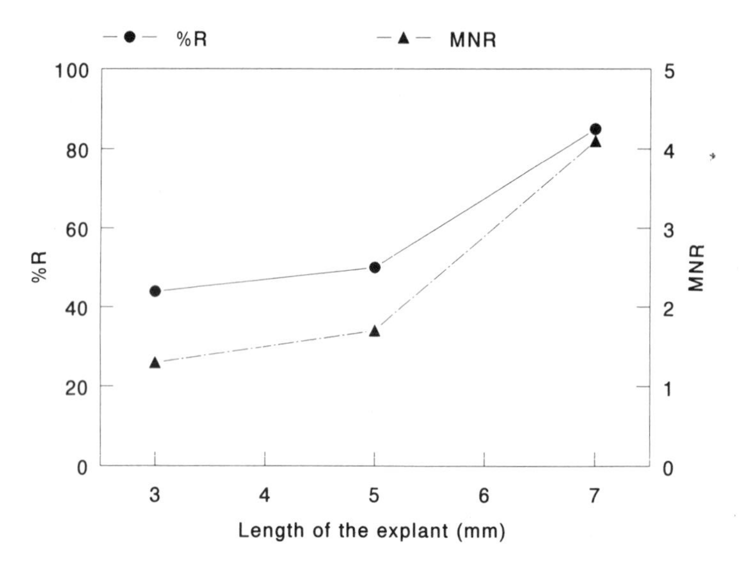

Figure 12. The effect of explant length on adventitious root formation. All explants were placed with their basal ends down.

DISCUSSION

Our experiments with rose clearly support the view that a complex of plant and environmental factors determines rooting in vitro. The results of our experiments with rose in vitro are largely in accordance with the older literature (Olieman van der Meer et al., 1971; Pierik and Segers, 1973; Pierik and Steegmans, 1975).

From literature it is quite clear that carbohydrates are a very important factor for rooting, generally requiring a relatively high sugar concentration, which is in agreement with our work. The remarkable difference in rooting in rose between glucose and sucrose was never described in literature, and we have no explanation for that phenomenon. Connor and Falloon (1993) concluded that the osmotic effect of sugar provides an important trigger for root initiation in *Asparagus*, whereas the nutritional effect of a high sucrose concentration stimulates the development of the storage roots.

In most reports auxin is promotive only during the first days after isolation. A 24-h pulse with auxin had a maximum effect on rooting of rose (De Klerk and ter Brugge, 1992), whereas Van der Krieken et al. (1993) showed that incubation of *Malus* stem explants for 3 days on a medium with IBA resulted in a better root formation as compared with 2 or 5 days.

The role of light/darkness during adventitious root formation is very puzzling, because not only inhibition of rooting by light was reported (Caboni et al., 1992; Drew et al., 1993; Druart et al., 1982; Hammerschlag, 1982; Karhu and Zimmerman, 1993; Olieman-van der Meer et al., 1971; Pierik and Steegmans, 1975; Van der Krieken et al., 1992; Rugini et al, 1993; Van der Krieken et al., 1992a; Zimmerman, 1984), but also promotion (Bertazza et al., 1995; Caboni et al., 1992; Gautheret, 1969; Jarvis and Shaheed, 1987; Vieitez et al., 1989; Walker et al., 1987). A reasonable explanation for the diverse effect of light/darkness is difficult. However, many reports concluded that the site of action of darkness/light was the basal part of the cutting, where root initials arise (Karhu and Zimmerman, 1993; Rugini et al., 1993; Vieitez et. al., 1989).

Dark treatment is particularly important at the beginning of rhizogenesis (De Klerk en ter Brugge, 1992; Drew et al., 1993; Druart et al., 1982; Hammerschlag, 1982; Pawlicki and Welander, 1992; Rugini et al., 1993; Van der Krieken et al., 1992). In stem explants of rose, light has a promotive effect on rooting in the first week after isolation (Fig. 7 and 8). It may be concluded that dark/light sensitivity for rooting occurs very early in the rooting process.

Since rooting is very often inhibited by light, it seems appropriate to summarize the explanations for this phenomenon. Dumas and Monteuuis (1995) and Hammerschlag (1982) suggest that reduction of the light at the base of shoots provides an environment conductive to the accumulation of photosensitive auxin and/or cofactors in the basal tissues. Druart et al. (1982) concluded that during the root induction phase of apple, darkness favoured the increase of the peroxidase activity, whereas during the root initiation phase the peroxidase activity decreased; this confirms the validity of peroxidase as a marker for rooting. Karhu and Zimmerman (1993) suggested that differences in phenolic metabolism exist between darkened and non-darkened apple microcuttings. Druart et al. (1982) concluded that phenol content can indeed be considered as a biochemical marker for rooting, since its variation was exactly the reverse of that of peroxidase activity. Hammerschlag (1982) concluded that the inhibitory effect of light may not be solely due to auxin breakdown, because rooting in the light was not promoted by increasing auxin levels in the medium. Van der Krieken et al. (1992a) found that root formation in apple in vitro occurred after incubation in the dark on a medium with IBA and riboflavin; they concluded that in the light (in a medium with riboflavin) 75% of the IBA was destroyed, whereas only 25% of the IBA was destroyed in the absence of riboflavin. Independent of the presence of riboflavin, no breakdown of IBA occurred in the dark, indicating that light and riboflavin stimulate the inactivation of IBA by oxidation.

The high rooting efficiency under red and white light, and the inhibition of rooting under far-red and darkness, suggest the involvement of the phytochrome system in rhizogenesis (Bertazza et al., 1995; Letouzé and Beauchesne, 1969; Rossi et al., 1993). Phytochrome induction of rhizogenesis may occur in part through modulation of auxin levels, as reported for phytochrome regulation of stem growth (Behringer et al., 1992).

In a few reports (Pierik and Steegmans, 1975; Van der Krieken et al., 1993; Pawlicki and Welander, 1992) it was shown that incubation of stem explants with basal ends up, resulted in an increase in the number of roots, whereas stem discs with basal ends down in the medium decreased rooting. Three hypothetical explanations can be given for the promotive effect of inverting explants on rooting (Pierik and Steegmans, 1975): (1) oxygen supply at the basal part is much better when the explants are inverted; (2) basipetally transported substances accumulate in inverted explants as no diffusion from the basal part into the agar medium can occur; (3) the normal auxin gradient from tip to base is preserved. The first explanation seems to be more likely, since regeneration of organs has a definite oxygen requirement and it is unlikely that rooting occurs under anaerobic conditions (basal ends down). The question whether substances beneficial for rooting diffuse into the agar (in explants with basal ends down), remains unanswered. Whether preservation of the auxin gradient in inverted explants takes place, needs further research.

ABSTRACT AND CONCLUSIONS

Factors affecting adventitious root formation in vitro were studied on stem explants of rose. Rooting at the basal ends of the stem segments was genotype dependent and occurred optimally in stem explants with their basal ends up. Continuous darkness was important for rooting, although light during the first week promoted rooting in comparison to continuous darkness. Good root formation occurred only when both sugar and macrosalts were present in the medium. Adventitious root formation was much better on a culture medium with glucose than with sucrose. The agar brand played a very important role during rooting. Auxin was an absolute requirement for rooting and was only necessary during the first 24 hours of in vitro culture.

ACKNOWLEDGMENTS

I gratefully ackowledge the help of the M.Sc. students H.M. Donkers, T.P. van der Heide, A.C. Vreden, C.W. Belgraver and O. Bicakci, which carried out many of the experiments with rose in vitro. Their results strongly contributed to the composition of this article. I appreciate the critical remarks of Mr. M.J. Schornagel. Thanks are also due to Mr. P. Sprenkels for the technical assistance in the research project with rose.

REFERENCES

Behringer, F.J., Davies, P.J. and Reid, J.B., 1992, Phytochrome regulation of stem growth and indole-3-acetic acid levels in the lv and Lv genotypes of *Pisum*. *Photochem. Photobiol.* 56:677.

Bertazza, G., Baraldi, R. and Predieri, S., 1995, Light effects on in vitro rooting of pear cultivars of different rhizogenic ability, *Plant Cell Tissue Organ Culture* 41:139.

Caboni, E., Boumis, G and Damiano, C., 1992, Effect of phenols, gibberellic acid and carbohydrates on the root ing of apple rootstock Jork, *Agronomie* 12:789.

Conner, A.J. and Falloon, P.G., 1993, Osmotic versus nutritional effects when rooting in vitro asparagus minicrowns on high sucrose media, *Plant Sci.* 89:101.

De Klerk, G.J. and Ter Brugge, J., 1992, Factors affecting adventitious root formation in microcuttings of *Malus*, *Agronomie* 12:747.

Drew, R. A., McComb, J.A. and Considine, J.A., 1993, Rhizogenesis and root growth of *Carica papaya* L. in vitro in relation to auxin-sensitive phases and use of riboflavin, *Plant Cell Tissue Culture* 33:1.

Druart, Ph., Kevers, C., Boxus, P., T. Gaspar, T., 1982, In vitro promotion of root formation by apple shoots through darkness. Effect on endogenous phenols and peroxydases, *Z. Pflanzenphysiol.* 108:429.

Dumas, E. and Monteuuis, O., 1995, In vitro rooting of micro-propagated shoots from juvenile and mature *Pinus pinaster* plants: influence of activated charcoal, *Plant Cell Tissue Organ Culture* 40:231.

Gautheret, R.J., 1969, Investigations on the root formation in the tissues of *Helianthus tuberosus* cultured in vitro, Am. J. Bot. 56:702.

Hammerschlag, F., 1982, Factors influencing in vitro multiplication and rooting of the plum rootstock Myrobalan (*Prunus cerasifera* Ehrh.), *J. Am. Soc. Hort. Sci.* 107: 44.

Jarvis, B.C. and Shaheed, A., 1987, Adventitious root formation in relation to irradiance and auxin supply, *Biol. Plant.* 18:100.

Karhu, S.T. and Zimmerman, R.H., 1993, Effect of light and coumarin during root initiation on rooting apple cultivars in vitro, *Adv. Hort. Sci.* 7:33.

Letouzé, R. and Beauchesne, G., 1969, Action d'éclairement monochromatiques sur la rhizogenèse de tissus de topi- nambour, *C. R. Acad. Sci. Paris* 269:1528.

Murashige, T. and Skoog, F., 1962, A revised medium for rapid growth and bioassays with tobacco tissue cultures, *Physiol. Plant.* 15:473.

Olieman-Van der Meer A.W., Pierik, R.L.M. and Roest S., 1971, Effects of sugar, auxin and light on adventitious root formation in isolated stem explants of *Phaseolus* and *Rhododendron*, *Meded. Fak. Landbouw Wet. Gent* 36:511.

Pawlicki, N. and Welander, M., 1992, The effects of benzyladenine and gibberellic acid on adventitious root forma- tion in apple stem discs, *Agronomie* 12:783.

Pierik R.L.M. and Segers, Th., 1973, In vitro culture of midrib explants of *Gerbera*: adventitious root forma- tion and callus induction, *Z. Pflanzenphysiol.* 69:204.

Pierik, R.L.M. and Steegmans, H.H.M., 1975, Analysis of adventitious root formation in isolated stem explants of *Rhododendron*, *Scient. Hortic.* 3:1.

Rossi, F., Baraldi, R., Facini, O. and Lercari, B., 1993, Photomorphogenic effects on in vitro culture of *Prunus* rootstock GF 655–2, *Plant Cell Tissue Organ Culture* 32:145.

Rugini, E., Jacoboni, A. and Luppino, M., 1993, Role of basal shoot darkening and exogenous putrescine treatments on in vitro rooting and on endogenous polyamine changes in difficult-to-root woody species, *Scient. Hortic.* 53:63.

Van der Krieken J.M, Breteler, H., Visser, M.H.M. and Jordi, W., 1992, Effect of light and riboflavin on indolebutyric acid-induced root formation on apple in vitro, *Physiol. Plant.* 85:589.

Van der Krieken, W.M., Breteler, H. and Visser, M.H.M., 1992a, The effect of the conversion of indolebutyric acid into indoleacetic acid on root formation of *Malus*, *Plant Cell Physiol.* 33:709.

Van der Krieken, W.M., Breteler, H., M.H.M. and Mavridou, D., 1993, The role of the conversion of IBA into IAA on root regeneration in apple: introduction of a test system, *Plant Cell Rep.* 12:203.

Vieitez, A.M., San-Jose, M.C. and Ballester, A., 1989, Progress towards clonal propagation of *Camellia japonica* cv. *Alba Plena* by tissue culture techniques, *J. Hort. Sci.* 64:605.

Walker, N., Jaques, R. and Miginiac, E., 1987, Action of light on rooting in vitro and acclimatization of *Sequoia sempervirens* to soil, *Acta Hort.* 212:289.

Zimmerman, R.H., 1984, Rooting apple cultivars in vitro: Interactions among light, temperature, phloroglucinol and auxin, *Plant Cell Tissue Organ Culture* 3:301.

38

ROOTING INDUCTION IN ENCAPSULATED BUDS OF M.26 APPLE ROOTSTOCK FOR SYNTHETIC SEED

Alvaro Standardi and Emanuele Piccioni

Istituto di Coltivazioni arboree
University of Perugia
Borgo XX Giugno, 74
06121 Perugia, Italy

INTRODUCTION

In 1993, Redenbaugh reported that the term "synthetic seed" (synseed) should be referred only to an encapsulated somatic embryo. In more recent years, synthetic seeds were defined as artificially encapsulated somatic embryos, shoots or other vitro-derived tissues that can be used for sowing under in vitro or ex vitro conditions (Aitken-Christie et al., 1995). It seems logical, therefore, that the terminology that was initially set up for somatic embryo synthetic seeds be now used for any kind of encapsulated explant. The term "conversion" was explained as growth and development of both shoot and root systems, with minimal swellings, callus production, etc. In one word, "conversion" expresses the production of a green plant with a normal phenotype from a synthetic seed (Redenbaugh et al., 1988; Redenbaugh, 1993). When the above described criteria are met, it seems justified that the term "conversion" can be reasonably used to describe the development of a full plantlet from any synseed, either made from a bipolar (somatic embryo) or an unipolar explant. Talking about synseeds, there are other terms that should instead be avoided, such as "germination", which is proper of the real seed, or "plantlet formation", which can generally be related to any way of producing a small rooted plant from micropropagated units (Debergh and Read, 1991). Furthermore, the term "regrowth" should be used to indicate any other vegetative act different from conversion that is performed by an encapsulated propagule after sowing, such as rooting or sprouting or emission of a leaf, etc.

Conversion is one the most important aspect of the synseed technology, and still one of the factors limiting its practical use (Kozai et al., 1991; Piccioni et al., 1995; Tautorus and Dunstan, 1995). In contrast to a somatic embryo, which is a bipolar structure and therefore a complete plantlet in itself (Redenbaugh, 1993), shoots and buds do not have root meristems, and they must regenerate roots in order to be able to convert (Bapat, 1993). Furthermore, synthetic seeds should be able to survive at least short and medium

Biology of Root Formation and Development, edited by Altman and Waisel.
Plenum Press, New York, 1997.

term storage, in order to be convenient in commercial uses (Piccioni et al., 1996). Different authors described how encapsulated buds of banana and mulberry produced whole plantlets without specific root induction treatments, and also after medium term cold storage (Bapat and Rao, 1990; Ganapathi et al., 1992; Bapat, 1993). Encapsulated buds of apple sown in vitro, instead, sprouted well also after medium term storage, without regenerating roots (Piccioni and Standardi, 1995; Piccioni et al., 1996). In another report, a protocol was described for root induction and encapsulation of buds of the apple clonal rootstock M.26, with 66% synseed conversion on agar medium (Piccioni, 1997). The same technique was used in the present work, in which two different substrata and two storage periods were compared in allowing conversion from root-induced encapsulated micropropagated buds of M.26.

EXPERIMENTAL

Microcuttings (3–4 mm-long uninodal cuttings with one bud) of the clonal apple rootstock M.26 (*Malus pumila* Mill.) were excised from micropropagated shoots at the end of a proliferation subculture. Stabilization of culture, medium composition and growth chamber conditions are described elsewhere (Piccioni, 1997). The microcuttings were subjected first to a root induction treatment, made of an auxin liquid treatment (24 hour in a 24.6 μM IBA -Indolebutyric acid- and 15 g/l sucrose solution, 100 rpm in darkness) and then to a root initiation treatment (6 days in darkness on a nutritive medium) before encapsulation. Encapsulation was made with sodium alginate, enriched with an artificial endosperm made of nutrients and growth regulators. Root initiation conditions and encapsulation protocol were as previously described (Piccioni, 1997). Part of the root-induced encapsulated microcuttings were stored in a refrigerator at 4°C for 60 days before sowing. Storage conditions have been described previously (Piccioni et al., 1996). Both stored and non-stored microcuttings were sown in an agarized medium and in a soil-mix one. Both media were previously enriched with inorganic and organic nutrients and sucrose, and then autoclave-sterilised at least 1 day before use. Sowing was made with respect to the microcutting polarity, with the base of the explant downward and keeping the explant upper half outside the sowing medium. Vessels were kept in the growth chamber. Viability (green appearance, with no necrosis or yellowing), regrowth (rooting or sprouting) and conversion rates were monitored after 60 days from sowing.

RESULTS AND DISCUSSION

M.26 synseeds sown directly after encapsulation without storage showed a different behaviour according to the sowing medium and to storage (Table 1). The soil mix supported only 40% viability in non-stored microcuttings, while stored ones gave much better results (Table 1). Since agar medium gave always 100% viability, it is to be supposed that the tested soil mix was not an optimal substratum for synseed survival and conversion, probably as far as nutrients and water availability were concerned. Visual observations after 30 days from sowing showed that nearly 90% of the explants on soil mix were viable. The subsequent decline of viability is probably related to nutrient deficiency in the medium. It seems evident that the artificial endosperm included in the alginate capsule was not able to support the explant viability for long periods. Therefore, a fast establishment of the root system and of a functioning photosynthetic leaf system are essential in allowing

Table 1. M.26 synthetic seed performance after 60 days from sowing, according to medium and cold storage*

Treatments							
Medium	Storage	Viability (%)	Conversion (%)	Sprouting (%)	Rooting (%)	Roots (n)	Fresh weight of plantlets (mg)
Agar	0 days	100.0	63.3	90.0	63.3	3.4	151.0
Agar	60 days at 4°C	100.0	83.3	93.9	83.6	3.5	179.0
Average		100.0 a	73.3 a	91.9 a	73.4 a	3.4 a	165.0 a
Soil mix	0 days	40.0	10.0	10.0	10.0	1.0	17.4
Soil mix	60 days at 4°C	90.4	31.1	88.5	31.1	2.2	49.5
Average		65.2 b	20.6 b	49.3 b	20.6 b	1.6 b	33.4 b

*Each value is the average of 4 repetitions, made of 10 synseeds each. Average data of each column followed by different letters are different, according to the Duncan's test, $\alpha<0.05$.

plantlet survival on non-agarized substrata. In these aspects, microcutting quality also played an important role: as reported next, stored explants sprouted and rooted more vigorously and quickly, maintaining higher viability on soil mix after 60 days (Table 1).

Conversion rates were very different according to medium composition and storage. Non-stored synseed conversion rates sown on agar medium confirmed previous results (Piccioni, 1997). On the contrary, non-stored units sown on soil mix had very low conversion rates. As well as viability, cold storage increased conversion rates both on soil mix and agar media (Table 1) (Fig. 1).

Sprouting rates were generally very high on agar medium, with very little differences between stored and non-stored microcuttings. A very high increase was instead de-

Figure 1. M.26 plantlet obtained from a synthetic seed sown on agar medium after 60 days of storage at 4°C. Note the capsule fragments (arrow) at the base of the explant (bar=1 cm).

tectable in sprouting rates in soil mix-sown synseeds after storage, confirming previous results (Piccioni et al., 1995).

Rooting confirmed the general trend of the other parameters. Agar medium allowed generally higher rooting rates, and cold storage enhanced rooting performances both on agar and soil mix media (Table 1). As for rooting quality, agar-sown converted synseeds had an average of 3.5 root/unit, with no differences between stored and non-stored microcuttings. Soil mix-sown converted synseeds, instead, produced 1 root/unit without storage and 2.2 root/unit after storage.

Fresh weight of the plantlet obtained from the M.26 synseeds were also very different between treatments, in agreement with the general results of the experiment (Table 1).

A relevant production of callus was often visible, regardless of treatments, on the microcuttings that did not sprout, but particularly on those that did not root. The callus was usually entrapped inside the alginate capsule.

The overall results of the experiments are very encouraging the use of micropropagated buds of M.26 for synthetic seed production. In fact, 83.3% conversion rate obtained on agar medium after cold storage (Fig. 1) is to be considered an optimal performance of these synseeds, definitely comparable to that of encapsulated somatic embryos (Fujii et al., 1990, 1992; Piccioni et al., 1996).

It is interesting to note that in all combinations the factor limiting conversion was always rooting. Rooting and conversion rates were always the same, because all rooted explants had also sprouted. The encapsulated buds already had an organized shoot meristem, which did not lose its sprouting potential during root induction, encapsulation and sowing. On the contrary, the root meristem had to be regenerated through direct rhizogenesis. In fact, sprouting always preceded rooting.

Cold storage already proved to enhance M.26 encapsulated microcuttings vigour and sprouting when the storage conditions provided to the synseeds were appropriate (Piccioni et al., 1996). This probably happens because cold temperatures help removing residual dormancy from the cultured tissues (Pierik, 1987). The positive effect of the cold on rooting may be also related to a dilution of the auxin effect on callus production. In fact, callus occurred mainly on the unrooted explants, and a negative interaction between rooting and callus growth was already reported (Piccioni, 1997). This is probably related to competition between the two regenerative pathways, in which the progression of the rooting process limits cell growth towards unorganised callus. In fact, different morphogenic pathways are mutually exclusive (Ammirato, 1985). Different induction treatments with lower auxin concentration and/or shorter exposure of the microcuttings to IBA should be performed.

Visual observations during the experiment seemed to show that apical buds sprouted and rooted at a higher rate and faster and more vigorously than axillary buds, showing eventually a much better conversion rate. Axillary and apical buds were randomly chosen and not kept apart during the experiment, so it is not possible to perform a real analysis of the data on this point, but at the end of the experiment it seemed very clear that callus production and insufficient rooting and/or sprouting was much more frequent in the axillary buds than in the apical ones. It is suggested, therefore, that further experiments be carried out distinguishing between the two explant types.

CONCLUSIONS

The results reported in the present work confirm the possibility of using encapsulated root-induced micropropagated buds of the M.26 clonal rootstock for synthetic seed production. Conversion rates of 83.3% and 31.1% on agar and soil mix media, respectively, were achieved. Even though the soil mix conversion rate is still low for practical use, these results are extremely encouraging. It must be outlined that the soil mix used was identical to those used in greenhouse for plant nursery. Apart from the sterile conditions of the experiment, therefore, conversion of the M.26 synseeds on soil mix could be considered very similar to a direct conversion *in vivo*, which is highly desirable (Debergh & Read, 1991, Fujii et al., 1992). Improving sowing and nursing techniques and the soil mix composition, especially in nutrient and water availability, will probably increase conversion rates up to acceptable standards. An appropriate microorganism control system would afterwards give the final step towards *ex vitro* use of M.26 synthetic seed.

ACKNOWLEDGMENTS

Research carried out with the financial support provided by the Italian National Council of Research (CNR), "Contratto integrato CNR/Università n.96.00036.PF01".

REFERENCES

Aitken-Christie J., Kozai T. and Smith M.A.L., 1995. Glossary. In: Automation and Environmental Control in Plant Tissue Culture. J. Aitken-Christie, T. Kozai & M.A.L. Smith (Eds.). Kluwer Academic Publishers, Dordrecht, The Netherlands: ix-xii.

Ammirato P.V., 1985. Patterns of development in culture. In: Tissue Culture in Forestry and Agriculture. Basic Life Sciences, Vol.32. R.R. Henke, K.W. Hughes, M.J. Constantin and A. Hollander (Eds.). Planeum Press, New York, NY, (USA): 9–29.

Bapat V.A., 1993. Studies on synthetic seeds of sandalwood (*Santalum album* L.) and mulberry (*Morus indica* L.). In: Synseeds: Applications of Synthetic Seeds to Crop Improvement. K. Redenbaugh (Ed.). CRC Press Inc., Boca Raton, Ca (USA): 381–407.

Bapat V.A. and Rao P.S., 1990. In vivo growth of encapsulated axillary buds of mulberry (*Morus indica* L.). Plant Cell, Tissue and Organ Culture 20: 67–70.

Debergh P.C. and Read P.E., 1991. Micropropagation. In: Micropropagation. Technology and Application. P.C. Debergh & R.H. Zimmerman (Eds.). Kluwer Academic Publishers, Dordrecht, (The Netherlands): 1–14.

Fujii J.A.A., Slade D., Aguirre-Rascon J., Ruzin S.E. and Redenbaugh K., 1992. Field planting of alfalfa artificial seed. In Vitro Cell. Dev. Biol., 28P: 73–80.

Fujii J.A.A., Slade D., Olsen R., Ruzin S.E. and Redenbaugh K., 1990. Alfalfa somatic embryo maturation and conversion to plants. Plant Science, 72: 93–100.

Kozai T., Ting K.C. and Aitken-Christie J., 1991. Considerations for automation of micropropagation systems. Trans. of the ASAE 35: 503–517.

Ganapathi T.R., Suprasanna P., Bapat V.A. and Rao P.S., 1992. Propagation of banana through encapsulated shoot tips. Plant Cell Rep., 11: 571–575.

Piccioni E. 1997. Plantlets from encapsulated micropropagated buds of M.26 apple rootstock. Plant Cell, Tissue and Organ Culture 47: 255–260.

Piccioni E. and Standardi A., 1995. Encapsulation of micropropagated buds of six woody species. Plant Cell, Tissue and Organ Culture, 42: 221–226.

Piccioni E., Falcinelli M. and Standardi A., 1995. La "germinazione" del seme sintetico di erba medica (Medicago sativa L.). Rivista di Agronomia, 29(4): 567–573.

Piccioni E., Standardi A. and Tutuianu V.C., 1996. Storage of M.26 apple rootstock encapsulated microcuttings. Advances in Horticultural Sciences 10: 185–190.

Pierik RLM, 1987. The influence of physical factors on growth and development. In: In Vitro Culture of Higher Plants. R.L.M. Pierik (Ed.) Martinus Nijhoff Publishers, Dordrecht, The Netherlands: 115–125.

Redenbaugh K., Fujii JA and Slade D., 1988. Encapsulated plant embryos. In: Biotechnology in Agriculture. Advances in Biotechnological Processes, vol.9. Avshalom Mizrahi (Ed.). Alan R. Liss inc., New York, NY, (USA): 225–248.

Redenbaugh K., 1993. Introduction. In: Synseeds: Applications of Synthetic Seeds to Crop Improvement. K Redenbaugh (Ed.) CRC Press Inc., Boca Raton, Ca (USA): 3–7.

Tautorus T.E. and Dunstan D.I., 1995. Scale-up of embryogenic plant suspension cultures in bioreactors. In: Somatic Embryogenesis in Woody Plants. Volume 1 - History, Molecular and biochemical Aspects, and Applications. S.M. Jain, R.K. Gupta, R.J. Newton (Eds.). Kluwer Academic Publishers, Dordrecht, The Netherlands: 265–292.

RELATING THE POST-HARVEST PHYSIOLOGY OF SWEET POTATO STORAGE ROOTS WITH STORABILITY FOR A RANGE OF EAST AFRICAN VARIETIES

Deborah Rees,[1] Adam Pollard,[1] Dominic Matters,[1] and Edward Carey[2]

[1]Natural Resources Institute, Chatham Maritime, Kent ME4 4TN,United Kingdom.
[2]International Potato Center, Regional Office, P.O.B. 25171, Nairobi, Kenya.

INTRODUCTION

Sweet potato (*Ipomoea batatas*) is very important as a staple crop in many developing countries. However, its potential is constrained by the perishability of the storage roots which limits the time that the crop can be kept in the ground after maturity and limits post-harvest shelf-life. Given the genetic diversity of sweet potato germplasm, there is great potential for improving storability of the root through breeding. Root composition and metabolism are key factors affecting rates of physiological deterioration and susceptibility to pests and diseases. The purpose of this study is to identify the key physiological factors associated with perishability, with the objective of determining selection criteria to be used within breeding programmes.

EXPERIMENTAL

In March 1995 and April 1996 sweet potatoes were grown in Kenya, air-freighted to the UK and maintained under simulated tropical storage conditions (27°C, R.H.>95%). Two varieties studied were perceived by Kenyan farmers and traders as being perishable, and two as having long-shelf life. During storage, measurements were made of respiration, rates of loss of dry and fresh weight, sugar content (hplc), rates of wound healing (staining for lignin) and sprouting.

RESULTS AND DISCUSSION

Physiological differences related to perishability were observed. Over the course of six weeks of storage, "perishable varieties", Kemb10 and SPK004, had respiration rates up to twice that of the "long-shelf-life" varieties Kemb36 and KSP20. This was reflected in the rate of loss of dry matter. Although more precise information about relative rates of deterioration in-country were lacking, on the basis of this study, KSP20 was identified as the best variety for long-term storage, having a low metabolic rate, high rates of wound healing and low sprouting. SPK004 was identified as the worst storing variety by the same criteria. KSP20 was distinguished from the other varieties by a low dry matter content, low sucrose levels and high monosaccharide (glucose and fructose) levels. This is contrary to previous observations on North American varieties that indicate that low dry matter is associated with high respiration rates and perishability (Kays 1985). This may be related to the fact that East African varieties tend to have higher dry matter contents than North American varieties. As sugar profile is obviously an indication of metabolic status, the relationship between sugar levels, metabolic rates, sprouting, wound healing etc. requires additional investigation. The possibility of using sugar profiles as a selection criterion for long-storing varieties should be considered, although it will be important to consider acceptability of taste by consumers.

ACKNOWLEDGMENTS

This work was supported by funding from the Overseas Development Administration.

REFERENCES

Kays, S.J., 1985, The physiology of yield in sweet potato. *in* Sweet Potato Products: A Natural Resource for the Tropics, J.C. Bouwkamp ed. CRC Press, Boca Rouge, Florida USA.

FACTORS CHARACTERIZING THE ROOTING POTENTIALITIES IN PEACH (*Prunus persica*, L. BATCH) CV FERTILIA I CUTTINGS

G. Bartolini,[1] G. Di Monte,[2] P. Pestelli,[1] and M. A. Toponi[2]

[1]Institute for Propagation of Woody Plants, CNR, Scandicci, Firenze. [2]Institute of Plant Physiology and Biochemistry, CNR, Area Ricerca, Roma.

The peach plant production by self-rooting of the leafless hardwood cuttings occurs in two steps: rooting of the cuttings by different periods of basal heating, followed by transplanting into soil. Some difficulties are encountered in obtaining good results because of the frequent failures after transplanting. The time of the year when the cuttings are collected appears to have the most influence, showing that a particular physiological state is able to favour the rooting and survival.

Therefore, the aim of this work was to investigate the relationship between the rooting and survival percentage and the soluble carbohydrate and putrescine availability both at the moment of cutting collection from the mother plant and during the rooting process up to the moment of transplanting.

Rooting experiments with cuttings made from September to December show values which reach 95% for the cuttings collected in mid-October and decrease rapidly to reach the value of almost 10% for cuttings collected in December.The rooting percentage was recorded after 60 days.

The survival percentage of the same cuttings transplanted after 20 or 60 days of basal heating reaches 80% for the cuttings collected in mid-October more quickly if the transplant is made after 60 days (roots are well developed), 15 days later if made after 20 days (roots not evident in the basal end of the cutting). The survival was recorded in July. Mid-October is the best period for both rooting and survival; before and after this time the values increase and decrese rapidly.

We thus studied the variations in soluble carbohydrate availability and polyamine amounts, particularly at the two principal moments of the propagation: collection of the cuttings from the mother plant and time of transplanting.

The values of soluble carbohydrate availability in the basal end of the cuttings (shoots of stock plant) before basal heating show that sorbitol is present in a large amount over the entire period. In mid-October a general increase can be noted, sight for glucose and fructose and sharper for sucrose. For polyamines the analyses made on the basal end of the same cuttings rarely show the presence of spermine and spermidine while a very significant presence of putrescine is revealed in the cuttings collected during the period September-October.

The analyses on the soluble carbohydrate amounts during the rooting period indicate that they decrease rapidly during the first 20 days of basal heating, (50%), to reach very small values on the 60th day.

These values represent the soluble carbohydrate amounts present in the basal end of the cuttings at the two moments of transplanting.

In the same samples the presence of spermine is very rare and while the spermidine appears after 5 days of basal heating the putrescine is always present showing significant amounts in the cuttings collected during the period September–October. The values of the three polyamines continue to be elevated until the 20th day of basal heating while in the material collected in October an increase in putrescine is present on 60th day. The tissue of the cuttings made from shoots of the stock plants does not show the same potentialities during the year and every year. The results obtained strengthen the hypothesis of a particular potential of the cutting tissues during the month of October.The variations on soluble carbohydrates and the presence of putrescine, considered a rooting marker, correlated to the primordia formation, indicate that in the shoots collected in mid-October the primordia are stimulated to develop into roots, and also that these conditions are able to induce its consolidation in the cutting tissue, without secondary formations or alterations, thus ensuring a good survival.

CHANGE OF FREE AMINO ACIDS IN ROOT AND SHOOT OF KOREA GINSENG (PANAX GINSENG) DURING EMERGENCE

Hoon Park,[1] Mee-Kyoung Lee,[1] Byung-Goo Cho,[1] and Byung-Ju Choi[2]

[1]Korea Ginseng and Tobacco Research Institute
Daejeon, Korea
[2]College of Industry and Science
University of Kongju
Yesan, Korea

INTRODUCTION

Korea ginseng (Panax ginseng CA. Meyer) is the representative herb of the oriental medicinal plants. The plant has a perennial root which grows very slowly and is generally harvested after 6 years. Since the ginseng plant is very sensitive to high fertility soils, no fertilizers are used. The heavier the root is the higher is the price per unit weight, in Korea ginseng. This is explained by the increase of the central portion (xylem-pith) as compared with the outer part (phloem-cortex) (Park and Lee 1989). The compactness of the central part is very important for quality assessment. There is almost no saponins in the central part but it contains a relatively large quantity of soluble proteins, although ginseng saponins have been considered as the major biologically active compound by most pharmacologists (Kim 1978). Thus, the present research, on nitrogen metabolism, is very important not only for the production of large roots but also for the search of new bioactive nitrogen compounds, e.g. peptides, which are assumed to be in the high quality roots. Only the amino acid content in ginseng roots was investigated before for pharmaceutical and nutritional purposes (Park et al., 1990) and for quality relation (Lee and Park 1987). There is no report on the amino acid composition in relation to nitrogen metabolism of ginseng. Since arginine constitutes more than 60% of the total free amino acid in ginseng roots, arginine seems to be a reserve nitrogen source in root for regrowth. For the elucidation of the nitrogen metabolism during regrowth, the time course of amino acid composition of various organs (root, rhizome, stem and leaf) was investigated during emergence and leaf unfolding.

Biology of Root Formation and Development, edited by Altman and Waisel.
Plenum Press, New York, 1997.

EXPERIMENTAL

Plant material : Four years old plants were taken, washed, separated into the central part (xylem-pith), rind (cortex-phloem-epidermis) of tap root, rhizome, stem and leaves. The plant material was freeze-dried and powdered. Free amino acid extraction : Ground plant material (0.2g) was soaked overnight in 50ml of 75% ethanol and extracted at 70°C for 30 minutes with reflux. The extract solution was filtered and concentrated in a flash-evaporator below 40°C. The concentrate residue of the aerial part extract was dissolved with 1.2ml distilled water and well mixed, with 0.1% TFA-30% methanol. Three ml of solution was passed through pretreated SEP-PAK C18 cartridge (Millipore 1984). The residue of root extract was dissolved with 2ml of 0.2M Na-citrate buffer (pH 2.2). Amino acid analysis: Sample solution was filtered through 0.45mm millipore filter and applied to Ultrapac-II cation exchange resin column and analyzed using LKB 4150 Alpha-Amino Acid Analyzer. Standard amino acids were obtained from Sigma Co. Ltd. and used for calibration.

RESULTS AND DISCUSSION

Change of total free amino acid content in root : The change of total free amino acid (TFA) content, before and after new shoot emergence, is shown in Fig. 1. TFA content in the pith-xylem greatly decreased first and slightly increased with shoot growth. Then it decreased to less than half at the end of leaf spreading (May 24). TFA in the phloem-cortex was slightly decreased first and then greatly increased with shoot growth. Then it decreased to about one quarter of the original value. Stem length was 2cm on April 19 and 6.7cm on April 27. It was 45cm when the leaves were fully spread. Total free amino acid content in the rhizome was much higher (about 100mmole/g d.w.) than in the roots before

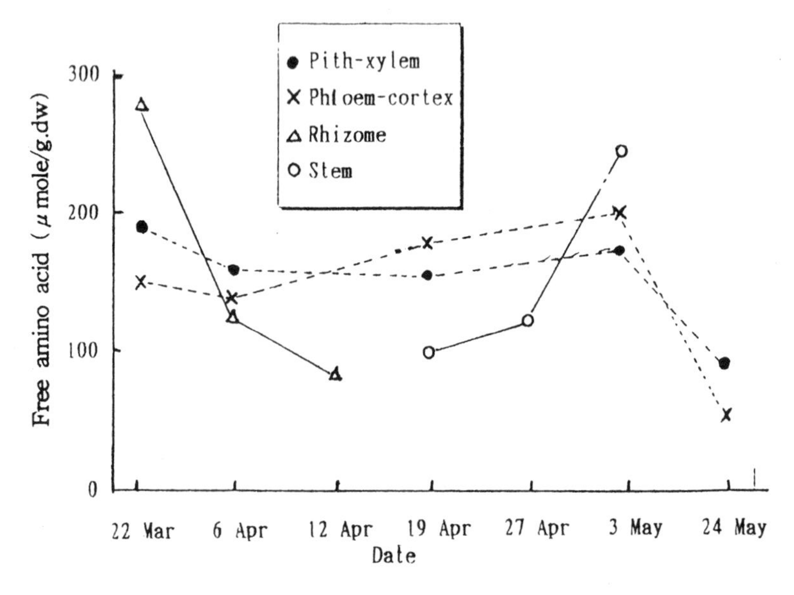

Figure 1. Time course of total free amino acids in roots, rhizomes and stems, of P. ginseng during new shoot emergence at 4th year.

Table 1. Changes of free amino acids in pith-xylem of P. ginseng root during emergence (mmole/g,d.w.)

	Date				
Amino acid	Mar. 22	Apr. 6	Apr. 19	May 3	May 24
Asp	13.28	11.10	7.44	3.39	2.13
Thr.*	25.12	18.58	21.57	16.69	7.37
Glu.	8.63	1.33	1.52	1.59	0.42
Pro.	2.11	2.24	1.60	2.69	T
Gly.	0.51	0.49	0.60	1.01	0.56
Ala.	3.32	2.03	1.61	1.56	2.72
Val.	1.33	1.34	1.37	1.83	0.58
Met.	0.53	0.53	0.49	0.77	T
Ile.	1.47	1.21	1.55	2.86	0.54
Leu.	3.08	3.06	3.55	6.40	0.86
Tyr.	1.52	1.43	1.68	4.58	1.26
Phe.	1.54	1.52	1.89	5.04	0.53
His.	10.07	13.27	10.26	20.79	18.74
Lys.	1.41	1.63	2.24	8.73	2.88
Amm.	15.47	8.20	10.19	14.26	9.87
Arg.	103.75	95.27	90.33	91.00	41.25
Total	194.62	163.21	157.86	183.41	89.69

Thr* : including serine; T : trace; Apr. 19 : emergence (2cm)

emergence. It rapidly decreased to much lower level than the roots. TFA content in stems increased faster than the decrease rate in the rhizome. Total amino acid included ammonia. Change of free amino acids in root : Seventeen amino acids were detected in the pith-xylem part (Table 1) and 16 in the phloem-cortex part (Table 2). Methionine was only measurable in the phloem-cortex part. Tryptophan was not measurable in both parts. There

Table 2. Changes of free amino acids in phloem-cortex of P. ginseng root during emergence (mmole/g d.w.)

	Date				
Amino acid	Mar. 22	Apr. 6	Apr. 19	May 3	May 24
Asp	11.02	9.37	13.17	4.80	1.63
Thr.*	15.03	10.70	23.24	20.86	4.08
Glu.	6.22	1.09	1.46	2.72	0.63
Pro.	2.88	2.14	1.86	1.37	T
Gly.	0.39	0.46	0.80	1.01	1.00
Ala.	3.25	1.88	2.68	1.58	1.60
Val.	0.95	0.99	2.06	2.58	T
Ile.	1.06	1.16	2.12	2.84	0.52
Leu.	2.19	2.60	4.05	5.24	0.92
Tyr.	0.95	1.98	1.54	3.93	0.83
Phe.	0.98	1.91	2.18	4.04	0.50
His.	7.88	7.72	13.30	10.67	8.71
Lys.	1.20	1.37	2.23	5.54	1.41
Amm.	6.15	5.60	12.57	25.49	6.75
Arg.	91.95	92.00	97.70	120.40	26.50
Total	152.09	140.97	181.71	214.84	55.07

Thr.* : including serine; T : trace; Apr. 19 : emergence (2cm)

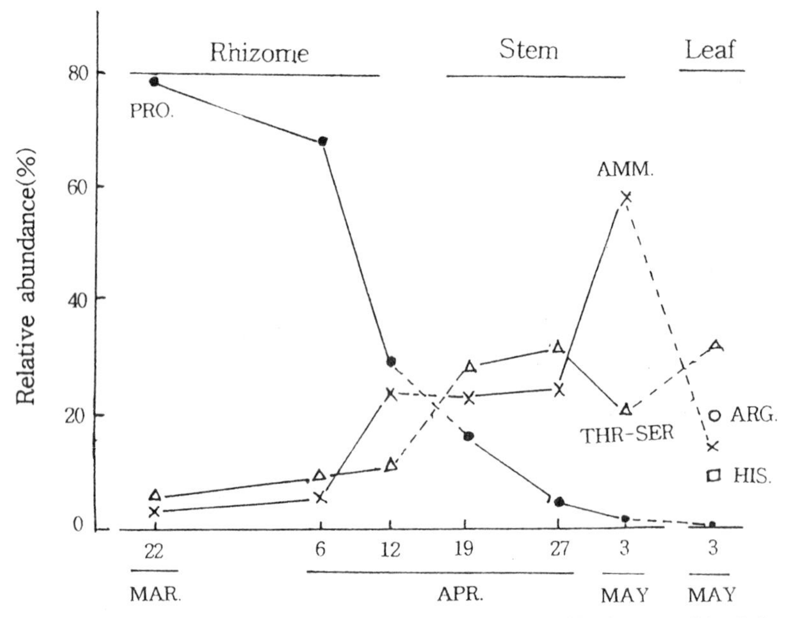

Figure 2. Chromatogram of free amino acids in phloem-cortex of P. ginseng on May 3rd.

were two unknown conspicuous peaks, just before that aspartic acid (Fig. 2). These unknowns are under investigation and are not included in the presented study. According to simple correlation amino acid composition was very similar in the two parts of the roots and differed between growth stages. This indicates that amino acid metabolism in roots depends on that of the shoot. Ginseng roots are the source organ for the supply of nitrogen to the shoots, until complete leaf spreading occurs in the middle of May. Five amino acids and ammonia in the pith-xylem were listed according to their abundance in Table 3. The similar trend was shown in the rind (Table not shown). Since ammonia content was considerably high, and increased with shoot growth, it was included for discussion with amio acids. These five major amino acids and ammonia contributed approximately 90% of the total free amino acids. Arginine was the most abundant one (46–55%) in all cases. When stem growth was most vigorous (May 3, Table 3), phenylalanine and tyrosine increased, indicating specific requirement of phenolic amino acid. Based on the changes of arginine

Table 3. Changes in relative abundance of the major free amino acids in pith-xylem of P. ginseng during emergence (% of total)

	1	2	3	4	5	6	Total
Mar. 22	Arg.	Thr.(+Ser)	Amm.	Asp.	His.	Glu	
	53.3	12.9	7.9	6.4	5.2	4.4	90.1
Apr. 6	Arg.	Thr.(+Ser)	His.	Asp.	Amm.	Leu.	
	58.4	11.4	8.1	6.8	5.0	1.9	90.6
Apr. 19	Arg.	Thr.(+Ser)	His.	Amm.	Asp.	Leu.	
	57.2	13.7	6.5	6.5	4.7	2.2	90.8
May 3	Arg.	His.	Thr.(+Ser)	Amm.	Lys.	Leu.	
	49.7	11.3	9.1	7.8	4.8	3.5	86.2
May 24	Arg.	His.	Amm.	Thr.(+Ser)	Lys.	Ala.	
	46.0	20.9	11.0	8.2	3.2	3.0	92.3

Table 4. Changes in free amino acids in rhizomes, stems and leaves of P. ginseng during emergence (mmole/g,d.w.)

	Rhizome			Stem+Leaf		Stem	Leaf
Amino acid	Mar. 22	Apr. 5	Apr. 12	Apr. 19	Apr. 27	Mar. 3	Mar. 3
Asp	5.98	5.67	2.75	3.72	2.06	3.04	2.70
Thr.*	14.80	11.85	8.95	28.22	38.63	52.27	15.05
Glu.	2.99	1.53	4.24	4.85	2.94	3.24	4.04
Pro.	228.95	87.41	23.42	15.89	4.59	3.94	T
Gly.	0.68	T	1.17	0.7	1.00	1.10	0.34
Ala.	2.55	1.80	2.91	5.02	7.27	7.75	3.20
Val.	T	T	T	1.32	2.27	1.64	T
Ile.	1.50	T	0.95	0.54	2.18	1.86	T
Leu.	1.17	T	1.13	1.00	3.93	3.04	T
Tyr.	T	T	2.30	1.43	3.64	4.25	T
Phe.	T	T	T	1.91	2.02	1.12	T
His.	6.54	5.61	9.76	7.65	12.16	15.18	4.52
Lys.	1.31	2.83	0.40	0.82	1.77	0.37	1.57
Amm.	9.90	6.83	19.21	21.87	28.89	150.57	6.76
Arg.	15.17	3.89	4.65	4.92	9.59	9.12	9.38
Total	291.54	127.43	81.89	99.01	122.96	258.50	47.77

Thr.* : including serine; T : trace

and ammonia, the differences between the central and the rind can not be explained by decomposition of arginine or by production of ammonia. Except for arginine, there were some changes in the order of abundance among four other amino acids and ammonia. Along with the shoot growth, histidine and lysine tended to increase while arginine decreased. Histidine reached 16~21% at the end of full growth period of the aerial part on May 24. Change of free amino acids in the rhizome and shoots : Free amino acid composition of rhizomes during emergence, is shown in Table 4. The peak of proline was espscially large in the rhizomes (Figure not shown) as compared with that of the roots (Fig. 2). The main amino acids in the rhizomes were proline (76.5%), arginine (7.7%) and threonine-serine (5.1%). The 4th abundant one is ammonia (3.4%) (Fig. 3). On April 12 just before emergence, proline remained the most abundant (28.6%). Other major amino acids were histidine (11.9%), threonine (7.6%) and arginine (5.7%). An interesting result

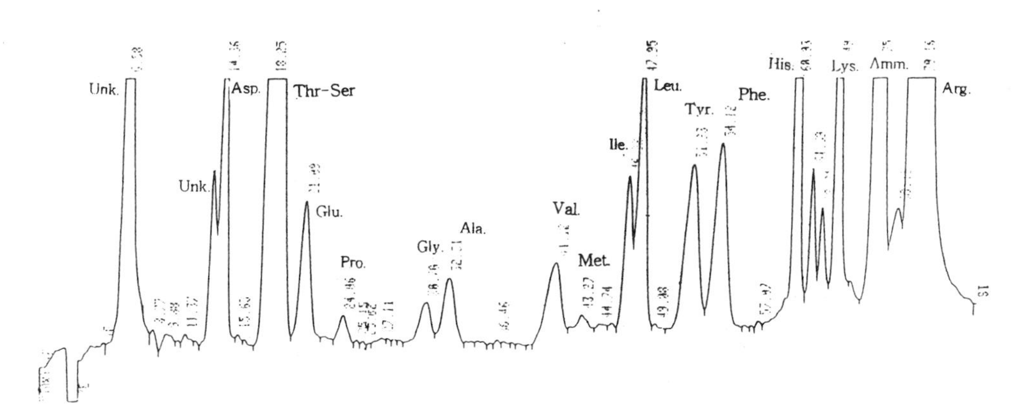

Figure 3. Time course of the major free amino acids in rhizomes and stems of P. genseng during emergence.

is the increase of ammonia. It constituted 23.5% of TFA and reached almost the amount of proline (Fig. 3). This may indicate that proline or arginine are decomposed to ammonium for translocation. In the growing shoot, stem length was 2.0cm on April 19 and 6.7cm on April 27. The shoot is mostly consisted of stem tissues. Thus, FA content could be compared with that of the stem on May 3rd. In the growing stem proline decreased rapidly while ammonium increased rapidly. When the stem was at the rapid growth phase ammonium content reached its highest values (Fig. 3). Threonine-serine was the second abundant acid and increased. TFA content of the leaves was one fifth of that of the roots and threonine-serine was most abundant (29.7%). Ammonium was the third in abundance (14.2%) and was followed by histidine (9.5%). When the change of the major amino acids in various parts of Korea ginseng during emergence was rearranged (Table not shown) it is very clear that each organ has its key amino acids and that they are arginine for roots, proline for rhizome, ammonium-threonine for stems and threonine for leaves.

CONCLUSION

The major free amino acid in roots was arginine whereas proline seems to be used for stem growth. Ammonia is supplied from the roots by decomposition of arginine. Threonine is the major amino acid in stems where ammonia content is higher than that of threonine. Threonine and arginine are major amino acids, in leaves. Threonine seems to be translocated from the roots to the leaves through the rhizomes and stems. Proline appeares to be translocated directly to the stem. Ammonia appeares to have a great role in amino acid metabolism, especially of arginine, during regrowth of the shoots of Korea ginseng.

REFERENCES

Kim, N.D. 1978 Ginseng pharmacology, In "Korean Ginseng" H.W. Bae ed. Korea Ginseng and Tobacco Res. Inst. p. 89

Lee, M.K. and Park, H. 1987, Free amino acids of xylem-pith in Panax ginseng root. Korean J. Ginseng Sci. 11:32 Millipore corporation 1985, Amino acid analysis system, operator's manual p. 3–2

Park, H. and Lee, M.K. 1989, The recent progress in the concept of biologically active substance in Panax ginseng in relation to cultivation and processing. Proc. 2nd. Int. Sym. on Recent Advance in Natural product Res., Natural product Res. Inst. Seoul. Nat. Uni. Seoul, Korea, p. 450

Park, H., Cho, B.J., and Lee, M.K. 1990 Nitrogen compound of Korea ginseng and their physiological significance. Korean J. Ginseng Sci. 14:317

THE ENHANCING METHOD OF GINSENOSIDE PRODUCTION FOR HAIRY ROOT MASS CULTURES OF GINSENG (*Panax ginseng* C.A. Meyer)

Baik Hwang,[1] Kyeong-Min Ko,[1] Deok-Chun Yang,[2] Ji-Chang Park,[2] and Kwang-Tae Choi[2]

[1]Department of Biology and HRC, Chonnam National University, Kwangju 500-757.
[2]Korea Ginseng and Tobacco Research Institute, Taejon 305-345, Korea.

Panax ginseng is an economically important medicinal herb used as a vitalizing and stimulating agent. The root has been used as a drug since ancient times and is expensive due to a long and complicated cultivation (Arya et al., 1991). Therefore, we have studied the production of the crude and/or the active ingredients, the saponin ginsenosides, by hairy root cultures. We investigated the two-step process to improve the growth and ginsenosides production by ginseng hairy root cultures in 3 L bioreactor with air bubble.

A hairy root clone of *Panax ginseng*, HRB-15 was cultured under various conditions with 3–5 L bioreactor to enhance growth and ginsenosides production. Two-step process of hairy roots culture, with yeast elicitation or without ammonium nitrate in the culture medium, was developed. Fifty μg of yeast elicitor per g fresh weight showed a synergistic effect on the ginsenoside synthesis of hairy roots, 20 days after culture. At that time, the ginsenoside content was 1.15%. However, growth of the hairy roots decreased by 21%, as compared with roots that was grown under dark conditions. Elimination of ammonium nitrate from the culture medium after 20 days of culture, reduced the content of total ginsenoside to 1.26%, but significantly raised that ginsenoside-Re and -Rd. Growth of the hairy roots decreased by 10% as compared with those grown under dark conditions.

By the presented system, we have demonstrated a unique two-step process of hairy root cultures to maximize the production of biomass and secondary metabolites. We found the procedure to enhance ginsenosides of production by growing hairy roots. These results show that there are many factors that are involved in the production of ginsenosides, with the use of large scale cultures of ginseng hairy roots being one of them.

ACKNOWLEDGMENTS

This research was supported by a grant from Hormone Research Center (HRC-96-2022) and '95 Post-doc. of KOSEF, Korea.

REFERENCE

Arya, S., Liu, J. R., and Eriksson, T., 1991, Plant regeneration from protoplasts of *Panax ginseng* through somatic embryogenesis, *Plant Cell Rep.*, 10: 277.

40

MORPHOLOGICAL STRUCTURED MODEL FOR HAIRY ROOT CULTURES

I. Berzin,[1] D. Mills,[2] and J. C. Merchuk[3]

[1]Department of Chemical Engineering
Ben-Gurion University of the Negev
Beer-Sheva, Israel
[2]Institute for Applied Biology
Ben-Gurion University of the Negev
Beer-Sheva, Israel
[3]Unit of Biotechnology
Ben-Gurion University of the Negev
Beer-Sheva, Israel

INTRODUCTION

The possibility of using hairy root (HR) cultures for producing secondary metabolites on a large scale has recently received a great deal of attention (Flores et al., 1987; Hamill et al., 1987; Rhodes et al., 1987; Scheidegger, 1990; Toivonen, 1993). However, in order to reliably design and scale-up HR culture systems, their growth kinetics must be understood and mathematically modeled (Flint-Wandel et al., 1993; Inomata et al., 1993; Kim et al., 1995; Taya et al., 1989). The shape of the HR batch growth curve has been described using Monod's equation (Hilton et al., 1988) or the logistic equation (Toivonen et al., 1990), both of which predict the change of total root biomass with time. However, published results (Croes et al., 1989; Yonemitsu et al., 1990; Flores, 1987; Aird et al., 1988) suggest that different HR morphologies result in dissimilar levels of secondary metabolite production. Thus two HR cultures with a similar biomass but different root architectures could have completely different product yields.

A morphological structured model would therefore be of great interest. Because HR cultures are highly branched and are not connected to aerial plant parts, morphological structured models of natural root systems cannot adequately describe their architecture. A kinetic model of branching growth of plant HRs published by Taya et al. (1989) assumes linear root extension, binary division of the root-tip meristem and a linear relation between biomass and root length. Although this model provides a close correlation with the biomass data of some HR cultures, its description of HR morphology and branching patterns does not conform to the findings obtained in our experimental system (Duran, 1993). HR laterals

Biology of Root Formation and Development, edited by Altman and Waisel.
Plenum Press, New York, 1997.

are formed in the zone behind the tip meristem, resulting in delayed generation of successive branches, as in natural plant roots (Rose, 1983). Here we present a morphological structured model for HR cultures, that allows detailed description of root architecture and evaluation of root biomass at any desired moment of growth in a batch culture.

EXPERIMENTAL

HR cultures of *Symphytum officinale* were maintained by subculturing a single root tip in 250 ml Erlenmeyer flasks containing 30 ml of liquid MS (Murashige and Skoog, 1962) medium with 3% sucrose. The tissue was incubated at 25°C in the dark at 150 rpm on a gyratory shaker. Every 3–4 days the fresh and dry weights of three root systems were measured. Their morphological characteristics (length and diameter of each lateral) were determined by image analysis using a desktop scanner (Pan et al., 1991) and NIH Image program. Sugar concentrations in the medium were also determined using colorimetric and enzymatic methods (Chaplin et al.).

THE MODEL

Definitions

1. HR systems are described hierarchically in terms of locus of origin of the laterals (Barley, 1970 ; Hackett et al., 1972; Lungley, 1973). The inoculum tip is designated as "zero-order", laterals branching off from inoculum as "first-order", laterals branching off from first-order laterals as "second-order", and so on. Only the first four orders (zero to third) are relevant to our HR system and will be considered here, but the model is capable of describing the growth of higher-order systems as well.
2. To distinguish chronologically between laterals belonging to the same order, each lateral is designated by a sequence of numerical indices indicating its chronological location; the first ("i-") index indicates order of branching among first-order laterals, the second ("j-") index shows order of branching among second order laterals, and the third ("k-") index shows order of branching among third-order laterals. Obviously, the total number of non-zero indices of a lateral corresponds to its order (e.g., a lateral designated "000" is zero-order, i.e., it is the inoculum). For example, a lateral designated "ijk" ($i,j,k \neq 0$) must be a third-order lateral; it must have been the kth lateral to branch off from a second-order lateral, which was the jth lateral to branch off from a first-order lateral, itself the ith to branch off from the inoculum.

Assumptions

Most of the following assumptions were tested by experiment and some are based on published data.

1. The HR system can be treated as a set of cylinders of decreasing diameter.
2. The HR laterals grow only in one dimension by extension at the root meristem (Taya et al., 1989). The diameter of a lateral of given order is considered constant.

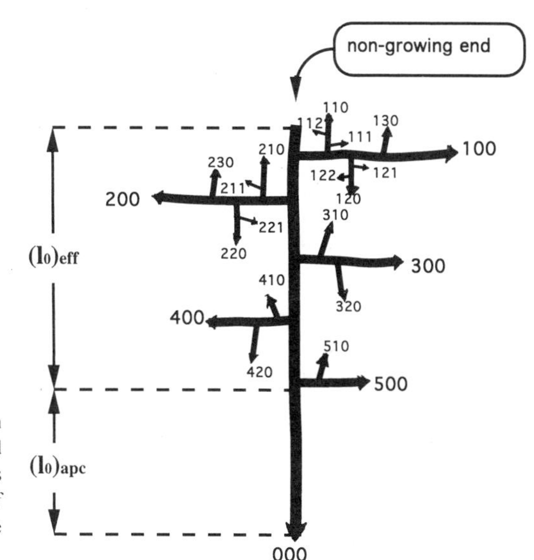

Figure 1. Components of the architecture of an idealized HR system. Each lateral is characterized by an index which specifies both its order and its chronological location. Each lateral is composed of branching and non-branching parts (illustrated here for the zero-order lateral).

3. Each HR lateral includes a non-branching part l_{apc} in which cells divide and differentiate, and a branching part l_{eff} which is called the effective part (Rose, 1983).
4. Along the effective part, branching occurs with a linear density q_g, which is characteristic of each order.
5. The ratio between HR biomass and volume (ρ_v) is constant throughout the root system and remains constant during entire culturing period .
6. The ratio between HR net water amount (wet weight minus dry weight) and biomass wet weight (W_c) is constant throughout the root system and remains constant during entire culturing period.
7. The biomass yield $Y_{x/s}$ is constant (Toivonen et al., 1990).

Morphological Structured Model for HR Growth

The elongation rate of each lateral in the HR system can be described by the logistic equation:

$$d(l_{ijk}) / d(t) = k_g \, l_{ijk} \, (1 - l_{ijk} / l_{ijk\text{-}max}) \tag{1}$$

where

$$l_{ijk\text{-}max} = Y_{l/s\text{-}g} \, S_{ijk\text{-}0} \tag{2}$$

This equation can be applied to each and every lateral of the HR system. It takes into account both the characteristics of the order and the changes in growth-medium composition. Thus it can describe the different elongation rates of two laterals of the same order that branched off at different times, and also the different elongation rates of two laterals of different order that branched off at the same time.

After integration eq. 1 becomes:

$$l_{ijk} = [l_{g\text{-}0} \, e^{k_g t_{ijk}}] \; / \; [1 - (l_{g\text{-}0} \; / \; l_{ijk\text{-}max}) \, (1 - e^{k_g t_{ijk}})] \tag{3}$$

The overall length of each order of laterals can be calculated as follows:

$$L_0 = l_{000} \tag{4}$$

$$L_1 = \Sigma\, l_{i00} \tag{5}$$

$$L_2 = \Sigma\Sigma\, l_{ij0} \tag{6}$$

$$L_3 = \Sigma\Sigma\Sigma\, l_{ijk} \tag{7}$$

The overall length of the HR system is therefore:

$$L = L_0 + L_1 + L_2 + L_3 = \Sigma\, L_g \tag{8}$$

Since the number of laterals of order (g+1) is proportional to the effective length of the laterals of order (g), the former can be calculated from:

$$n_{g+1} = q_{g+1}\,(L_{eff})_g \tag{9}$$

and the overall number of laterals is:

$$N = \Sigma\, n_g \tag{10}$$

Using the typical diameter of each order, its volume can be calculated as follows:

$$V_g = \pi\, D_g^{\,2}\, L_g / 4 \tag{11}$$

and the overall volume of the HR system is therefore:

$$V = \Sigma\, V_g \tag{12}$$

The biomass can be calculated using the volumetric density:

$$X_d = \rho_v\, V \tag{13}$$

$$X_w = X_d / (1 - W_c) \tag{14}$$

Changes in the sucrose concentration can be calculated using the yield factor ($Y_{x/s}$), which is the ratio of biomass formed to substrate consumed.

$$\Delta S = \Delta X_d / (Y_{x/s}\, V_b) \tag{15}$$

RESULTS AND DISCUSSION

Testing the Model

The first step was to check the reliability of the assumptions. Most of the latter were validated by experiments (Figures 2–5).

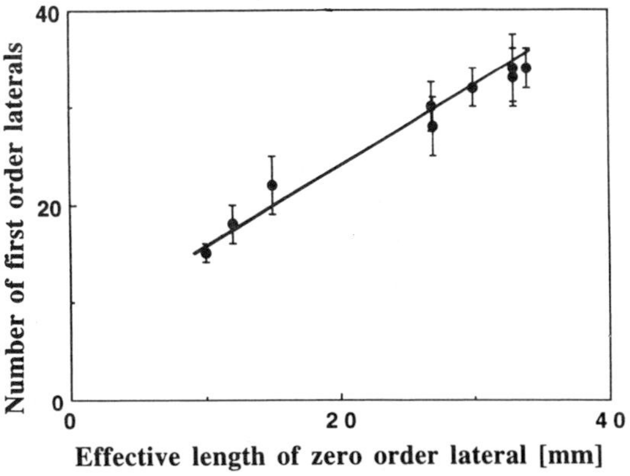

40

20

0

0 20 40

Number of first order laterals

Effective length of zero order lateral [mm]

Figure 2. Number of first-order laterals as a function of the effective length of zero-order laterals. The slope is the first-order branching frequency.

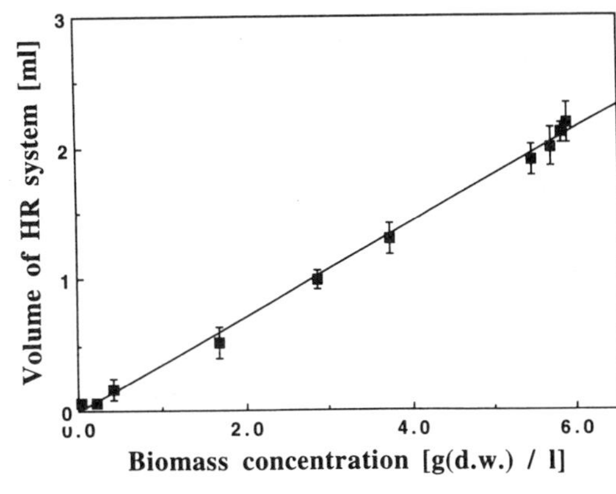

Figure 3. Volume of the root system as a function of dry weight. The straight line indicates a constant density of HR.

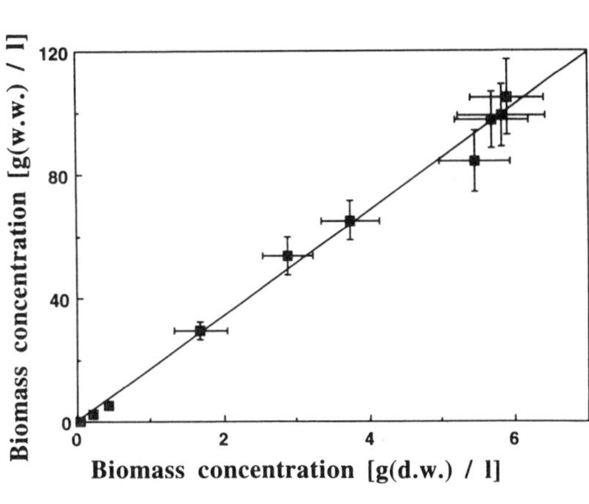

Figure 4. Relationship between fresh weight and dry weight. The straight line indicates a constant water content.

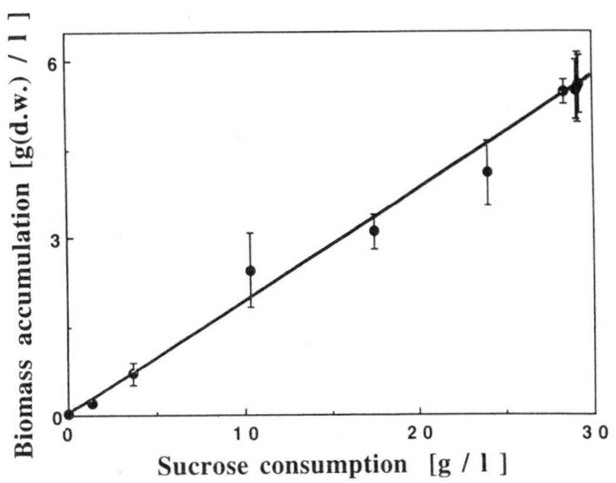

Figure 5. Relationship between dry weight formation and sucrose consumption during HR cultivation.

The above equations were used to write an iterative computer program for calculating changes in the number, length, volume and dry and wet weight of HR laterals of each order and of the overall HR system during the growth period. Changes in sugar concentration in the growth medium were also calculated. The structure of the simulation program is presented in Fig. 6. The predicted length and biomass are compared with the experimental data in Figs. 7 and 8.

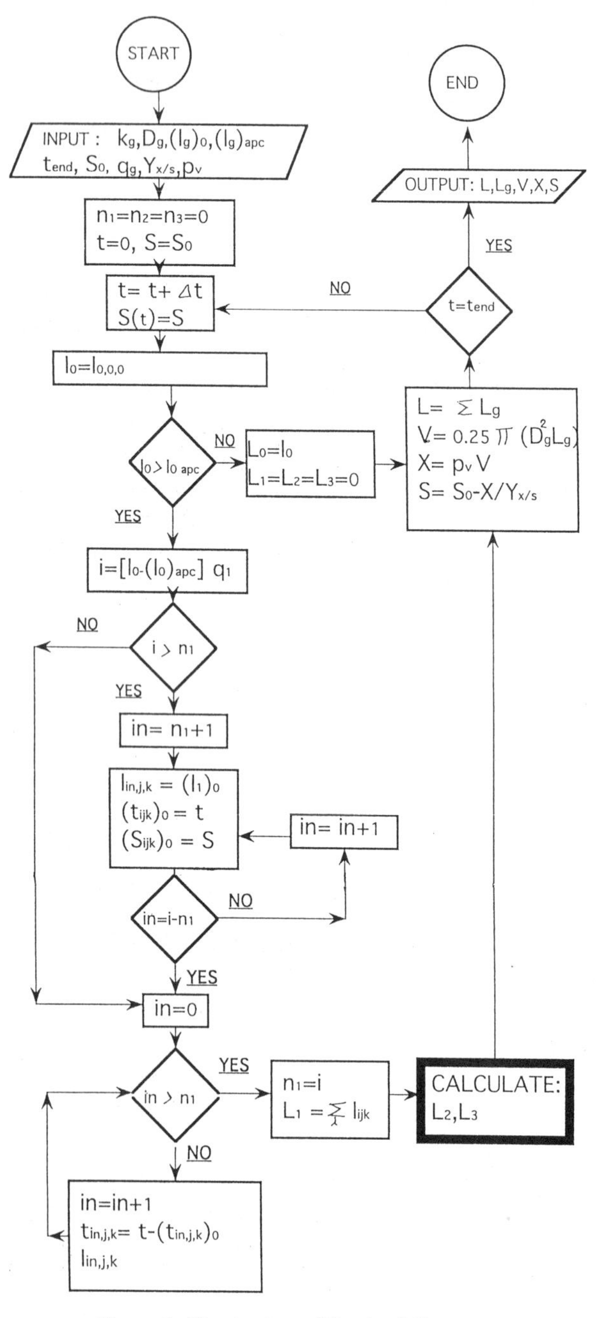

Figure 6. The structure of the simulation program.

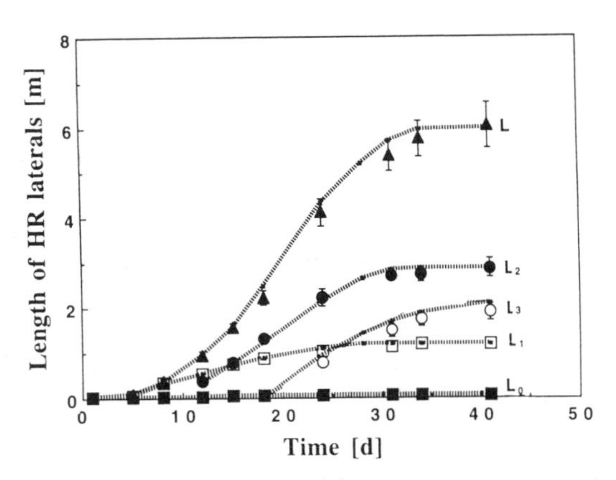

Figure 7. Lengths of each order and of overall HR system versus time of cultivation. The dots represent experimental data and the curves show the results of the simulation.

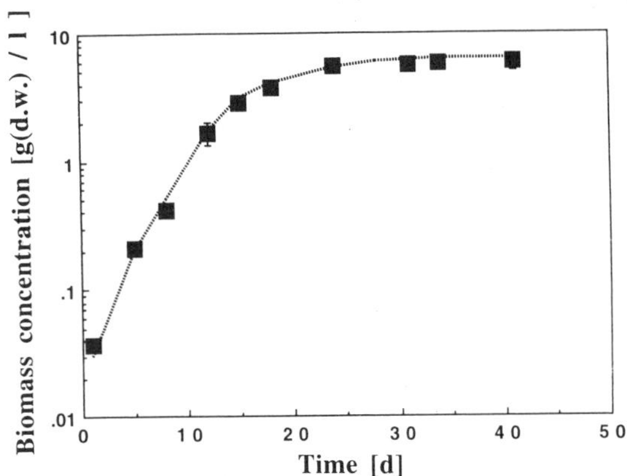

Figure 8. Biomass (dry weight) of the HR system versus time of cultivation. The dots represent experimental data and the curve shows the simulated results.

The values predicted by the model are in fair agreement with the experimental data. The model presented here leaves some questions unanswered. Due to the limited information given in the literature regarding the morphology of HR cultures, our model was confined to simulating HR culture in *Symphytum officinale* and has yet to be tested with other HR cultures. As mentioned above, the cultures were initiated with a single tip, which is not a realistic practice on a large scale. Therefore, the influence of the inoculum size (number of tips) should be considered. Although most of the parameters and constants used in this model, have a biological meaning which is quite easy to measure or estimate, finding expressions connecting these quantities would make the model easier to use. Finally, scaling-up HR cultures will have to await elucidation of the correlations between HR morphology and secondary metabolite production.

CONCLUSIONS

A morphological structured model of HR cultures has been proposed and found to satisfactorily simulate the growth characteristics and architecture of HR cultures of *Symphytum officinale*. Additional aspects should be considered in order to predict a closer fit to HR morphology and correlation with secondary metabolites production.

ACKNOWLEDGMENT

We wish to express our especial gratitude to Mrs. Shivta Wenkart for sharing her experience and knowledge with us.

NOMENCLATURE

D- diameter of lateral [mm]; k- parameter of the logistic equation [d^{-1}]; l- length of lateral [mm], l_0- initial length of lateral [mm]; l_{apc}- length of non-branching area [mm]; l_{eff}- length of branching area [mm]; l_{max} - maximum length of the lateral; L- overall length of laterals [mm]; n- number of laterals of a given order [-]; N- overall number of laterals [-]; q- branching frequency [mm^{-1}]; S- sucrose concentration [g/l]; S_0- sucrose concentration at initiation [g/l]; t- time from inoculation [d]; V-volume of laterals [ml], W_c- water content of HR [-], X_d- dry biomass concentration [g/l]; X_w- wet biomass concentration [g/l]; Yl/s- length yield factor [-], Yx/s- biomass yield factor [-]; ρ_v- density of HR. Subscript: g- index for order of HR laterals (g=0,1,2,3); ijk- chronological index for HR laterals.

REFERENCES

Aird, E.L.H., Hamill, J.D., Robins, R.J.R., and Rhodes, M.J.C., 1988, Cromosome stability in transformed hairy root cultures and the properties of variant lines of Nicotiana rustica hairy roots. in: "Manipulating Secondary Metabolism in Culture" R.J. Robins, and M.J.C. Rhodes eds. pp.137–144, Cambridge University Press. Cambridge.

Barley, K.P., 1970, The configuration of the root system in relation to nutrient uptake. Adv. Agron. 22:159–201.

Croes A.F., Van der Berg, A.J.R., Bosveld M., Breteler, H., and Wuiems, G.J., 1989, Thiophene accumulation in relation to morphology in roots of Tagetes patula. Planta, 179: 43–50.

Chaplin, M.F. and Kennedy, J.F., 1986, "Carbohydrate analysis, a practical approach" p. 6, 2nd edition, IRTL Press Limited, Oxford.

Doran, P.M.,1993, Production of chemicals using genetically transformed plant organs. Adv. Biochem. Eng. Biotechnol 48: 115–168.

Flint-Wandel, J., and Hjortso, M., 1993, A flow cell reactor for the study of growth kinetics of single hairy roots. Biotechnol. Tech. 7: 447–452.

Flores, H.E., 1987, Use of plant cells and organ culture in the production of biological chemicals. in: "Biotechnology in Agricultural chemistry" H.M. LeBaron, R.O. Mumma, R.C. Honeycutt, and J.H. Duesing eds. pp. 66–86. American Chemical Society Symposium, Series 334. Washington DC.

Flores, H.E., Hoy, M.W., and Pickard, j.j., 1987, Secondary metabolites from root cultures. Trends Biotechnol 5: 64–68.

Hacket, C., and Rose, D.A., 1972, A model of extension and branching of seminal root of barley, and its use in studying relations between root dimensions. Aust. J. Biol. Sci 25:669–679.

Hamill, J.D., Parr, A.J., Kim, S., Hopper S., and Hjortso M., 1995, Hairy root growth models: effect of different branching patterns. Biotechnol. Prog. 11:178–186.

Lungley, D.R., 1973, The growth of root systems - a numerical computer simulation model. Plant and Soil 38: 145–159.

Murashige, T., and Skoog, F. A., 1962, A revised medium for rapid growth and bio assays with tobacco tissue cultures. Physiol. Plant. 15:473–497.

Pan W.L. and Bolton, R.P., 1991, Root quantification by edge discrimination using a desktop scanner. Agronomy J. 83: 1047–1052.

Rhodes, M.J.C., Robins, R.J. and Walton, N.J., 1987, New routes to plant secondary products. Bio/Technology 5: 800–804.

Rhodes, M.J.C., Robins, R.J., Hamill, J.D., Parr, A.J., and Walton, N.J., 1987, Secondary product formation using Agrobacerium rhizogenes-transformed "hairy root" cultures. TCA Newsl. 53: 2- 15.

Rose, D.A., 1983, The description of the growth of root systems. Plant and Soil 75: 405–415.

Scheidegger, A., 1990, Plant biotechnology goes commercial in Japan. Trends Biotechnol 8: 197–198.

Taya, M., Kino-Oka, M., Tone, S., and Kobayashi, T., 1989, A kinetic model of branching growth of plant hairy roots. J. Chem. Engl. Japan 22: 698–700.

Toivonen, L., 1993, Utilization of hairy root cultures for production of secondary metabolites. Biotechnol. Prog. 9: 12–20.

Toivonen, L., Ojala, M., and Kuppinen, V., 1990, Indole alkaloid production by hairy root cultures of cetharanthus roseus: growth kinetics and fermentation. Biotech. Lett. 12:519–527.

Yonemitsu, H., Shimomura, K., Satake, M., Mochida, S., Tanaka, M., Endo, T., and Kaji, A., 1990, Lobeline production by hairy root cultures of Lobelina inflata. Plant Cell Rep. 9: 307–310.

41

AEROPONICS

A Search for Understanding Roots

Amram Eshel and Yoav Waisel

Department of Plant Sciences
Tel Aviv University
Tel Aviv, Israel 69978

It is well accepted that plant shoots form specific shapes and develop, at maturation, into predicted contours and sizes. This was thoroughly studied and amply reported in the literature. Certain aspects of the functional significance of these shapes with regards to their ability to sustain mechanical strains and to capture light, were analyzed in detail (Niklas, 1992). However, when the plants' "Hidden Half", is discussed, rather little is known, regarding the regulation of root architecture and its functional significance (Lynch and Nielsen, 1996).

The first question that comes to mind, when comparing root and shoot architecture is whether root systems tend to develop into a predictable composition and construct. So far there is no definite answer to this question, in spite of the fact that plant roots have been studied intensively by scientists for many years (Cannon, 1911; Weaver, 1919; Kutschera, 1960; Fitter, 1996). This poor state of our scientific knowledge results from two compounding factors—firstly, the opaque medium that the roots grow in, does not allow for direct observation of intact developing root systems. Secondly, the roots are very sensitive to their immediate environmental conditions, and any non-uniformity in the solid or liquid medium will bring about changes in root development.

There are indications, from excavated roots of a few plant groups, e.g., Cacti, or a few woody species, that a tendency to form root systems with specific shape does exist (Kutschera, 1960; Nobel, 1996). But for most other plants there simply is not enough information to indicate this way or another. In any case, the factors that are involved in regulating the development of the forms that were observed have not been discerned.

The second question which needs to be answered regards the quantitative relationships between the various constituents of each root system. It is reasonable to assume, that relationships between various parts of the root system are similar to those that exist among branches of the crown. Source-sink interactions that are mediated by plant hormones presumably take part in determining the joint function and development of the roots into one integral system (Eshel and Waisel, 1996). However, we do not have enough data to sub-

Biology of Root Formation and Development, edited by Altman and Waisel.
Plenum Press, New York, 1997.

stantiate or refute these assumptions, especially with regards to root systems of perennial plants. In order to tackle such questions we have to be able to manipulate intact root systems and follow their responses to our treatments, under constant and uniform conditions, and deviate from there to determine those factors that play a role in shaping the root system.

In order to understand fully the function-structure relationships within one root system, we should characterize the physiological processes that control the multiform appearance and function of the various root types which comprise every root system. The performance of the entire root system is an integration of the contributions of its multiple components. Each component operates according to its specific characteristics and to the conditions it is subjected to. There is multitude of evidence that individual components of any root system have their own unique characteristics (Waisel and Eshel, 1991; Eshel and Waisel, 1996).

The following questions still have to be answered:

1. What switches the initiation and growth of lateral roots on and off?
2. What determines their growth pattern and life span?
3. Is the direction of their growth predetermined?
4. What are the correlations between various root types?

In the past, root research has been frequently hampered by technical and by methodological difficulties. Most roots are hidden underground, therefore, they must be excavated to enable treatment, observation and analysis. However, it is practically impossible to dig out plant roots without destroying their delicate root hairs, without damaging the fine roots and without breaking some of the larger ones. Thus, functional analysis of such roots are doomed to fail, even before the first experiment has started, because of the inherent misfunction of injured roots. Moreover, whole systems, of soil grown roots, cannot be continuously inspected and successively observed, because the developmental changes are concealed by the medium. Thus, the main drawback of investigating excavated roots, for physiological and developmental studies, is that excavation excludes any return to the original physiological status. They present a one time event which cannot be reconstructed again. Thus, research of dynamic processes, that is based on such a methodology yields some information regarding the location of various roots, but is superficial when the developmental or functional aspects of root behavior are considered.

An alternative approach is to study roots through observation windows in root-boxes, large rhizotrons, or through minirhizotron transparent tubes (Box, 1996). Such studies are focused on the few roots that happen to grow between the bulk of the soil and the impermeable window. However, the question whether such roots truly represent the bulk of the root system remains open. These roots comprise only a small fraction of the whole root system that is not subjected to the physical forces of the soil particles, and to the flow and diffusion of substances (liquid solution and gases). They are not exposed to the temperature regime that characterizes the bulk soil, but are in a rather undefined set of conditions. Roots that reach the observation windows change their growth direction, and probably also their growth rate. In any case, their environmental conditions are different from those of all the other constituents of the root system. It is therefore difficult to conclude from such studies much about the "natural" behavior of the roots in the soil.

What root research needed badly was a system that enables the investigation of whole root systems of fully grown plants, in a setup where the roots can be continuously studied for prolonged periods. Hydroponics can be used for such purposes to a limited extent. However, because of the practical limitations of the volume of water-filled containers, it has been used for investigations of young and small plants only.

Aeroponic rhizotrons overcome most of the described difficulties, and provide exactly the type of research facility needed (Zobel et al., 1976; Zsoldos et al., 1987; Waisel and Breckle, 1987; Young and Young, 1991; Weathers and Zobel, 1992; Waisel, 1996). The roots of plants grown in aeroponic rhizotrons, are allowed to develop when suspended in moist air, and sprayed intermittently by a fine mist of a nutrient solution. The conditions in the aeroponic chamber are uniform and can be accurately controlled and monitored. The lack of large residual volumes of solution or of thermal inertia allow for quick and accurate manipulation of the experimental conditions. These include composition of the nutrient solution, rate of spraying (intensity of water supply), composition of the gaseous environment and of the temperature. The scientist's imagination and budget constrains are the only limitations to the size of plants that can be accommodated in such facilities or to the sophistication of their employment and control systems.

Aeroponic rhizotrons have several unique advantages:

1. They can be constructed in various sizes and can fit the specific requirements of most experimental projects.
2. They enable continuous observations of each of the roots, and the repeated measurement of their development and function, under uniform environmental conditions.
3. They enable measurements and manipulation of the growing root axes of perennial plants through several annual cycles.
4. They enable distinction between various growth patterns of roots, and comparison between the behavior of different roots, of a similar status or similar location in one root system.
5. They enable sampling or treatment of selected roots, without causing damage to the others.
6. They provide information regarding the growth potential of roots, rather than measurement of growth under the limitations imposed by the mechanical impedance of a specific soil or by the aeration characteristics of certain media.
7. Aeroponics is a research system where roots are subjected to a uniform environment without limitations of nutrient supply and without the formation of depletion zones. Roots can be grown under what seems to be optimal conditions (high [O_2], lack of mechanical impedance, optimal water conditions, variable or constant mineral composition etc.).
8. When carefully treated, roots of aeroponically grown plants seem to be free of diseases or pests, so common in roots grown in other research installations.
9. Root measurement in an aeroponic system is non-destructive. It enables frequent observation and recording of each component of the root system, rather than that of few of them, of unknown rank and status.
10. The use of aeroponics enables repeated measurements of the activity of same roots, and by that reduces the sampling errors.
11. Aeroponics constitutes a uniform environment, allowing the scientist to alter rapidly the nutrient milieu, and the gas composition of the atmosphere around the roots, without even touching them.

Aeroponic systems have been used in several laboratories for studies of various aspects of root structure and function. For review see Waisel (1996).

Such a discussion must not overlook the possible weak points of aeroponic systems. The roots growing in such systems are not exposed to the effects of mechanical contacts with a solid component of their environment. We know very little with respect to the importance of such contact. What processes are affected? Is there a threshold value for strength and duration? Where is the site of signal perception, if and when it takes place? Is the winding growth pattern of roots an inherent characteristic expressing a search for a proper contact, or is it an induced trait? Does it affect the physiological properties of the root? Since roots grown in aeroponics are not supported by a solid matrix or float in liquid, they are subjected to gravitational strains. How does such pulling strain affect the root behavior? Certain indications for the different properties of aeroponically grown roots were reported in the studies of Zsoldos et al. (1987) and Szabo-Nagy et al. (1994) who compared nutrient uptake by aeroponically grown and hydroponically grown plants. The conclusion of these studies was that the aeroponically grown roots lacked a certain metabolic component of the potassium uptake system. Such questions and many more, should be investigated in the future and answered, by comparing the behavior of plants grown under different experimental conditions.

SYNOPSIS

Aeroponics overcomes many of the difficulties that hamper the use of other growth installations for root research. It is used for observations of root growth without the mechanical, physical and electrochemical limitations which are imposed by the soil or by an aqueous environment. Root measurement, in such a system is not destructive, it enables frequent and repetitive observations of each of the roots of whole root systems. Aeroponics gives a uniform environment to all the roots of the system, with the possibility to alter the nutrient milieu, and the gas composition of the atmosphere around the roots, very rapidly.

REFERENCES

Box, J.E. Jr., 1996, Modern methods for root investigation. *in*: "Plant Roots: The Hidden Half, second edition, revised and expanded", Y. Waisel, A. Eshel and U. Kafkafi, eds., Marcel Dekker Inc. New York. pp.193–237.

Cannon, W.A., 1911, Root habits of desert plants. Carnegie Inst. Wash. Publ. No. 131.

Eshel, A. and Waisel, Y., 1996, Multiform and multifunction of various constituents of one root system. *in*: "Plant Roots: The Hidden Half, second edition, revised and expanded", Y. Waisel, A. Eshel and U. Kafkafi, eds., Marcel Dekker Inc. New York. pp. 175–192

Fitter, A., 1996, Characteristics and functions of root systems. *in*: "Plant Roots: The Hidden Half, second edition, revised and expanded", Y. Waisel, A. Eshel and U. Kafkafi, eds., Marcel Dekker Inc. New York. pp. 1–20.

Kutschera, L., 1960, Wurzelatlas mittleuropaischer Acker- und Kulturpflanzen. DLG Verlag, Frankfurt am Main.

Lynch. J., And Nielsen, K.L., 1996, Simulation of root system architecture. *in* "Plant Roots: The Hidden Half, second edition, revised and expanded", Y. Waisel, A. Eshel and U. Kafkafi, eds., Marcel Dekker, Inc., New York, pp.247–257.

Niklas, K.J., 1992, Plant biomechanics: An engineering approach to plant form and function. Chicago Univ. Press, Chicago, 607 p.

Nobel, P.S., 1996, Ecophysiology of roots of desert plants, with special emphasis on agaves and cactai. *in*: "Plant Roots: The Hidden Half, second edition, revised and expanded", Y. Waisel, A. Eshel and U. Kafkafi, eds., Marcel Dekker, Inc., New York, pp. 823–844.

Szabo-Nagy, A., Abdulai, M.D., and Erdei, L., 1994, Morphological and physiological differences between aeroponically and hydroponically grown sunflower plants. *Acta Biol. Hung.* 45:25–37.

Waisel, Y., 1988, Aeroponic observatories: useful tools for teaching and for long term studies of root behavior. Proc ISSR meeting, Uppsala, 4:27.

Waisel, Y., 1996, Aeroponics: A tool for root research. *in*: "Plant Roots: The Hidden Half, second edition, revised and expanded", Y. Waisel, A. Eshel and U. Kafkafi, eds., Marcel Dekker Inc., New York. pp. 239–245.

Waisel, Y., and Breckle, S.W., 1987, Differences in responses of various radish roots to salinity. *Plant Soil* 104:191–194.

Waisel, Y. and Eshel, A., 1991, Multiform behavior of various constituents of one root system. *in*: "Plant Roots: The Hidden Half", Y. Waisel, A. Eshel and U. Kafkafi, eds., Marcel Dekker, Inc., New York, pp. 39–52.

Weathers, P.J., and Zobel, R.W., 1992, Aeroponics for the culture of organisms, tissues and cells. *Biotech. Adv.* 10:93–115.

Weaver, W.J.E., 1919, the ecological relations of roots. *Carnegie Inst. Wash. Publ. No.* 286.

Young, W.M., and Young, S.Y., 1991, Basic study on a new soilless culture. II Effects of oxygen level in rhizosphere on the physio-ecological characteristics of tomato in aeroponics. *J. Kor. Soc. Hort. Sci.* 32:434–439.

Zobel, R.W., Del Tredici, P. and Torrey, J.G., 1976, Methods for growing plants aeroponically. *Plant Physiol.* 57:344–346.

Zsoldos, F., Vashegyi, A., and Erdei, L., 1987, Lack of active K^+ uptake in aeroponically grown wheat seedlings. *Physiol. Plant.* 71:359–364.

42

MOSSES AS PETRI DISH MODELS FOR UNDERGROUND FORAGING

Tsvi Sachs

Department of Botany
Hebrew University
Jerusalem 91904
Israel

INTRODUCTION

Various mosses have been grown on agar in Petri dishes (Allsop and Mitra, 1956; Bopp, 1952, 1962, 1992; Duckett, 1994). As could be expected for autotrophic organisms, they require no additions other than salts containing the known essential ions. Individual plants spread through the substrate by means of protonemata, reaching considerable distances relative to their size. Protonemata of neighboring plants, however, do not mingle (Bopp, 1952). This avoidance reaction was observed for members of the same clone and thus, presumably, the very same genotype. The influence of protonemata on one another was not related to a depletion of the nutrients from the substrate but rather to an inhibitory substance which was extracted from the medium, though it has not yet been purified (Bopp and Klein, 1963; Klein, 1967).

Thus physiological work points to the potential of mosses as experimental systems. Though many don't penetrate the substrate, others are among the simplest organisms to make their living as land plants, taking advantage of the proximity of the two contrasting habitats, the air and the soil. Below the surface, however, they have filaments that are only a single cell in diameter—protonemata—which are clearly not homologous to roots or any other multicellular organ. Yet the similrity of functions suggests that a comparison of protonemata with roots might offer important insights.

Hence the questions addressed here are: How extensive is the system of protonemata when the mosses are in soil rather than agar? Does mutual inhibition play a role in these conditions? How do the protonemata react to soil heterogeneity, such as stones or other physical obstacles? Remarkably enough, there seems to be no information about these questions. Even the mutual avoidance of the protonemata of neighboring mosses in agar has apparently not been demonstrated for mosses growing in soil. With one exception (Watson, 1981), there has been no discussion of its ecological role. In more general terms, "there is a need to reconsider the importance of protonemata" (Duckett and Matcham,

Biology of Root Formation and Development, edited by Altman and Waisel.
Plenum Press, New York, 1997.

1995). One reason for this past neglect might be that in many mosses the protonemata are restricted to early stages, following spore germination. The absence of protonemata in the mature state, however, is far from being a general rule.

A note about terminology. There are varied cellular filaments on individual moss plants, and they appear to carry out different roles. Spores germinate and develop filaments called protonemata. Similar if not identical filaments are formed on mature plants, especially, but not only, during regeneration ("secondary protonemata"). These filaments have been referred to as chloronemata when they are green and presumably photosynthetic, rhizoids when they anchor plants and caulonemata when they are dark in color and spread through the substrate. It is the caulonemata that can form buds which develop into additional green plants (Bopp, 1962, 1992). The same basic filamentous structure can thus differentiate in various ways (Duckett, 1994) which often intergrade or overlap. This differentiation, interesting though it is, was not the topic here, and for the purpose of the present study the general term protonema (plural protonemata) will suffice.

EXPERIMENTAL

Funaria hygrometrica plants were originally grown from spores supplied by Dr. I. Herenstadt. The mosses were then reproduced by regeneration from one clone. The substrates used were 1.5% agar, sand and soil. The sand originally came from a shore locality, and was washed a number of times by being boiled in de-ionized water. The soil, treated in the same way, was Terra Rosa from Giv'at Ram campus, Jerusalem. Hoagland nutrient solution (Hewitt, 1963) was added as specified. No sterility precautions were taken beyond the initial autoclaving of the agar and the boiling of the sand and soil. Infestation by fungi occurred only on agar and infestation by algae was rare. Dishes in which it occurred were discarded.

The closed Petri dishes were kept at all stages in a growth room with a 12hr photoperiod. The light, 180 μmol m^{-2} s^{-1}, was provided by white fluorescent tubes and an equal wattage of incandescent bulbs. Temperatures during lighted and dark hours were 24±2°C and 17±1°C, respectively. The dishes were not sealed and were opened at least once a week. De-ionized water was added as needed to keep the dishes moist, but with no free water.

Protonemata development in sand and soil was observed by flooding the dishes with de-ionized water and picking up the green plants with fine forceps. These plants could be transferred to a separate dish with water in which the sand (but not the soil) was washed off with no damaged to the protonemata. Microscope slides of regions of special interest were prepared with no staining.

RESULTS AND DISCUSSION

Funaria protonemata as an Experimental System

Funaria hygrometrica plants with 3–4 leaves were cut and placed on sand with 0.2 strength nutrient solution (Fig. 1A). Within 3d they formed many protonemata. These grew in all directions; within 14 days they reached the length of 18mm, about 6 times the height of the plants (Fig. 1B). By this time the thicker filaments had turned dark brown. The mat formed by the protonemata growing in sand covered the entire area surronding

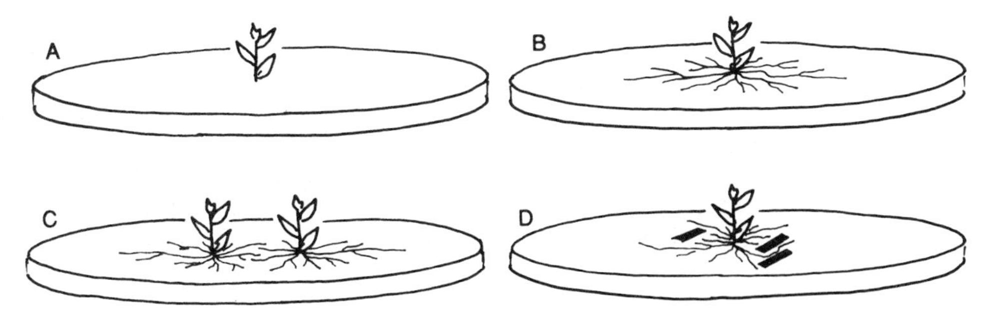

Figure 1. A. A cut moss plant with three expanded leaves, about 3mm high, placed in sand in a closed Petri dish (not drawn to the same scale). B. Within two weeks this plant regenerated a mat of protonemata, closely covering a circular area up to 18mm in diameter. C. Two plants formed independent protonemata systems that came into close contact but did not penetrate one another's territory. D. Protonemata often stopped just short of glass obstacles placed in the sand. They bent around such obstacles (on the right) when this did not lead to their meeting other protonemata of the same or of other plants.

the plants. The distances between the protonemata did not exceed 0.2mm, and did not vary much (precise measurements were not possible, because removing the sand disturbed the protonemata). This system was three dimensional—protonemata occurred at more than one level—but it penetrated the sand to a depth of only a few mm. Its individual protonemata were not straight, though they kept their general direction without meeting their neighbors and without overlaps or crosses.

Mosses could thus be valuable for the study of plant responses to soil heterogeneity. They are small and grow well at low light intensities. These traits allow for work in closed Petri dishes, maintained in relatively large numbers in uniform, controlled conditions. Relative to the size the plants the foraging through the soil extends to considerable distances and a macroscopic area is thoroughly covered. This foraging could be made essentially two-dimensional using a shallow layer of sand. This greatly simplifies a complicated story.

Protonemata differ from roots in being cellular filaments rather than organs. They form a thick matt, in which no local source of nutrients could be missed. Protonemata resemble root hairs, known to exhibit developmental plasticity (Bates and Lynch, 1996) and they also resemble mycorrhiza, from which they differ in being integral parts of the moss plant.

Development in Different Conditions

Though the form of the moss plants was basically similar on sand, agar and soil, there were differences. There were more branched green filaments—chloronemata—when the mosses were grown on agar. Branching of protonemata within the substrate, however, was more pronounced in sand and most pronounced in soil, where its detailed development could not be followed. The length of the protonemata was certainly greater in washed sand, with no nutrient solution, than in soil (18 rather than 12mm in 14d). Soil is the richer substrate, so the mosses were therefore grown on sand with different concentrations of a the nutrient solution. The development of the mosses, both above and within the substrate, was a function of the concentration of the nutrient solution. Some influence was noticed even at 0.3% of the full prescribed strength. The optimal response was around 20% and even full strength Hoagland solution caused no observable damage. The distances covered by the protonemata were greater on washed sand with no added nutrients than on nutrient concentrations that were optimal for overall development (4 rather than 3mm in 5d).

Thus the rate at which the protonemata spread was not a simple function of plant development. This means that in spite of, or perhaps because of, their simple structure mosses are remarkably plastic in their developmental reactions to substrate conditions. In adverse conditions there was greater growth in length, which increased the chances of encountering favorable patches of soil. In the following the emphasis is on the details of this plasticity as it applies to individual protonemata rather than to the plant as a whole.

The Influence of Neighboring Plants

Growth of protonemata in all directions and the occupation of a circular area around the original plant occurred in a homogeneous substrate. The situation was quite different when a number of individuals were planted close together. The protonemata systems of different plants reached direct contact, but they were easily separated when the sand was flooded and the mosses lifted (Fig. 1C). Washing the sand confirmed that the domain of an individual plant was closely occupied by its protonemata. Invasions by protonemata of the territories that had been taken over by neighboring moss plants did occur, but they were relatively rare. This was true even when the mosses were planted only 3mm from one another.

The avoidance reaction was consistent and pronounced even when it greatly constrained the development of individual plants. An individual surrounded by other plants made do with a limited area, rather than invade neighboring territories. Some protonemata did form, however, when the plants were planted within an occupied region and the only alternative to invasion was no development and, presumably, death. There was no similar avoidance of the roots of plantlets of *Kallanchoë* growing in the same Petri dishes.

Avoidance was not a function of the substrate. It was noticed in agar, where it had been previously described (Bopp, 1952; Bopp and Klein, 1963). In our growing conditions it was more obvious and repeatable when the mosses were grown in sand, and this was true regardless of the addition of nutrient solutions. Neighbor avoidance also occurred when the mosses grew in soil.

The interface between the protonemata of two neighbors was studied in detail. The protonemata of the neighboring plants came into close contact and no straight line separated their domains (Fig. 1C). Only on rare occasions were protonemata seen to turn away from the contact region. However, once they met the protonemata did not continue developing and there was no deep intermingling of the two systems. This developmental arrest was also expressed by the protonemata that had stopped growing remaining thin and maintaining their light rather than deep brown color.

These responses are presumably the same phenomenon that had been demonstrated for mosses growing in agar (Bopp and Klein, 1963). Use of different nutrient concentrations and the extraction and partial identification of an inhibitory substance (Klein, 1967) demonstrated that an actual secretion is involved, not the depletion of resources by the protonemata. In the present experiments, too, mosses with washed sand, with no added nutrients, on one side and soil in which another moss was growing on the other, developed more protenmatrra into the sand than the soil.

The secreted substance could be involved in allelopathic relations. It is, however, active in the relations between plants that originated in the same clone and can be assumed to be genetically identical. It can be expected to be ecologically significant: since the development of the mosses is generally clonal, there should be a high chance of a growing moss encountering other individuals with the same genetic constitution. In this context observations on the reactions of mosses of other species would be critical. The problem has been that these don't grow in sand in the conditions used here; preliminary observations of the growth of plants of different species in soil indicate, but as yet do not prove, that the avoidance reaction is species specific. It is thus possible that the ecological role of the secreted substance is not so much in relation to neighbors as in the promotion of optimal foraging by a clone and even by an individual plant (Bopp, 1952, 1962). The growth of protonemata is not along straight lines, and the inhibitory signal may be a mechanism that prevents their running into each other and covering the same ground many times.

Local Responses to Physical Obstacles

The observations of the dramatic responses to neighboring plants to one another raised the question of the effects of non-living obstacles. Stones, for example, can be common obstacles in natural conditions and a prominant expression of substrate heterogeneity. They could be expected to force protonemata to change their course and, more generally, to perturb the organization of the protonemata of individual plants. Natural obstacles were therefore simulated in the Petri dishes by carefully placed fragments of microscope cover slips.

Glass obstacles necessarily modified protonemata development (Fig. 1D). A key observation appeared to be that the protonemata generally (but not invariably) stopped at a short distance from the actual obstacle. As in the case of encountering neighbors, they remained thin and did not turn dark. There were two significant exceptions to these generalizations. Protonemata growing above the sand, in humid air, hit the glass and actually stuck to it before their growth ceased. In the sand there were protonemata that changed direction in the immediate vicinity of the obstacle. Their growth continued at an angle that generally enabled them to avoid any direct contact with the glass. Such changes in the direction of growth were observed where they did not entail direct contact with other protonemata, of either same plant or of other plants. This meant that the plants were able to invade a zone between two obstacles placed close to one another.

It is possible that these observations are another of expression of the very same mechanism that enables the protonemata to avoid one another. The protonemata might stop before an obstacle because of the reduced diffusion and thus accumulation of the inhibitory substance. This interpretation is supported by the changes in the course of protonema growth where there were mechanical obstacles, but no neighbors (Fig. 1D). In such cases changes in the course of protonema development could be due to low inhibitor concentrations on one sides of the obstacle. This is in accordance with the observations that where neighboring plants met development ceased locally, with no adverse effect on the moss plants as a whole. A general suggestion would be that the secretion of the inhibitory substance is a means by which an individual protonema obtains information about its environment and chooses its course. Thus a system of external rather than internal interactions (Sachs, 1991) appears to be employed in the patterning of the protonemata system.

Could anything similar have a role in root development? Because of dimension differences, a role for a diffusing substance is probably less likely for roots than for protonemata. However, the possibility that a root might secrete a substance that regulates its own development still deserves the consideration it has not received. A local, specific inhibitory effect of roots on the growth of roots of other plants of the same species has been detected (Mahall and Callaway, 1991, 1996). Further, allelopathic substances, generally studied in relation to seed germination or the overall development of neighboring plants, can have an effect on the direction of root growth (Paradales et al., 1992). An emphasis of the effects of allelopathic substances on the growth of individual roots seems to be called for.

CONCLUSIONS

Though the protonemata of mosses are strands of single cells that are quite different from roots, they could be a model system, raising questions concerning the ways plants might cope with soil heterogeneity. Moss protonemata were inhibited from invading the territory of another moss, even when the two plants were genetically identical. This is due to a substance (or substances) secreted to the immediate environment. The protonema also avoided stones and other inanimate obstacles. They grew around them only when such growth did not involve coming close to other protonemata, even ones belonging to the very same plant. It is likely that they were able to avoid obstacles by means of the very same mechanism used to detect neighbors. As suggested by Bopp (1952, 1962), a secreted substance might play a role not only in sensing the environment but in the organization of protonemata of an individual moss.

ACKNOWLEDGMENTS

I thank Dr. I. Herenstadt and Professor C. C. Heyn for discussions and generous help with the literature and plant identification and Dr. M. Hassidim for technical help. At a crucial stage of this work Professor M. Bopp helped with information, an essential discussion, and reprints. Dr. H. During and Prof. J. G. Duckett helped with literature about moss development.

REFERENCES

Allsopp, A., and Mitra, G. C., 1956, The heterotrichous habit in the protonema of Bryales. Nature 178: 1063.

Bates, T. R., and Lynch, J. P., 1996, Stimulation of root hair elongation in Arabidopsis thaliana by low phosphorous availability. Plant Cell Env. 19: 529.

Bopp, M., 1952, Entwicklungsphysiologische Untersuchungen an Laubmoosprotonemen. Zeit. Bot. 40: 119.

Bopp, M., 1962, Development of the protonema and bud formation in mosses. J. Linn. Soc. (Bot.) 58: 305.

Bopp, M., 1992, The morphogenetic system of the moss protonema. Cryptogamic Bot. 3: 3.

Bopp, M., and Klein, B., 1963, Versuch zur analyse Protonemaentwicklung der Laubmoose. I. Endogene Wahcstumregulatoren im Protonema von *Funaria hygrometrica*. Portugaliae Acta Biologica. A-7: 95.

Duckett, J. G., 1994, Studies of protonemal morphogenesis in mosses. V. *Diphyscium foliosum* (Hedw.) Mohr (Buxbaumiales). J. Bryology, 18: 238.

Duckett, J. G., and Matcham, H. W., 1995, Studies of protonemal morphogenesis in mosses VII. The perennial rhizoids and gemmiferous protonema of *Dicranella heteromalla* (Hedw.) Schimp. J. Bryology 18: 407.

Hewitt, E.J. 1963, Nutrition in plants in culture media. In: Steward, F. C. (ed.) Plant Physiology, Vol 3: 97, Academic Press, N. Y.

Klein, B. 1967. Versuche zur Analyse der Protonema entwicklung der Laubmoose IV. Der Endogene Factor H und seine Rolle bei der morphogenese von *Funaria hygrometrica*. Planta, 73: 12–27.

Mahall, B. E., and Callaway, R. M., 1991. Root communication among desert shrubs. Proc. Nat. Acad. Sci. (Wash.) 88: 874.

Mahall, B. E., and Callaway, R. M., 1996. Effects of regional origin and genotypes on intraspecific root communication in the desert shrub *Ambrosia dumosa* (Asteraceae). Am. J. Bot. 83: 93.

Paradales, J. R. Jr., Kono, Y., Yamaguchi, A., and Iijima, M., 1992, Seminal root growth in Sorghum (*Sorghum bicolor*) under allelopthic influences from residues of Taro (*Calocasia esculenta*). Ann. Bot. 69: 493.

Sachs, T., 1991, Pattern Formation in Plant Tissues, Cambridge Univ. Press, Cambridge.

Watson, M. A., 1981, Chemically mediated interactions among juvenile mosses as possible determinants of their community structure. J. Chem. Ecol. 7: 367.

43

RESOLVING CIRCUMNUTATIONAL AND GRAVITROPIC MOVEMENTS OF ROOT TIPS

Amram Eshel

Department of Botany
The George S. Wise Faculty for Life Sciences
Tel-Aviv University, Tel-Aviv 69978
Israel

1. INTRODUCTION

The movements of growing roots allow them to find the path of least resistance through soil and maximize the utilization of soil resources by proper deployment of members of the root system in the rooting volume. Circumnutational and gravitropic movements of root tips are important components of these processes. Darwin and Darwin (1880) coined the terms "circumnutation" and "geotropism" for the phenomena described earlier by Sachs and others. They studied these phenomena and showed each to occur separately in growing radicles of seedlings of a number of species. They also recognized the adaptive significance of the circumnutations in the penetration of the soil. The adaptive significance of gravitropic responses of plants roots and shoots was discussed in detail by Barlow (1995).

The two phenomena were associated with one another and thus circumnutations were attributed by many scientists to "overcompensation" of gravitropic response (cf. Johnsson, 1979). In more recent literature the circumnutations, also referred to as "spontaneous oscillations", were regarded as noise that interferes with measurements of gravitropic curvature of radicles (Barlow and Rathfelder, 1985; Nelson and Evans, 1986; Selker and Sievers, 1987). Barlow et al. (1991) compared the curvature of root tips displaced from the vertical at various degrees. Only those displaced between 10° and 40° continued to show oscillatory circumnutations. Roots positioned at 90° returned to the vertical direction smoothly revealing no nutations. In a later study (Barlow et al., 1993) a root positioned initially at 98° from vertical also showed some oscillations about the smooth bending curve.

The role of hormones, especially auxin, in the gravitropic curvature of roots was studied extensively (cf. Pilet, 1996). Evans (1991) suggested that differential distribution of auxin across the root, that brings about differences in growth and gravicurvature, is fol-

Biology of Root Formation and Development, edited by Altman and Waisel.
Plenum Press, New York, 1997.

lowed by an adaptation process whereby the auxin sensitivity of cells in the two sides of the elongation zone changes. During the later phases the growth pattern begins to reverse, and as a result the cell elongation rates do not simply come back to the prestimulation value. Instead, they overshoot dramatically thereby leading to oscillations.

Legue et al. (1992) who studied movements of lentil roots on a clinostat concluded that oscillations of the root that were not limited by gravity could be responsible for the spontaneous curvature. Later, experiments performed during NASA Spacelab missions have shown that circumnutations were initiated and continued in microgravity conditions. It was therefore concluded that gravitropic-overcompensation cannot be regarded as the exclusive explanation for circumnutational oscillations. Brown (1993) proposed a new model based on the assumption that while a root bends, cells on the convex side of the bend peripheral are stretched, and those on the concave side are compressed. Relief of these strains leads to movement of the tip to the other side. Baluska et al. (1996) measured differences in cell elongation between outer and inner cortex at both sides of gravistimulated roots, that should inevitably lead to development of inner tension. They proposed that this tension plays a role in gravitropic root bending.

Brown (1993) concluded his review stating that research on circumnutation is still largely in the exploratory stage and for most objectives further descriptive studies of kinematics of circumnutations will be needed. Studies of gravitropism in recent years were done by a few laboratories, which specialized in lapse photography and special video digitizing equipment (e.g. Ishikawa et al., 1991). The aim of the current report is to describe utilization of computerized video techniques, that became readily available in recent years, for the studies of circumnutational and gravitropic movements of root tips.

2. EXPERIMENTAL

Experiments were conducted with corn and with *Arabidopsis*. Seedlings of corn were germinated for two days in vertical rolls of moist germination paper. During experiments, the seedlings were fixed at various angles deviating from the vertical, in the space between moist germination paper and a plate of glass. The experiments were conducted at 22–25°C under constant light supplied by an incandescent bulb. Preliminary experiments have shown that graviresponse in this genotype of corn was insensitive to light.

Root tip movements were recorded by a home video cam-corder (Panasonic NV-S700EN) for 6h. Later, the video tape was played back in the same cam-corder connected to an S-video port of a Silicon Graphics Indy workstation. The video images were digitized using Silicon Graphics Iris standard "Snapshot" software tool. The resulting series of images were combined to a computerized video movie either by Silicon Graphics Iris "Moviemaker" or by Adobe "Premiere 3.0" running on a PowerMac.

The position of the root tip during all stages of bending was measured on the computer image using Silicon Graphics "ImageView". The radius of curvature of the bending root and the length of the arc were calculated using the formulae described in Fig. 1.

Seeds of *Arabidopsis thaliana* were sown on MS-sucrose-agar (Sigma M-9274) plate and grown for two weeks in a growth chamber with 10h photoperiod and day/night temperature regime of 22°C / 17°C. For the experiment, the plates were turned 90° and fastened to a stage of an inverted microscope that was laid on its side. A CCD video camera (Sony DXC-151P) was attached to the microscope and images were recorded every 5 min as computer files on a 486 PC. Calculations of curvature and processing of the video movie were done by software as described above.

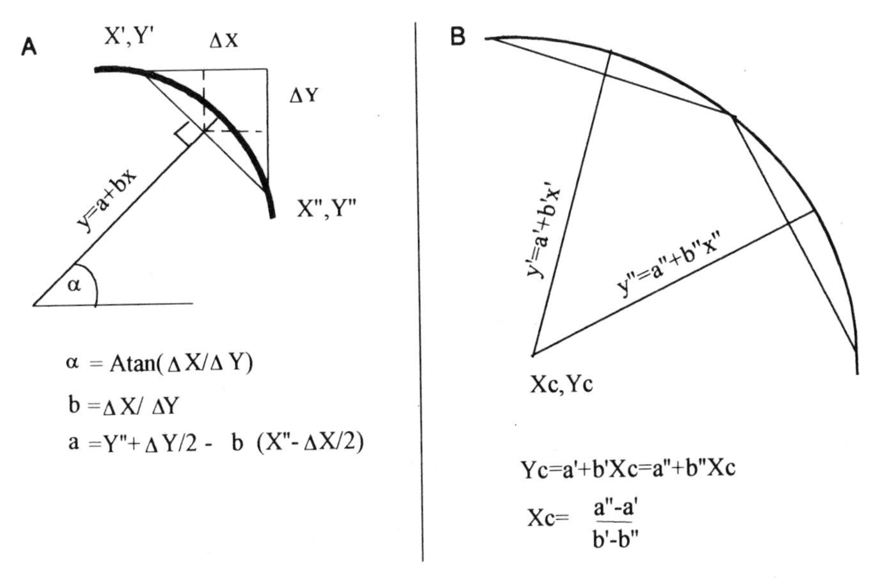

Figure 1. Finding the radius of curvature and arc length of a curved root from coordinates of three points along its central line: A. Calculating the equation that describes the perpendicular bisector of a chord connecting two points. B. Finding the coordinates of the center of curvature by intersection of the two chord bisectors.

3. RESULTS AND DISCUSSION

An example of the gradual change in root curvature with time is shown in Fig. 2. Other roots exhibited similar curves, undulating about the smooth hyperbola that describes the gravitropic curvature. Such undulations with a period of *ca.* 40 min were specified as circumnutational by Antonsen et al. (1995). These authors revealed the movements by

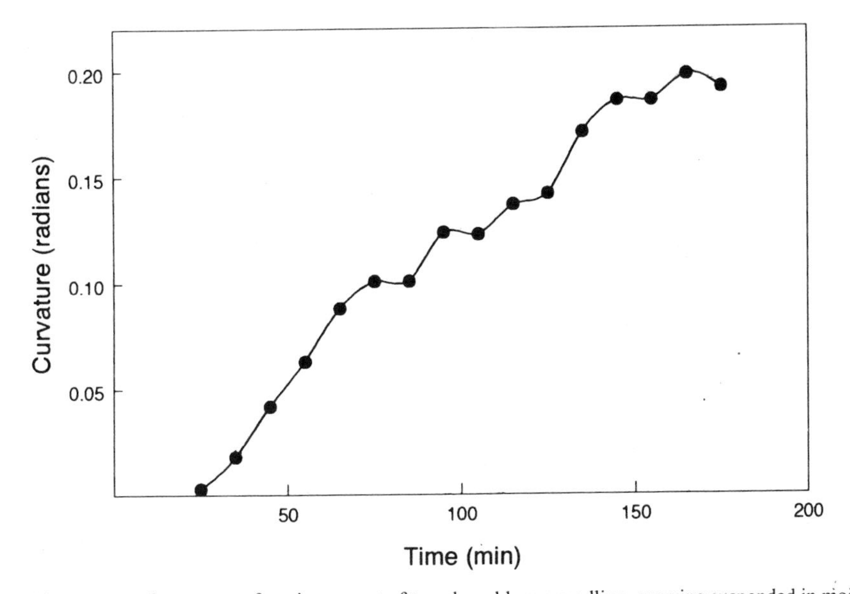

Figure 2. Time course of curvature of a primary root of two-day-old corn seedling, growing suspended in moist air.

applying complex signal processing techniques to data collected in Spacelab on graviresponding lentil roots.

Barlow (1992) presented theoretical analysis for the various concurrent movements of root tips. The differential distribution of waves of growth, that proceed along the roots, in one dimension was postulated to account for pendulum-like nodding of the root tip. The existence of multiple sensors and multiple response zones that are responsible for the various movements which take place concomitantly in the same root were also postulated. In a later review (Barlow et al., 1994), the short-period movements were named micronutations, and the authors proposed that they are distinct from the long period (1–8 h) circumnutations.

Careful observation of the video movie revealed that the circumnutations are due to transient variations in the growth of the most distal cells in the elongation zone, whereas the gravitropic bending is due to changes which take place in other parts of the elongation zone, further away from the tip. It therefore concurs with Barlow's hypothesis that different groups of cells are involved in the two movements of the root tip which take place concomitantly. It is proposed that the circumnutational movements result from rapid reversible changes in cell volume related to plasmamembrane-associated cation pumps.

Graviresponse of *Arabidopsis* roots growing on the surface of the agar plate or inside the plate was followed. In both cases the roots responded gravitropically and did not exhibit any nutations. The graviresponse of roots growing on the surface of the agar was much faster. The lower partial pressure of oxygen inside the agar might be the cause of that difference. Similarly, Hoson et al. (1996) reported that the graviresponse of certain species is abolished altogether upon submergence. The *Arabidopsis* plants used were two weeks old, and this might account for the lack of nutations. It was already noted by Darwin and Darwin (1880) that this type of movement is much more evident in young seedlings. Only the apical part of the elongation zone participated in the graviresponse of the *Arabidopsis* root. In the basal part of the elongation zone, there was no difference between cell extension at the two sides. As a result, the trajectory followed by the apex of the graviresponding root was a combination of horizontal extension and gravitropic curvature (Fig. 3).

4. CONCLUSIONS

Computerized video is becoming readily available, and it will enable detailed studies of root movements, to be performed on a wide variety of plants. The complex movements of root tips of graviresponding corn and *Arabidopsis* roots were studied. The results

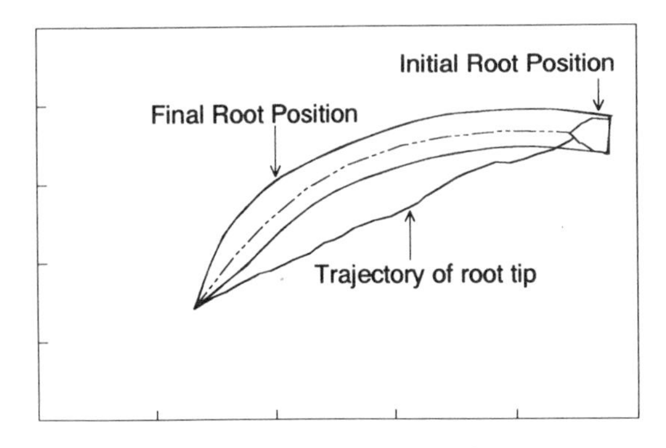

Figure 3. Initial and final positions of a graviresponding *Arabidopsis* root growing on the surface of a vertical agar plate. The growth was followed for 4h after the root had been challenged at 90°. The root tip itself followed during that time the trajectory marked by the solid line.

are in agreement with the assumption that differences in elongation of cells at various positions *along* the elongation zone, as well as *across* it, are responsible for these complex movements.

ACKNOWLEDGMENT

The author expresses his sincere thanks to Dr. P.W. Barlow for his discussion and commnets to the manuscript.

REFERENCES

Antonsen, F., Johnsson, A., Perbal, G., and Driss-Ecole, D., 1995, Oscillatory growth movements of roots in weightlessness. *Physiol. Plant.* 95:596–603

Baluska, F., Hauskrecht, M., Barlow, P.B., and Sievers, A., 1996, Gravitropism of the primary root of maize: a complex pattern of differential cellular growth in the cortex independent of the microtubular cytoskeleton. *Planta* 198:310–318

Barlow, P.W., 1992, A conceptual framework for investigating plant growth movements, with special reference to root gravitropism, utilizing a microgravity environment. *Microgravity Q.* 2:77–87

Barlow, P.W., 1995, Gravity perception in plants: a multiplicity of systems derived by evolution? *Plant, Cell Env.* 18:951–962

Barlow, P.W., Brain, P., Butler, R., and Parker, J.S., 1991, Modelling the growth response of gravireacting roots. *Aspects of Appl. Biol.* 26:221–225

Barlow, P.W., Parker, J.S., Butler, R., and Brain, P., 1993, Gravitropism of primary roots of *Zea mays* L. at different displacement angles. *Ann. Bot.* 71:383–388

Barlow, P.W., Parker, J.S., and Brain, P., 1994, Oscillation of axial plant organs. *Adv. Space Res.* 14:(8)149-(8)158

Barlow, P.W., and Rathfelder, E.L., 1985, Distribution and redistribution of extension growth along vertical and horizontal gravireacting maize roots. *Planta* 165:134–141

Brown, A.H., 1993, Circumnutations: From Darwin to space flights. *Plant Physiol.* 101:345–348

Darwin, C.A., and Darwin, F., 1880, The Power of Movement in Plants. John Murray, London.

Evans, M.L., 1991, Gravitropism: Interaction of sensitivity modulation and effector redistribution. *Plant Physiol.* 95:1–5

Hoson, T., Kamisaka, S., and Masuda, Y., 1996, Suppression of gravitropic responses of primary roots by submergence. *Planta* 199:100–104

Ishikawa, H., Hasenstein, K.H., Evans, M.L., 1991, Computer, based video digitizer analysis of surface extension in maize roots. Kinetics of growth rate changes during gravitropism. *Planta* 183:381–390

Johnsson, A. 1979 Circumnutation, *in*: "Physiology of Movements" W. Haupt, and, M.E. Feinleib eds. vol. 7 Encyclopedia of Plant Physiology NS. Chapter 5.3.1 pp. 627–645 Springer, Heidelberg.

Legue, V., Driss-Ecole, D., and Perbal, G., 1992, Cell cycle and cell differentation in lentil roots grown on a slowly rotating clinostat. *Physiol. Plant.* 84:386–392

Nelson, A.J., and Evans, M.L., 1986, Analysis of growth patterns during gravitropic curvature in roots of *Zea mays* by use of a computer based video digitizer. *Plant Growth Regul.* 5:73–83

Pilet, P.-E., 1996, Root growth and gravireaction: A reexamination of hormone and regulator implications. *in*: "Plant Roots: The Hidden Half" Y. Waisel, A. Eshel, and U. Kafkafi eds. Chapter 15 pp. 285–305. 2nd edition, Marcel Dekker, NY.

Selker, J.M., and Sievers, A., 1987, Analysis of extension and curvature during the graviresponse in *Lepidium* roots. *Amer. J. Bot.* 74:1863–1871

44

THE MUTUAL DISTRIBUTION OF COMPETING ROOT SYSTEMS

A Stationary Model

Ariel Novoplansky[1] and Dan Cohen[2]

[1]Ben-Gurion University of the Negev
The Jacob Blaustein Institute for Desert Research
Sede-Boker Campus, 84990 Israel
[2]Department of Evolution, Systematics and Ecology
The Silberman Institute of Life Sciences
The Hebrew University of Jerusalem, 91904 Israel

INTRODUCTION

The spatial distribution of plant roots in the soil is a complex dynamic process that changes in response to the changing morphological, physiological and ecological state of plants and in response to the spatial and temporal distribution of soil resources (Caldwell and Richards 1986, Brisson and Reynolds 1994). The changes in soil resources represent the balance between inputs and losses which include both physical and biotic factors. The depletion of resources by roots is a major feedback process in the dynamics of resources in the environment (e.g. Russel and Newbould 1969, Caldwell 1987).

In this paper we present a simple model that attempts to characterize the stationary evolutionary stable (ESS) spatial distribution of the root systems of adjacent plants. Such a distribution is the result of competition for nutrients between the plants under unrealistic conditions of a uniform constant supply of a limiting nutrient. The model offers equilibrium distributions of root systems that may be an approximation of an idealized stable community of long-living fully-grown perennial plants or trees. Although such a stationary model is rather unrealistic in describing dynamic root systems under field conditions which are invariably changing, it can serve as an important tool in predicting possible trends and traits which arise from *violations of its assumptions*, especially those which are related to the dynamics of both resource availability and plant development in space and time. The value of such an unrealistic model lays in its power to generally predict the degree of overlapping of real root systems according to the degree to which they deviate from its basic assumptions.

The observed distributions of root systems indicate a very large variation in the degree of overlap between neighboring plants (Cannon 1911, Biswell 1935, Rogers and

Biology of Root Formation and Development, edited by Altman and Waisel.
Plenum Press, New York, 1997.

Head 1969, Atkinson et al. 1976, D'Antonio and Mahall 1991, Brisson and Reynolds 1994). Our main objective is to explain and predict the distribution of roots within the root system of each individual plant, and the degree of overlap between the root systems of neighboring plants, as a function of the properties of the distances between their shoot bases, their sizes and their nutrient uptake and requirements. The model and its analysis are comparable to the modeling of the ESS of the home ranges of foraging animals competing for spatially distributed resources (e.g. Milinsky and Parker 1991).

THE MODEL

General

We construct and present a simple model of the dynamic processes and equilibrium states of both the soil nutrients and the lateral root distribution of one plant or of two neighbouring plants. The model is based on the following simplified assumptions:

a. Root growth is an increasing function of the fitness gain per unit of nutrient uptake.
b. There is a uniformly distributed constant input of one limiting nutrient, a constant proportional loss of the nutrient by leaching, and negligible long range diffusion.
c. Nutrient uptake per unit root biomass is an increasing function of nutrient concentration, and possibly also of the demand function for this nutrient.
d. Transport and maintenance costs of the roots increase with the distance from the base of the shoot. Maintenance costs also include the turnover of small roots and root hairs.
e. Under constant conditions, and when all the plants have reached their final mature size, it is assumed that the spatial distribution of the root systems of all the plants have reached a constant stationary equilibrium state.
f. The root system of an individual plant without competition will stop growing when the net gain per unit root growth is zero or negative for all its components.
g. The equilibrium lateral limit of the root system of a single plant should be at the distance at which the distance cost exactly balances the fitness gain by nutrient uptake at the equilibrium nutrient concentration that is reached without root growth.
h. The root systems of competing neighboring plants should also stop growing when the net gain per unit root growth is zero or negative for all the components in the root systems of all the plants.

These conditions imply that:

a. The equilibrium density of roots will be higher near the base of the plant.
b. The equilibrium nutrient concentration will be lowest near the base of the plant because of the stronger depletion.
c. The equilibrium lateral extent of the root systems of single plants of different species will be an increasing function of the efficiency of the uptake and utilisation functions of the nutrient by these species.

The Basic Equations of the Mathematical Model

The uptake rate function U(Y) of the nutrient Y per unit root is:

$$U(Y)\frac{U_m Y}{(K+Y)} \quad (1)$$

where U_m is the maximal rate, and K is an apparent Michaelis-Menten constant of the uptake process. We also derive the inverse function Y(U):

$$Y(U)\frac{UK}{(U_m - U)} \quad (2)$$

At the steady state in any small soil volume, uptake must be equal to the difference between input and loss, i.e.

$$U(Y)\,R = INPUT - LOSS\ Y \quad (3)$$

where R is the amount of roots per unit soil volume, and LOSS is the loss coefficient for Y, caused by leaching and decomposition.

E.g. if there are no roots, R = 0, and the steady state nutrient level is then:

$$Y = \frac{INPUT}{LOSS} \quad (4)$$

In the presence of plant roots, we combine equations (1) and (3) to get the equilibrium Y ^ as the solution to the equation:

$$\frac{Y\wedge}{(K+Y\wedge)} = \frac{(INPUT - LOSS\ Y\wedge)}{(U_m R)} \quad (5)$$

i.e. the quadratic equation

$$LOSS\ Y^2 + Y(K\ LOSS + U_m R - INPUT) - K\ INPUT = 0 \quad (6)$$

If the LOSS parameter is relatively small and can be neglected, we get that:

$$Y\wedge \approx \frac{K\ INPUT}{(U_m R - INPUT)} \quad (6.1)$$

Root growth by any individual plant at any small soil volume is expected to stop when the nutrient concentration has been depleted to such a low level that the *additional* fitness contributed by any further root growth W is zero or negative. Such an equilibrium distribution of roots can be reached only after all the plants have stopped growing and have reached their final adult size. The equilibrium distribution of the root system of any one plant as a function of the *distance* from the base of the plant must therefore satisfy the equation:

$$MAX_R\ (W^*)\ \text{at}\ U(Y(X))\,F - M - TX = 0 \quad (7)$$

for each part of the root systems of any plant at a distance X from the center of the root system at the base of the plant. F is the conversion coefficient of nutrient uptake to carbon units, M is the maintenance energy per unit root, and T is the transport energy in carbon units per unit root per unit distance.

The spatial equilibrium distributions of the root systems of isolated plants, and of all the competing plants, are given by the simultaneous solutions of equation (7) for the root systems of all the plants, and of equations (5) or (6) for the steady-state distribution of the nutrient concentration in the soil, taking into account the total nutrient uptake by the roots of all the competing plants in any given soil volume.

Solving the Mathematical Equations

At the steady states for both nutrient level and root growth, we get from (3) and (7) that the steady state level of the nutrient at any distance X satisfies the following equation:

$$U(Y(X))\,R(X) = (\text{Input} - \text{LOSS}\;Y(X) \tag{8}$$

while zero growth rate of the roots satisfies the following equation:

$$U(Y(X))\,F = (T\,X + M) \tag{9}$$

Combining (8) and (9), we get for the equilibrium optimal amounts of roots:

$$R^*(X) = \frac{(\text{INPUT} - \text{LOSS}\;Y(X))\,F}{(T\,X + M)} \tag{10}$$

which is a decreasing function of X, reaching zero at a distance X_L, with

$$R(X > X_L) = 0, \quad Y(X > X_L) = \frac{\text{INPUT}}{\text{LOSS}}$$

From Equation (7) we get that at the limit of the root system of a single plant:

$$X_L = \frac{(U(\text{INOUT}/\text{LOSS})F - M)}{T} \tag{11}$$

that is,

$$X_L = ([(U_m\,(\text{INPUT/LOSS}) / (K + (\text{INPUT/LOSS}))]\;F - M) / T \tag{12}$$

From (10), we get that at the base of the plant, i.e. at X= 0,

$$R(X=0) = \frac{(\text{INPUT} - \text{LOSS}\;Y(X=0))F}{M} \tag{13}$$

However, since U(Y(X=0)) = M/F,

$$Y(X=0) = \frac{K\,M/F}{(U_m - M/F)} \tag{14}$$

from which we can get the expression for R(X=0).

Over the whole range of $0 < X < X_L$, we therefore get both:

$$Y(X) = \frac{K(TX+M)}{(U_m F-(TX+M))} \tag{15}$$

$$R(X) = \frac{F/(TX+M)(INPUT-(LOSS\ K\ (TX+M)))}{(U_m F-(TX+M))} \tag{16}$$

See Fig. 1a-b for schematic representation of the radial distributions of Y(X) and R(X) for a single plant.

If LOSS is relatively small, it can be neglected in the expressions for R(X) when the root density R(X) is large near the base of the plant. We then get approximately:

$$R(X) \approx \frac{INPUT\ F}{(TX+M)} \tag{17}$$

One Dimensional Spatial Distribution of the Root System of Two Competing Plants

Consider two plants, A and B, separated by a distance D. The equilibrium distributions of the root systems of the two neighbouring plants, and of the nutrient concentration

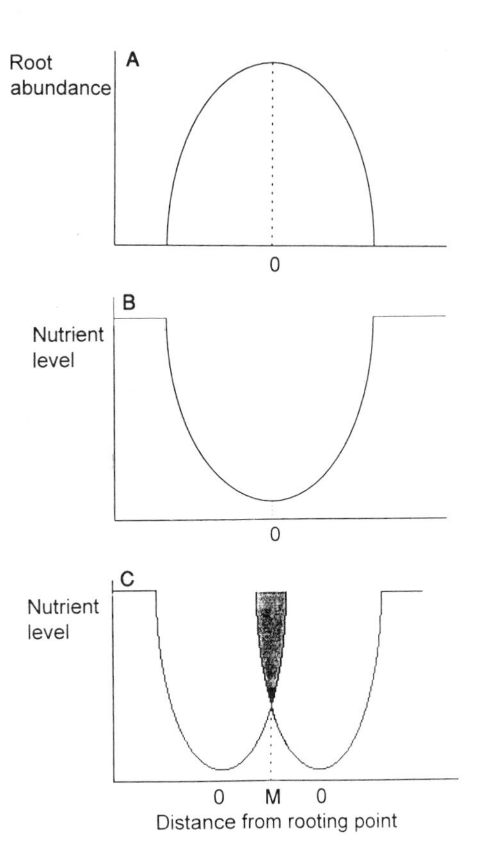

Figure 1. Root abundance (a), soil nutrient level for a single (b) and for two neighbouring plants (c). The shaded area represents the range in which both plants have negative fitness. A sharp demarcation boundary is expected at M, at the competitive equilibrium.

along the distance X from A to B (Fig. 1c) can be found by the simultaneous solution of the equilibrium conditions for the root systems of the two plants.

Plants of different species, or plants of the same species with different physiological, morphological and ecological states, may differ in a number of their characteristic parameters which influence the spatial distribution of their root systems.

To simplify the derivations, we assume that LOSS is negligible relative to INPUT and uptake. Using this simplification, we combine equation (3) for both plants, and arrive at the equilibrium:

$$R_A(X)\,U_A(Y(X)) + R_B(X)\,U_B(Y(X)) = \text{INPUT} \tag{18}$$

The conditions for the equilibrium ESS condition for zero growth for both plants when R_A (X)>0 and R_B(X)>0 respectably are:

$$\text{MAX}_{RA}(W_A{}^*(X)),\ \text{i.e., when } 0 = U_A(Y\,(X))\,F_A - T_A X - M_A \tag{19}$$

$$\text{MAX}_{RB}(W_B{}^*(X)),\ \text{i.e., when } 0 = U_B(Y\,(X))\,F_B - T_B(D - X) - M_B \tag{20}$$

Let us consider first the symmetric case, with identical A and B parameters for the two plants. In this case:

$$(R_A(X) + R_B(X))\,U(Y(X)) = \text{INPUT} \tag{21}$$

The conditions for the root growth equilibrium are :

$$0 = U(Y(X))\,F - T\,X - M \tag{22}$$

$$0 = U(Y(X))\,F - T\,(D - X) - M \tag{22.1}$$

and from (21),

$$U(Y(X)) = \frac{\text{INPUT}}{(R_A\ (X)) + R_B(X))} \tag{23}$$

Equations (22) and (22.1) for WA*(X) = 0 and WB*(X) = 0 can be satisfied together for (R_A(X) > 0 and R_B(X) > 0 only when X = D/2. The equilibrium solution in the symmetrical case must therefore be:

$$R_A{}^*(X > D/2) = 0$$

$$R_B{}^*(X < D/2) = 0 \tag{24}$$

with no overlapping between the spatial distributions of the root systems of the two plants (Fig. 1c). Because of the symmetry, $R_A(X) = R_B(D-X)$. The distribution of the roots and the nutrient satisfies the same equations as for the single plant, i.e.

$$R_A(X < D/2) = \frac{\text{INPUT F}}{(TX + M)} \tag{25}$$

$$Y_A(X < D/2) = \frac{K(TX+M)}{(UMF-(TX+M))} \tag{26}$$

with identical symmetrical expressions for $R_B(X > D/2)$, obtained by replacing X with D–X in equations (25) and (26).

When plants A and B differ in their fitness and cost parameters, and in their uptake rates, the ESS conditions when $R_A(X) > 0$ and $R_B(X) > 0$ respectably can be described more generally by solving the explicit equations:

$$MAX(W_A{}^*(X)),\ \text{i.e.,}\ 0 = U_A(Y(X))\,F_A - T_AX - M_A \tag{27}$$

$$MAX(W_B{}^*(X)),\ \text{i.e.,}\ 0 = U_B(Y\,(X))\,F_B - T_B(D - X) - M_B \tag{28}$$

If the uptake parameters of the two species are identical, then

$$(T_AX + M_A) / F_A = (T_B(D - X) + M_B) / F_B \tag{29}$$

$$X_{AB} = \frac{(F_AM_BF_BM_A + F_AT_BD)}{(T_AF_B + T_BF_A)} \tag{30}$$

independently of the uptake parameters, or of the input.

If the uptake functions differ by a constant C, such that $U_A = C\ U_B$, then at the ESS for non-overlapping $R_A(X) > 0$ and $R_B(X) > 0$,

$$\frac{T_AX_{AB} + M_A}{F_A} = \frac{C(T_B(D - X_{AB}) + M_B)}{F_B} \tag{31}$$

and $R_A\ (X > X_{AB}) = 0$ and $R_B(X < D - X_{AB}) = 0$, where:

$$X_{AB} = \frac{(CF_AM_A - F_BM_A + CF_AT_BD)}{(CF_AT_B + F_BT_A)} \tag{32}$$

independently of the K parameters of uptake and of the INPUT.

In general, since ESS $Y_A(X)$ is always an *increasing* function of X for plant A alone, and $Y_B(X)$ is always a *decreasing* function of X for plant B alone, The boundary X_{AB} is the solution to $Y_A(X) = Y_B(X)$, if they intersect over the range $0 < X < D$.

DISCUSSION

Our equilibrium model for the spatial distributions of root systems and nutrient levels in the soil makes several significant testable predictions:

1. The density of roots of an isolated plant is expected to be highest near the center of the root system and to gradually decrease to zero toward the outer limit (Fig. 1a). Accordingly, the nutrient concentration is expected to be lowest near the center of the root system, and to increase toward the periphery and reach the

steady constant level beyond the root system (Fig.1b). The root systems of isolated plants are expected to be radially symmetric in homogeneous environments. The lateral extent of the roots is expected to increase with increasing fitness gain per unit nutrient uptake, and to decrease with increasing costs of transport and maintenance of the roots. Additionally, the roots are expected to extend to greater distances into patches with higher nutrient input as the payoff of proliferating there is obviously greater (e.g. Drew and Saker 1975, Crick and Grime 1987, Campbell and Grime 1991, Gersani and Sachs 1992, Caldwell et al. 1992).

2. A sharp demarcation boundary is expected between the root systems of adjacent plants at the competitive equilibrium, where the roots of both plants will contribute negative fitness if they grow into the area occupied by their neighbours (shaded area in Fig. 1c). The position of the demarcation line between any two plants should be at the intersection between the equilibrium nutrient concentration functions of the two plants if they were isolated, and be in the middle between identical plants. The nutrient concentrations will have a maximum at these boundaries (Fig. 1c).
3. An inferior competitor will be eliminated if it is close enough to a superior competitor, when the lowest equilibrium nutrient concentration at its base is higher than the equilibrium nutrient concentration maintained by its neighbour at the same location (Fig. 2). Even a small competitive advantage will cause the elimination of the inferior competitor if they are spaced very close together. This consequence is very similar in its nature to the predictions of other mechanistic models of population-level nutrient competition (Tilman 1982, 1990). On the other hand, inferior competitors can coexist with superior competitors if they are sufficiently widely spaced (e.g. Heywood and Levin 1986)
4. The root systems of coexisting competing plants are expected to occupy areas of non-overlapping irregular polyhedrons, which satisfy the conditions of a stationary distribution of nutrient concentrations over the whole area, and of zero net growth for all the plants.

The natural distribution of root systems is expected to fit more closely these predictions under conditions that approximate more closely the main assumptions of the model: i.e. i) the temporal constancy of the input and of the loss coefficients of one limiting nutrient, and ii) the long term stationary root systems. These assumptions may hold reasonably well in a community of long living plants, with a constant input of one major limiting nutrient. The assumptions are also expected to better represent the distribution of natural root systems of competing plants of the same or of ecologically-equivalent species that have very similar physiological performance functions.

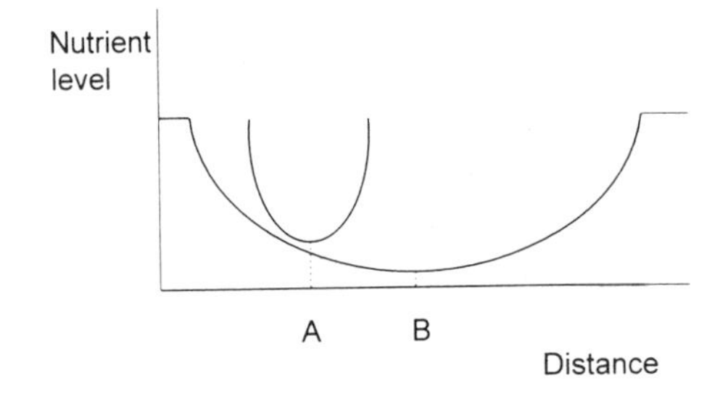

Figure 2. Equilibrium nutrient levels of a dominant (B) and a subordinate (A) plants. At the competitive equilibrium species B will competitively exclude species A.

Deviations from the predictions of the model can suggest the need for violations or modifications of the assumptions of the model. Several such predictions can be derived from some more realistic assumptions:

1. Fuzzy boundaries and partial overlapping are expected between root systems of plants of the same species when resource input varies, e.g. when it decreases or increases with time, it arrives in pulses, or is unpredictable. Although recent studies are touching on the problem of resource dynamics in time and space (e.g. Crick and Grime 1987, Campbell and Grime 1991, Bilbrough and Caldwell 1997), we know of no study that followed the degree of root intermingling under pulsed resource availability.
2. Fuzzy boundaries and partial overlapping are expected between root systems of plants of different species if they have different performance characteristics for different soil resources ("niches" *sensu* Whittaker 1975), Austin (1990).
3. Growing root systems of isolated plants are expected to approach an equilibrium quite slowly, depending on the cost of construction of the roots. Growing root systems of neighbouring plants are expected to grow beyond their eventual stable boundaries and temporarily overlap, while nutrient concentrations are above the levels that stop root growth.
4. A strong persistence of existing "over extended" roots is expected because the fixed cost of initial growth is often "expensive" and only partially recoverable (Novoplansky 1996). Thus, changes in the "overlapping zone" and "dynamic boundaries" between neighbouring plants are expected to be slow, especially if a considerable part of already existing roots is structural and long living. The major roots of trees or large shrubs are essentially permanent. In this case, the large scale features of the roots are expected to persist almost indefinitely at their initial locations, while the smaller roots and especially the extremely small roots that decay faster, will be replaced by new roots which will be situated in the most favorable locations nearer the plant center or the major roots, because the transport cost is probably greater in thin than in large roots.
5. Large persistent root systems are thus expected to develop sharper boundaries very gradually, and their shapes and locations will represent to a large extent the accidental pattern of their early development. For example, small scale patchiness of nutrient distribution in the soil during early growth, or some localized damage, can bias early root patterning. This will dictate a particular spatial distribution of the large root branches that will be maintained almost indefinitely in the adult plant, and will determine to a large extent its overlapping and boundaries with other root systems. The proposed mechanism for the origin and persistence of irregular patterns of root systems and their boundaries can be tested by tracing the location and branching patterns of a regenerating root system after the removal of roots of various sizes, ranks and locations. Such removal is expected to cause vigorous regeneration, with gradual plastic responses to the existing conditions at the time of regeneration, without necessarily regenerating the initial spatial distribution of the roots.
6. "War of attrition" (sensu Maynard-Smith 1982) is expected between overlapping root systems when the local nutrient concentration is depleted below its zero utility level. This is because persistence in spite of temporary losses might be more beneficial in the long run than an immediate retreat following a temporary loss. The time until withdrawal should be positively correlated with the unrecoverable investment in the relevant roots.

7. The growing root systems of annuals and seasonal root systems of perennials, are expected to have the greatest tendency for overlapping between nearby root systems. This is because these root systems can never reach a stationary level, and the nutrients are depleted continuously as the root systems grow. In such situations, a selective advantage is expected for an opportunistic uptake of the nutrients from the soil regions with higher nutrient concentrations. The degree of overlapping is also expected to be much stronger in annuals because the distances between neighbouring plants change every year, and cannot be predicted in advance. In contrast, in long living perennials with constant distances and locations, there may be feedback information from year to year about the availability of nutrients in different places and directions. Such information could in principle regulate the growth of competing root systems to increase efficiency and reduce wasteful overlapping and competition.
8. Strongly overlapping root systems are expected with initially high nutrient levels because this allows a faster uptake from temporarily high nutrient levels in the less depleted soil volume.
9. The absolute cost of the distance from the base decreases in smaller root systems. Thus, the *absolute* overlapping of the boundaries between the root systems of neighboring plants is expected to be negatively correlated with plant size.

This study is only touching on some ecological aspects of root intermingling. Obviously, a wide range of questions remains open in this rather neglected topic. For example, do plants use signals (e.g. R/FR, Solangaarachchi and Harper 1989, Novoplansky et al. 1990, Novoplansky 1991, or nutrient gradients), to estimate the presence and the quality of their neighbors or use any kind of self vs. non-self recognition mechanisms as was anecdotally mentioned by Mahall and Callaway 1991, 1996)? Such mechanisms or even changes based on the *probabilities* of encountering foreign vs. self roots would increase the efficiency of the root systems, and would be selected for.

CONCLUSIONS

The spatial distribution of plant roots in the soil is expected to depend on the ecological and the physiological state of the plants and on the distribution of soil resources in space and time. A simple model which characterizes the steady state ESS spatial distribution of the root systems of neighbouring long-lived plants predicts a sharp boundary between the root systems of adjacent plants, at which the roots of both plants contribute negative fitness if they grow into the area occupied by their neighbours. The boundary between the root systems is at the position where the stationary nutrient concentration of the neighbouring plants would be equal if they were isolated. A plant belonging to a species with a more efficient root system will eliminate a neighbouring plant belonging to a less efficient species if it lowers the nutrient concentration in the soil below the stationary maintenance level of the less efficient competitor.

Shifting and fuzzier boundaries with overlapping between neighbouring root systems are expected however if any of the assumptions of the model are violated, e.g.: when the resource supply is non-uniform, the life-time of the plants is limited (e.g. annual), or the root systems are not stationary, e.g. grow seasonally or there are more than one limiting soil factor, with different utilisation functions for different species.

The dynamics of growing or shrinking root systems, and of changing boundaries, is expected to be slower when the unrecoverable cost of root construction is higher.

ACKNOWLEDGMENTS

We thank Beata Oborny, Tsvi Sachs and Simon Barak for the insightful discussions and comments on early versions of this manuscript.

REFERENCES

Atkinson, D., Naylor, D. and Goldrick, G.A., 1976, The effect of spacing on the apple root system. Hort. Res. 16:89.

Austin, M.P., 1990, Community theory and competition in vegetation, *in*: "Perspectives on Plant Competition", J.B. Grace and D. Tilman, eds., Acad. Press, San Diego.

Bilbrough, C.J. and Caldwell, M.M., 1997, Exploitation of springtime ephemeral N pulses by six Great Basin plant species, Ecology 78:23.

Biswell, H.H., 1935, effects of environment upon the root habits of certain deciduous forest trees, Bot. Gaz. 96:676.

Brisson, J. and Reynolds, J.F., 1994, The effect of neighbours on root distribution in a creosotebush (*Larrea tridentata*) population, Ecology 75:1693.

Caldwell, M.M., 1987, Competition between root systems in plant communities, *in*: "Root development and Function", P.J. Gregory, J.V. Lake and D.A. Rose, eds., Cambridge University Press, Cambridge.

Caldwell, M.M., Dudley, L.M. and Lilieholm, B., 1992, Soil solution phosphate, root uptake kinetics and nutrient accuisition: Implications for a patchy soil environment, Oecologia 89:305.

Caldwell, M.M. and Richards, J.H., 1986, Competing root systems: morphology and models of absorption, *in*: "Economy of Plant Form and Function", T. Givnish and R. Robichaud, eds., Cambridge University Press, Cambridge.

Campbell, B.D. and Grime, J.P., 1991, A trade-off between scale and precision in resource foraging, Oecologia 87:532.

Cannon, W.A., 1991, The root habitat of desert plants, Pub. Carnegie Inst. of Washington 131:1.

Crick, J.C. and Grime, J.P., 1987, Morphological plasticity and mineral nutrient capture in two herbaceous species of contrasted ecology, New Phytol. 107:403.

D'Antonio, C.M. and Mahall, B.E., 1991, Root profiles and competition between the invasive exotic perennial, *Carpobrotus edulis*, and two native shrub species in California coastal scrub, Am. J. Bot. 78:885.

Drew, M.C. and Saker, L.R., 1975, Nutrient supply and the growth of the seminal root system in barley. II. Localized, compensatory increases in lateral root growth and rates of nitrate uptake when nitrate supply is restricted to only part of the root system. J. Exp. Bot. 26:79.

Gersani, M. and Sachs, T., 1992, Development correlations between roots in heterogeneous environments, Plant Cell & Env. 15:463.

Heywood, J.S. and Levin, D.A., 1986, Interactions between seed source, planting arrangement, and soil treatment in determining plant size and root allocation in *Phlox drummondii*, Oecologia 68:285.

Mahall, B.E. and Callaway, R.M., 1991, Root communication among desert shrubs. Proc. Nat. Acad. Sci. 88:874.

Mahall, B.E. and Callaway, R.M., 1996, Effects of regional origin and genotype on inrtaspecific root communication in the desert shrub *Ambrosia dumosa* (Asteraceae), Am. J. Bot. 83:93.

Maynard-Smith, J., 1982, "Evolution and the Theory of Games", Cambridge University Press, Cambridge.

Milinsky, M. and Parker, G.A., 1991, Competition for resources, *in*: "Behavioural Ecology", J.R. Krebs and N.B. Davies, eds., Blackwell Sci. Pub., Oxford.

Novoplansky, A. 1991, Developmental responses of *Portulaca* seedlings to conflicting spectral signals, Oecologia 88:138.

Novoplansky, A., 1996, Developmental responses of individual *Onobrychis* plants to spatial heterogeneity, Vegetatio 127:31.

Novoplansky, A. Cohen, D. and Sachs, T., 1990, How *Portulaca* seedlings avoid their neighbours. Oecologia 82:490.

Rogers, W.S. and Whittington, W.J., Factors affecting the distribution and growth of roots of perennial woody species, *in*: "Root Growth", W.J. Whittington, ed., Butterworths, London.

Russell, R.S. and Newbould, P., 1969, The pattern of nutrient uptake in root systems, *in*: "Root Growth", W.J. Whittington, ed., Butterworths, London.

Solangaarachchi, S.M. and Harper, J.L., 1989, The growth and asymmetry of neighbouring plants of white clover (*Trifolium repens* L.), Oecologia 78:208.

Tilman, D., 1982, "Resource Competition and Community Structure", Princeton Univ. Press, Princeton.

Tilman, D., 1990, Mechanisms of plant competition for nutrients: The elements of a predictive theory of competition, *in*: "Perspectives on Plant Competition", J.B. Grace and D. Tilman, eds., Acad. Press, San Diego.

Whittaker, R.H., 1975, "Communities and Ecosystems", 2nd ed. Macmillan, New-York.

FRACTAL BRANCHING: A MODEL FOR BRANCHING ROOT SYSTEMS

Louise Y. Spek[1] and Meine Van Noordwijk[2]

[1]Theoretical Biology, Utrecht University. Padualaan 8, 3584 CH Utrecht, The Netherlands.
[2]International Center for Research on Agroforestry (ICRAF), P.O. Box 161, 16001 Bogor, Indonesia.

Observations on the branching of trees made Leonardo da Vinci to claim, that the cross section area of the main stem is equal to the sum of the cross section areas of tree branches. In a formal expression for bifurcation this means (the model rule):

$$D^2 = \alpha \cdot (d_1^2 + d_2^2) \text{ (when a=1.0) (after Mandelbrot, 1983)} \quad (1)$$

D and (d_1, $d_{2)}$, represent diameter values before and after branching, respectively. The diameter exponent being 2, is close to that of botanical trees. The parameter a is a proportionality factor, and might be connected to physiological and/or morphological characteristics of root-systems. Changes in α rather strongly influence the pattern formation. Changes in the diameter ratio between d_1 and d_2, such as changes in the diameter exponent, influence the pattern formation. Our assumption is that Eq. 1 might be applicable to root-systems. (Van Noordwijk et al., 1994).

Repeated application of this model rule (Eq. 1), with changes in parameter values, give rise to a great diversity of root-like patterns, with fractal characteristics. Visualization of these patterns are realized in a three dimensional graphical display by PLUTON, a Molecular Graphics Program. (Spek, 1996). See Figure 1. Spreading and morphology of root-systems can be studied roundabout the display.

The great diversity of root-like patterns arising by repeated application of a simple branching rule (Da Vinci) and the visualization of these patterns in a three dimensional display (by PLUTON, offers interesting tools to study root-systems.

ACKNOWLEDGMENTS

With gratitude to Dr. A. L. Spek for essential help, to Prof. Dr. P. Hogeweg and Prof. Dr. W. Drenth, Utrecht University.

REFERENCES

Mandelbrot, B.B., 1983, The fractal geometry of nature. W.H. Freeman and Company New York.

Spek, A.L., 1996, PLUTON, a Molecular Graphics Program for display and analysis of crystal and molecular structures. Utrecht University.

Van Noordwijk, M., Spek, L.Y. and De Willigen, P., 1994, Proximal root diameter as predictor of root size for fractal branching models. I. Theory. *Plant Soil*, 64:107.

Figure 1. A three dimensional display by PLUTON of a calculated root-like pattern. The ratio of the diameter values after branching is: $d_1:d_2 = 1:3$ (or vice versa). $D_{start}=40$. $a=1.8$. Diameters in the displayed pattern are derivatives of the diameters as calculated in Eq. 1.

A ROOT MODEL USED IN METHODICAL RE-EVALUATION OF REGR-PROFILE DETERMINATION TECHNIQUES

Nirit Bernstein and Winfried S. Peters

Institute of Soils and Water, Volcani Center, POB 6, Bet Dagan, 50 250, Israel.

The characterization of spatial growth patterns is a pre-condition for the causal analysis of growth processes. The distribution of growth intensities in space are best expressed as Relative Elemental Growth Rates, (REGR). A multitude of alternative methods to determine REGR-profiles has been described, but a quantitative, comparative analysis is still lacking to date. A central problem for a methodological re-evaluation is that empirical REGR-profiles will always be estimative, since an accurate determination would require actual measurements of velocities at one instant of time. In principle, a comparison of estimates cannot establish a judgment concerning their relative accuracies; this can only be achieved by comparison of the estimates to a known REGR-profile. Therefore an assessment of the reliability of methods of REGR-profile determination must necessarily be based on a theoretical profile.

We created a model REGR-profile by a simple third-order polynomial. The coefficients were adjusted to render the model similar to empirical REGR-profiles from maize primary roots. The Displacement Velocity (DV; i.e. the velocity by which points at a certain position move away from the apex) was calculated as the integral of the REGR-profile. Growth trajectories were determined as the integral of the reciprocal DV function; these describe the change of position of points moving along the organ with time. The trajectories were used to calculate the changes in length of segments defined by pairs of points.

This methodical equipment enables us to perform theoretical marking experiments. In such experiments, which were often employed in the determination of REGR-profiles, points are marked along a growing organ, and either their DV or the length changes of their intervals are monitored.

Our theoretical experiments can be performed under varying conditions, e.g. with varying initial segment lengths, or durations of measurement. Existing methods of REGR-profile determination can be applied to the "experimental" data. The resulting profiles can be quantitatively compared to the original one. The "experimental" data are free of numerical errors and statistical variance, and are thus perfectly suited to evaluate systematic errors introduced by the particular method.

The results obtained so far prove the usefulness of our approach in the identification, quantification, and causal analysis of methodical errors. For example, we could demonstrate that segmental relative growth rates RGR) are an excellent tool for the establishment of REGR-profiles, provided that consistent arithmetical methods are used. This insight questions the necessity of advanced, complicated numerical methods favored by currently accepted theory.

A detailed description of the here described study is currently in *Plant Physiology* 113: 1395–1404.

45

A 1996 OVERVIEW OF THE RESEARCH INTERESTS FROM THE JERUSALEM SYMPOSIUM

Thomas Gaspar

University of Liege
Institute of Botany
B 22, Sart Tilman
B-4000 Liege, Belgium

Plants are our obligatory partners in life. They relieve us of carbon dioxide and provide us with the necessary oxygen. They are the source of our food and of many medicines. They constitute our ornamental landscape, and with their industrial exploitation with many purposes (wood, fibers, gums, etc.), they are integral part of our evoluting eco–socio-economical society.

Plants have unique characteristics which make them superior organisms: their capacity to fix solar energy, their plasticity in growth and development (number of organs, size, and time of sexualisation not predetermined), which reflects an intense capacity in sensing the environment factors, and which is made possible by their primary meristems, their organogenic totipotency which is the ability to initiate *de novo* primary meristems and to orient them in specific organo- and morpho-genic programs. Roots are such organs which can be formed *de novo*.

It is an evidence that there is no whole plant development without the roots which play essential roles (fixation, nutrition, hormonal control, allelopathy, etc.). These few indications stress the importance of a symposium on the biology of root formation and development.

Communications and research interests manifested at the Jerusalem symposium can be grouped around the formation, characteristics and biology of three root types: the seed-embryo-borne roots, the (adventitious) untransformed shoot-borne roots, and the mutants- and genetically transformed -borne roots. The main points discussed with these roots are reported in the table.

What to conclude? that we continue to be amazed in front of the diversities of root morphologies and their undefinite capacity of adaptation under extreme conditions of viability. Substantial progresses have been made in the understanding of these adaptations, mostly at the physiological level, but we still are far away from the fine diverse cell signalling systems involved. The same can be said concerning the regeneration of roots

Biology of Root Formation and Development, edited by Altman and Waisel.
Plenum Press, New York, 1997.

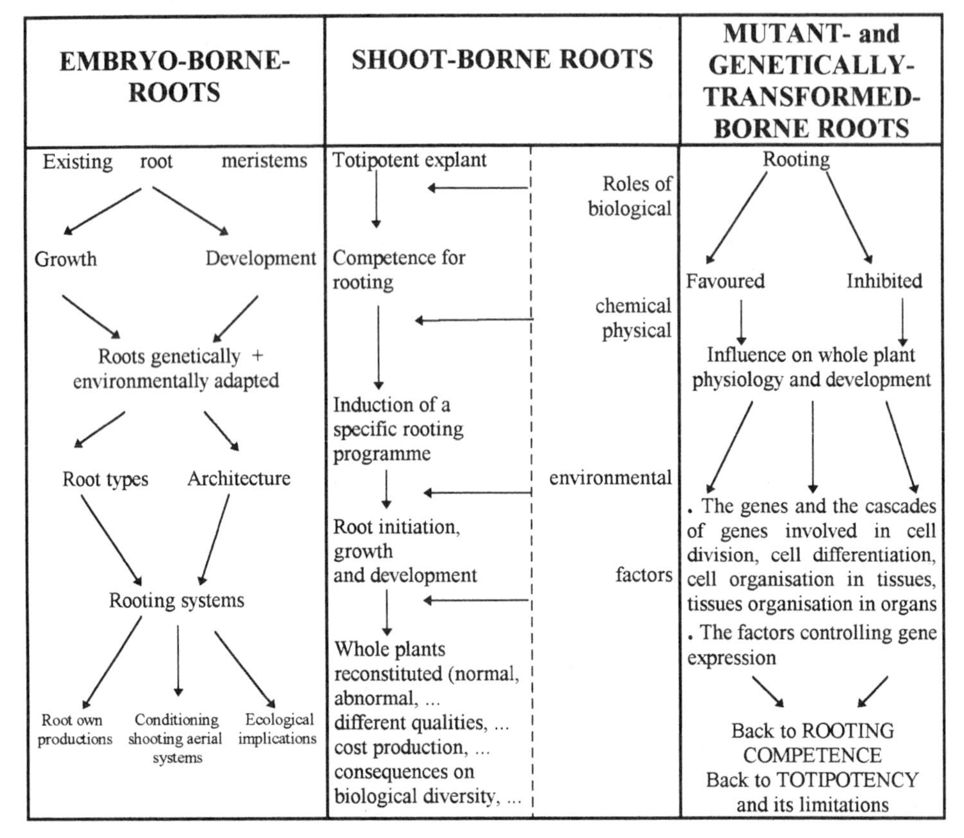

Figure 1. The biology of the three types of roots discussed at the 1996 Jerusalem symposium.

where we control organogenesis, that is a developmental process, with "growth" substances. As an example to show how our approach still is global, we rarely discriminate between the factors involved in cell division, cell differentiation, organisation of the differentiating cells into tissues, and organisation of the tissues into organs. Fortunately, attention has been drawn to selected new mutants with known gene alterations and known "abnormal" behaviours or responses at specific steps of developmental pathways. Genetic engineering is in parallel providing us with new experimental systems. Developing molecular biology with new powerful techniques offers new tools for identifying the genes involved. The next symposia unavoidably will go further with cascade events in gene actions, their interplays and the control mechanisms. We already expect from the future larger applications of root biology and plant totipotency but the researches probably will also show the limitations of the biotechnological manipulations.

INDEX OF PLANT GENERA AND PLANT COMMON NAMES

SUBJECT INDEX